Microbiological analysis of red meat, poultry and eggs

Related titles:

Improving the safety of fresh meat
(ISBN-13: 978-1-85573-955-0; ISBN-10: 1-85573-955-0)
It is widely recognised that food safety depends on effective intervention at all stages in the food chain, including the production of raw materials. Contaminated raw materials from agricultural production increase the hazards that subsequent processing operations must deal with, together with the risk that such contamination may survive through to the point of consumption. This collection provides an authoritative reference summarising the wealth of research on reducing microbial and other hazards in raw and fresh red meat.

Food safety control in the poultry industry
(ISBN-13: 978-1-85573-954-3; ISBN-10: 1-85573-954-2)
Consumers' expectations about the safety of products such as poultry meat and eggs have never been higher. The need to improve food safety has led to renewed attention on controlling contamination at all stages of the supply chain from 'farm to fork'. This collection reviews the latest research and best practice in ensuring the safety of poultry meat and eggs, both on the farm and in subsequent processing operations.

Emerging foodborne pathogens
(ISBN-13: 978-1-85573-963-5; ISBN-10: 1-85573-963-1)
Developments such as the increasing globalisation of the food industry, constant innovations in technologies and products, and changes in the susceptibility of populations to disease, have all highlighted the problem of emerging pathogens. Designed for microbiologists and QA staff in the food industry, and food safety scientists working in governments and academia, this collection discusses ways of identifying emerging pathogens and includes chapters on individual pathogens, their epidemiology, methods of detection and means of control.

Details of these books and a complete list of Woodhead's titles can be obtained by:

- visiting our web site at www.woodheadpublishing.com
- contacting Customer Services (e-mail: sales@woodhead-publishing.com; fax: +44 (0) 1223 893694; tel.: +44 (0) 1223 891358 ext. 130; address: Woodhead Publishing Ltd, Abington Hall, Abington, Cambridge CB1 6AH, England)

Microbiological analysis of red meat, poultry and eggs

Edited by
G. C. Mead

CRC Press
Boca Raton Boston New York Washington, DC

WOODHEAD PUBLISHING LIMITED
Cambridge England

Published by Woodhead Publishing Limited, Abington Hall, Abington
Cambridge CB21 6AH, England
www.woodheadpublishing.com

Published in North America by CRC Press LLC, 6000 Broken Sound Parkway, NW, Suite 300, Boca Raton, FL 33487, USA

First published 2007, Woodhead Publishing Limited and CRC Press LLC

British Library Cataloguing in Publication Data
A catalogue record for this book is available from the British Library.

Library of Congress Cataloging in Publication Data
A catalog record for this book is available from the Library of Congress.

Woodhead Publishing ISBN-13: 978-1-84569-059-5 (book)
Woodhead Publishing ISBN-10: 1-84569-059-1 (book)
Woodhead Publishing ISBN-13: 978-1-84569-251-3 (e-book)
Woodhead Publishing ISBN-10: 1-84569-251-9 (e-book)
CRC Press ISBN-13: 978-1-4200-4397-6
CRC Press ISBN-10: 1-4200-4397-8
CRC Press order number: WP4397

The publishers' policy is to use permanent paper from mills that operate a sustainable forestry policy, and which has been manufactured from pulp which is processed using acid-free and elementary chlorine-free practices. Furthermore, the publishers ensure that the text paper and cover board used have met acceptable environmental accreditation standards.

Typeset by SNP Best-set Typesetter Ltd., Hong Kong
Printed by TJ International Limited, Padstow, Cornwall, England

Contents

Contributor contact details

(* = main contact)

Editor
Professor G. C. Mead
Consultant
17 Harbutts
Bathampton
Bath
Somerset
BA2 6TA
UK

email: geoffmead@hotmail.com

Chapter 1

Dr C. de W. Blackburn
Safety & Environmental Assurance Centre (SEAC)
Unilever Colworth
Colworth House
Sharnbrook
Bedfordshire
MK44 1LQ
UK

email: clive.blackburn@unilever.com

Chapter 2

M Fogden
Medvek Consullltancy Ltd
3 Kipling Drive
Towcester
Northants
NN12 6QY
UK

email: medve@medvek.com

Chapter 3

Iain D. Ogden
Applied Food Microbiology Group
Dept. Medical Microbiology
University of Aberdeen
Foresterhill
Aberdeen
AB25 2ZD
UK

email: i.ogden@abdn.ac.uk

Chapters 4 and 7

Professor G. C. Mead
Consultant
17 Harbutts
Bathampton
Bath
Somerset
BA2 6TA
UK

email: geoffmead@hotmail.com

Chapters 5 and 9

Dr J. E. L. Corry
Department of Clinical Veterinary Science
University of Bristol Langford
Bristol
BS40 5DU
UK

email: janet.corry@bristol.ac.uk

Chapter 6

Dr C. O. Gill
Agriculture and Agri-Food Canada Lacombe Research Centre
6000 C & E Trail
Lacombe
Alberta
Canada
T4L 1W1

email: gillc@agr.gc.ca

Chapter 8

J. F. Rigarlsford
John Rigarlsford & Associates Ltd
Gold Hill Lodge
Gold Hill
Tansley
Derbyshire
DE4 5FG
UK

email: john@rigarlsford.com

Chapter 10

E. de Boer
Food and Consumer Product Safety Authority (VWA)
Regionale Dienst Oost
PO Box 202
7200 AE Zutphen
The Netherlands

email: enne.de.boer@vwa.nl

Chapter 11

Dr S. M. Russell* and Dr P. C. Vasavada
218 Poultry Science Building
University of Georgia
Athens, GA
30602-2772
USA

email: srussell@uga.edu

Chapter 12

Prof. M. Uyttendaele* and Prof. J. Debevere
Laboratory of Food Microbiology and Food Preservation
Department of Food Safety and Food Quality
Faculty of Bio-science Engineering
Gent University
Coupure Links 653
B-9000 Ghent
Belgium

email: johan.debevere@ugent. bemieke.uyttendaele@ugent.be

Chapter 13

R. Wood
Food Standards Agency
c/o Institute of Food Research
Norwich Research Park
Colney
Norwich
NR4 7UA
UK

email: roger.wood@foodstandards.gsi.gov.uk

Introduction

Some readers may wonder why the food commodities covered in this volume are confined to red meat, poultry and eggs. The answer is simply that these foods are, or have been, major global causes of foodborne human disease, and all are relatively susceptible to microbial growth and spoilage. Red and white meats are traditionally associated with food poisoning that arises mainly from mishandling of the meat in the kitchen. In England and Wales, for example, there were more than 0.6 million cases of foodborne illness that were attributable to red meat and poultry during 1996–2000, with 305 deaths. The principal causative agents were *Campylobacter* and *Salmonella* spp. (Health Protection Agency data). Shell eggs, on the other hand, were long regarded as a safe food to eat, even when raw or only lightly cooked and consumed by vulnerable groups in society. That viewpoint had to be modified, when strains of *S.* Enteritidis emerged in the 1980s with the capability of infecting the reproductive tract of the laying hen, because a small, but significant, proportion of shell eggs was then contaminated internally with *Salmonella* and there followed a pandemic of human salmonellosis. Thus, eggs had rapidly become one of the commonest sources of *Salmonella* outbreaks in many countries, a situation that took some years to show any real improvement – and could, conceivably, happen again!

Microbiological analysis has a lengthy history as a means of monitoring the microbial quality and safety of foods, whether in relation to guidelines, product specifications or legally enforceable standards. Following the more recent development and gradual implementation of a risk-based, preventative approach to food safety control, microbiological testing of foods has a further role to play and, in Europe, new microbiological criteria are being introduced that will encompass the food commodities considered here.

It seemed timely, therefore, to review the current analytical position and speculate about future developments.

In a series of chapters that have been written by international experts, the key aspects of microbiological analysis are described and discussed. Some, such as sampling methods and use of faecal indicators, are aimed specifically at the foods in question, while others have a wider relevance, including current approaches to testing of foods, detection and enumeration of pathogens and spoilage organisms, and microbial identification techniques. Attention is also given to the validation of analytical methods and Quality Assurance in the laboratory, both of which will have a considerable impact on future laboratory practices. Because of their present importance to the food industry, there are additional chapters on current and developing legislation in the European Union and the significance of *Escherichia coli* O157 and other VTEC.

The book is not intended as a bench manual, but aims to be an up-to-date reference work on what is clearly an important and dynamic area of food microbiology. As such, it will allow those responsible for product quality and safety to be fully informed about the issues involved in an area that is so crucial to the functioning of the food industry. The book will also provide students of food science and researchers with a scientific overview of the analytical field.

In my role as editor, I am indebted to all the other contributors for their diligence and hard work throughout the publication exercise. Thanks are also due to the publisher for continuing support and encouragement, and to my wife, Valerie, for her unfailing help in dealing with the manuscripts. It has been a pleasure to work with all those concerned.

1

Microbiological testing in food safety and quality management

C. de W. Blackburn, Unilever Colworth, UK

1.1 Introduction

There has been an inexorable move towards a Quality Assurance (QA) approach in the control of microbiological hazards in food, with the wide adoption of Hazard Analysis Critical Control Point (HACCP), HACCP-based approaches and pre-requisite programmes (PRPs) as preventative management systems (Blackburn, 2003). This has meant a change in the scientifically flawed approach of using microbiological testing as the sole means of ensuring that microbiological hazards are under control. Microbiological testing is now becoming integrated within these preventative management systems, and it can have a number of roles in monitoring, validation and verification. In addition, microbiological testing may be required to demonstrate compliance with microbiological criteria (whether standards, guidelines or specifications) and in the investigation of a suspected breakdown of process control.

Microbiological safety and quality of food are often separated and, in many cases, there is a clear reason for doing so: most foodborne pathogens will not be responsible for spoiling food (unsafe food may appear organoleptically unchanged) and most spoilage microorganisms are not pathogenic. However, the cases that illustrate safety and spoilage as being a continuum (e.g. the growth of spoilage bacteria or moulds that, under certain circumstances, can lead to safety hazards – the formation of biogenic amines and mycotoxins, respectively) together with the lack of a clear distinction between the safety and quality of food by the consumer, which is often mirrored in the eyes of the law, goes to emphasise the importance of managing safety and quality in an integrated way (Blackburn, 2006).

Food safety and quality need to be managed across the entire food supply chain, often referred to as 'farm-to-fork', 'plough-to-plate' or 'stable-to-table', and controlling microbial hazards in primary production can help to reduce the burden of preservation at subsequent stages of the supply chain and the risk of harm to the consumer. This is particularly important since it has been estimated that 75 % of emerging pathogens are zoonotic and that zoonotic pathogens are twice as likely to be associated with emerging diseases (Taylor *et al.*, 2001). It has been suggested that consumer, demographic and environmental trends and changes in farming practices and food manufacturing are likely to lead to an increased risk of foodborne illness in the future (Armstrong *et al.*, 1996; Käferstein and Abdussalam, 1999). This, in the context of an incredibly competitive global industry, where there is continual pressure to reduce costs, places great importance on the management of food safety and quality. The aim of this chapter is to consider food safety and quality management systems and the role that microbiological testing plays therein.

1.2 Control systems used in the food industry

1.2.1 Quality Control and Quality Assurance

Quality Control (QC) and Quality Assurance (QA) are two different approaches to delivering safety; both systems share tools, but the emphasis is very different (Table 1.1). Both approaches are legitimate, but they need totally different organisations, structures, skills, resource and ways of working (Kilsby, 2001). QC is a reactive approach influenced by the pressures in the external world. In a QC organisation, the emphasis is on measurement, which needs to be robust and statistically relevant, and the focus is on legal and commercial issues. In contrast, QA is a preventative approach driven by the company's internal standards. The emphasis is on operational procedures, which must be robust and reviewed regularly, and the focus is on the consumer.

There are several problems associated with relying on testing for product safety assurance (van Schothorst and Jongeneel, 1994). In order to apply

Table 1.1 A comparison of Quality Assurance and Quality Control

Factor	Quality Assurance	Quality Control
Approach	Preventative	Reactive
Reliance for delivering safety	Central standards and processes	Measurement
Focus	Consumer	Legal and commercial issues

Source: Kilsby (2001).

any statistical interpretation to the results, the contaminant should be distributed homogeneously throughout the batch. Since microbiological hazards are usually distributed heterogeneously, this means that there is often a major discrepancy between the microbiological status of the batch and the microbial test results (ICMSF, 1986). Even if the microbial distribution is homogeneous, testing a sufficient number of sample units for all the relevant hazards to obtain meaningful information may still be prohibitive. Perhaps most significantly, microbiological testing detects only the effects and neither identifies nor controls the causes.

As a consequence, there has been an inexorable move from QC to QA in the management of microbiological hazards in food, with the focus on preventative control measures rather than finished product testing. Although microbiological analysis has subsequently borne the brunt of much denigration, it still has a vital role to play as part of a QA framework, albeit with a shift in application and emphasis.

1.2.2 Product design

Food product design can be defined as the process and formulation factors intended to give the product its typical characteristics and allow it to meet customer expectation. Microbial contamination can then be defined as the presence of types or numbers of microorganisms not envisaged in the product design.

In food manufacture, the overriding microbiological concern is that of safety. Safety assurance is best obtained by focusing on 'safety by design', with a combination of formulation and processing conditions to ensure that pathogenic microorganisms are controlled (in the design). Provided that the necessary PRPs are in place, then the HACCP system is used to ensure that the safe design is implemented and that 'operational safety' is maintained. However, from a business perspective, the control of spoilage is also important and the use of a 'stable by design' approach (Blackburn, 2006) and implementation by means of HACCP-like principles, together with all the associated PRPs, can also be harnessed to help manage food quality.

1.2.3 Hazard Analysis Critical Control Point (HACCP)

HACCP is a food safety management system that uses the approach of identifying, evaluating and controlling hazards that are significant for food safety. HACCP was originally designed for manufacturing environments and it has been standardised (Codex Alimentarius Commission [CAC], 1997) and widely adopted (Mayes and Mortimore, 2001; Mortimore and Mayes, 2002). The HACCP process comprises seven principles, which are further broken down into tasks (Table 1.2).

Although it is widely accepted that HACCP is the most effective means of producing safe acceptable food, the classical HACCP approach is not

Table 1.2 HACCP principles and tasks required for their application

Principle	Task
	1. Assemble HACCP team 2. Describe product 3. Identify intended use 4. Construct flow diagram 5. On-site confirmation of flow diagram
1. Conduct a hazard analysis	6. List all potential hazards associated with each step, conduct a hazard analysis, and consider any measures to control identified hazards
2. Determine the Critical Control Points (CCPs)	7. Determine Critical Control Points
3. Establish critical limit(s)	8. Establish critical limits for each CCP
4. Establish a system to monitor control of the CCP	9. Establish a monitoring system for each CCP
5. Establish the corrective action to be taken when monitoring indicates that a particular CCP is not under control	10. Establish corrective actions
6. Establish procedures for verification to confirm that the HACCP system is working effectively	11. Establish verification procedures
7. Establish documentation concerning all procedures and records appropriate to these principles and their application	12. Establish documentation and record keeping

Source: CAC (1997).

fully applicable to other parts of the food supply chain. However, increasingly the HACCP approach is being modified to encompass the entire farm-to-fork continuum (Fig. 1.1). In the EU, this is being driven by legislation that has made HACCP mandatory in all food production and processing businesses (post-primary production) from 1 January 2006.

In the food service sector, adaptations of the HACCP approach in the form of a more 'user-friendly' format have been attempted and this has led to the development of Assured Safe Catering (ASC), Systematic Assessment of the Food Environment (SAFE) and generic HACCP approaches (Griffith, 2002). In response to the changes in European legislation, the UK Food Standards Agency (FSA) have produced an information pack 'Safer food better business' to help small catering businesses, such as restaurants, cafés and food takeaway establishments. The FSA Scotland has also developed 'CookSafe' which is designed to help catering businesses understand and implement a HACCP-based system.

Although primary production is currently exempt from the mandatory HACCP requirements of the European Commission, HACCP principles

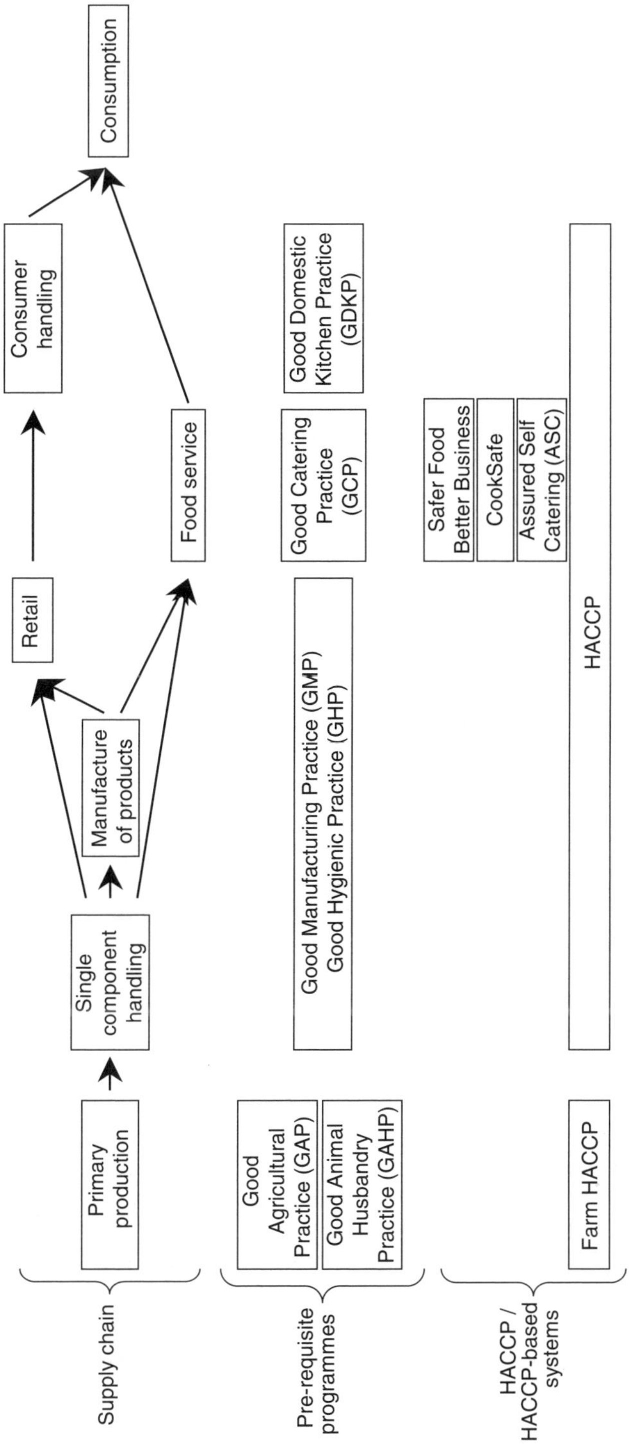

Fig. 1.1 Examples of management systems used in the food supply chain.

have been applied to farm production, and potential benefits for improving the health status of livestock, for reducing or controlling foodborne pathogens and for QA have been reported (Johnston, 2002). Hazard analysis and the identification of potential control measures can certainly be applied to the farm situation, but the identification of true CCPs is problematic, because the effects are often not quantitative and hence critical limits and a relevant monitoring system are difficult to establish (Maunsell and Bolton, 2004). However, a HACCP-based approach to food safety at the farm level can identify good farming practices (GFPs), which can be implemented in a similar way to CCPs (Maunsell and Bolton, 2004).

The use of HACCP and/or HACCP-based approaches has been applied to the meat and poultry industries. In the USA, there has been widespread recommendation for federal meat and poultry inspection to reduce its reliance on organoleptic inspection of carcasses and move to prevention-oriented systems based on public health risk (Cates *et al.*, 2001). Resulting HACCP-based slaughter inspection models have been developed, and plants operating under these models have been shown to maintain or even improve food safety, as measured by results of *Salmonella* and *Escherichia coli* testing (Cates *et al.*, 2001).

HACCP studies have indicated that feed is a critical point in the production of eggs free of pathogenic microorganisms (Cabo Verde *et al.*, 2004), and shell-egg washing is normally defined as a CCP in the commercial shell-egg washing and grading processes used in some countries (Srikaeo and Hourigan, 2002).

In addition to the use of HACCP or HACCP-based approaches across the food supply chain, these principles need to be implemented for new food products during the transition from product development to manufacture (from ‘concept-to-consumer’). To this end, the HACCP process often starts with a product/process concept, where design control points (DCPs) rather than CCPs are identified.

HACCP is targeted primarily at safety, but the same principle can be applied to the control of microbiological spoilage. In Australia in the 1990s, the recognition that customers expect safe food but discriminate on quality, when making their buying decision, led to the development of two key, voluntary, third-party certified standards focused on using HACCP principles for both quality and safety (Peters, 1998). Both standards were developed in 1995 after significant research into customer expectations, and small-to-medium size business development capabilities. Customer-defined product specifications became the key to developing these HACCP-based QA standards. The HACCP principles are used to identify quality control points (QCPs) and quality points (QPs) in the process. However, it should be noted that one of the causes attributed to failure to implement HACCP successfully has been the tendency to include quality as well as safety issues (Mortimore and Mayes, 2001).

Regardless of whether microbiological spoilage hazards are considered during the HACCP study, their control will need to be considered prior to

HACCP implementation. Indeed, microbiological control based on safety alone could well lead to the proliferation of spoilage microorganisms. In many cases, the critical limits of CCPs may need to be more severe to control microbial spoilage hazards, as compared with safety hazards alone. A good example is in the setting of heat processes, where a minimum treatment is likely to be defined by the most heat-tolerant pathogen that needs to be controlled, but relevant spoilage organisms often have greater heat resistance, thus requiring higher temperatures and/or longer treatment times (Gaze, 2005).

1.2.4 Pre-requisite programmes

It is generally agreed that the most successful implementation of HACCP is done within an environment of well-managed PRPs (Mortimore and Mayes, 2002). This is highlighted by the fact that the confusion surrounding the relationship between PRPs and HACCP is considered to be one of the causes of failure in successful HACCP implementation (Mortimore and Mayes, 2001). Safety and quality control points identified during the HACCP study should be separated in order to deal with the non-safety points through PRPs and the safety-critical points through CCPs (Suwanrangsi, 2001).

Although definitions vary, the concept of PRPs does not differ significantly from what may be termed good manufacturing practices (GMPs). GMP is concerned with the general (i.e. non-product specific) policies, practices, procedures, processes and other precautions that are required to consistently yield safe, suitable foods of uniform quality. Good hygienic practice (GHP) is the part of GMP that is concerned with the precautions needed to ensure appropriate hygiene and, as such, tends to focus on the pre-requisites required for HACCP (Table 1.3). Although GMP cannot substitute for a CCP, collectively it can minimise the potential for hazards to occur, thus eliminating the need for a CCP. The implementation of effective GMP will control 'general' or 'establishment' hazards, many of which would include potential spoilage microorganisms that would otherwise have to be controlled by a CCP. Failure to have GMP in place will inevitably lead to a large number of CCPs in the HACCP plan, covering both 'general/establishment' hazards and product-specific ones.

QA processes should prevent microbiological contamination, but these systems are never perfect all of the time, and sometimes there is a failure of control (or a failure to recognise hazards and implement the necessary control) at a particular stage in the food supply chain. Therefore, the possibility of unsafe or spoiled product reaching the market must be addressed and post-launch management systems need to be in place to deal with these situations. This could include monitoring systems, e.g. customer complaints, traceability and recall procedures (Venugopal *et al.*, 1996; EC, 2002).

Prerequisite programmes are also recognised for other parts of the food supply chain (Fig. 1.1). For catering/food service businesses, good catering

Table 1.3 Examples of good hygienic practices (GHPs)

Area, practice or procedure	Scope
Food manufacturing premises	Hygienic design and construction
Machinery	Hygienic design and construction Proper use Maintenance
Cleaning and disinfection procedures	Efficacy Frequency
General hygienic and safety practices	Microbial quality of raw materials Supplier Quality Assurance Hygienic operation of each process step Hygiene of personnel Food hygiene training Pest control Water control Air control Product rework procedures Product recall Waste management Labelling and traceability systems Transportation

Source: Brown (2002); Mortimore and Mayes (2002).

practice (GCP) is analogous to GMP (Griffith, 2002). With regard to primary production, good animal husbandry practices (GAHP) (Hafez, 1999; Johnston, 2002; CAST, 2004), good agricultural practices (GAP) (FDA, 1998) and good farming practices (GFP) (Maunsell and Bolton, 2004) have been documented.

The importance of GAHP being implemented pre-harvest to control *Salmonella* and *Campylobacter* within the poultry industry has been highlighted (Hafez, 1999). This requires effective hygiene measures, particularly cleaning, applied to poultry houses, feed mills and hatcheries and to the catching and transport of live poultry. To reduce carcass contamination during processing GMPs, particularly thorough schedules for cleaning and disinfection of equipment and plant, are also essential (Hafez, 1999).

1.2.5 Risk assessment

In a food safety context, the formalised meaning of risk assessment has evolved primarily from the CAC definitions (FAO/WHO, 1995), where risk assessment is the primary science-based part of risk analysis, dealing specifically with condensing scientific data to an assessment of the human health risk related to the specific foodborne hazard, and risk analysis according to these definitions also comprises risk management and risk communication (Schlundt, 2000). Microbiological risk assessment approaches have been utilised primarily by regulatory bodies and researchers in order to determine the best and/or most effective risk management options (Gale, 1996; Lawrence, 1997; Gofti *et al.*, 1999; Schlundt, 2000; Kelly *et al.*, 2003; McLauchlin *et al.*, 2004.) However, the food and beverage industry is beginning to apply these risk assessment approaches in order to help better manage microbial pathogens (Membré *et al.*, 2005, 2006; Syposs *et al.*, 2005). Risk assessment approaches have particular value when a risk management decision is required related to a critical and complex food safety issue where there may be a high degree of uncertainty and variability in the relevant information and data. For example, risk assessment outputs have enabled decisions to be made about heat process optimisation (Membré *et al.*, 2006) and shelf-life determination (Membré *et al.*, 2005). Risk assessment approaches are not without disadvantages in that they can be time consuming and require a great deal of detailed knowledge, as well as considerable skill, to implement.

1.2.6 Microbiological hazard management

Ross and McMeekin (2002) represented the potential interplay of the fundamental elements of pathogen management: the building blocks of scientific/industry knowledge; risk assessment as a decision support tool; and HACCP as the mechanism for translating quantitative, risk-based, food safety strategies into practical pathogen management systems to achieve the overall objective of 'safe' food. Blackburn (2006) went on to propose that microbial quality (spoilage) management could also be transposed onto this model, with the overall objective being extended to 'safe and stable' food and, in addition to Food Safety Objectives (FSOs) as targets, this could be extended to the concept of an acceptable level of spoilage (e.g. spoilage/defect/failure rate). Scientific and industry knowledge would be required to identify both the relevant microbial safety and spoilage hazards and the possible means for their control. HACCP and/or HACCP-like principles, together with associated PRPs, would be the mechanism for achieving this objective, with CCPs in combination with QCPs to control the safety and spoilage hazards. The decision-making tools linking scientific/industry knowledge and HACCP/PRPs would include risk assessment (Ross and McMeekin, 2002), as well as the more commonly used challenge testing, shelf-life assessments and predictive models (Blackburn, 2006).

1.3 Role of microbiological testing

1.3.1 Challenge testing and predictive models

Challenge testing and shelf-life assessments are often required for determining the safety and/or stability of a food product (CCFRA, 2004a). In essence, microbiological challenge testing involves the inoculation of a food with specific hazardous organisms and monitoring their growth, survival or death during storage under a defined set of conditions and/or after specific process steps. This type of test can be helpful in determining the ability of a food to support the growth of pathogens and in the validation of processes that are intended to deliver a defined degree of lethality against a target organism (IFT, 2001). However, there are a number of important factors that must be considered when designing and implementing a challenge test, including: selection of appropriate challenge organisms; inoculum level; duration and number of analyses; packaging and storage conditions; methodology; and data interpretation (Vestergaard, 2001).

Shelf-life assessment is more frequently associated with the use of 'naturally' contaminated food samples in order to determine the extent to which shelf-life is limited by the growth of spoilage microorganisms. Although from a practical perspective the assessment of safety and stability/shelf-life can be, and may have to be, determined separately, it is important that the results are combined so that decisions regarding the food product formulation, processing and storage conditions are made with all the relevant information.

This type of microbiological testing is expensive, time consuming and very product/process-specific. Therefore, it may have to be repeated, if the product and/or process is modified. These factors have been some of the main drivers for developments in the field of predictive microbiology, the concept and history of which have been reviewed in detail by McMeekin *et al.* (2002). Mathematical microbiology models can help describe the growth, survival and death of microorganisms in food, as affected by the intrinsic factors (characteristics of the food, e.g. pH, a_w, preservatives) and extrinsic factors (characteristics of the environment, e.g. temperature, gas atmosphere). In addition to the numerous predictive microbiology models that have been published, several software systems incorporating microbiology models have been produced, some of which are commercially or freely available (Blackburn, 2003).

Predictive models have the potential for a range of safety and spoilage applications, including shelf-life determination and extension, distribution and storage condition assessment, product formulation and re-formulation, process design, risk assessment, GMP, HACCP and as an alternative or adjunct to challenge testing (Vestergaard, 2001). However, the extent to which the application of predictive models can be relied upon has been the subject of considerable debate. The US Department of Agriculture Food Safety and Inspection Service (USDA FSIS) has gone so far as to say that 'it is not possible or appropriate to rely solely upon a predictive modelling program to determine the safety of foods and the effectiveness of processing systems' (FSIS, 2002). FSIS also state that predictive models do not replace the need for challenge testing. The Institute of Food Technologists (IFT) take a more balanced view, highlighting the value of combining predictive models with challenge tests and the potential for using challenge-test data for model development and validation (IFT, 2001). Taking a pragmatic approach, it is clear that predictive models can provide a powerful source of information and a tool for its practical application, whilst not completely negating the need for microbiological testing. Utilising predictions requires a considerable amount of knowledge of the food, the process, the microorganism of concern and the model itself, and therefore models do not replace the judgement of a trained and experienced microbiologist.

Developments in information technology have also led to the construction of microbiological expert systems. Originating from the artificial intelligence field of research, expert systems are essentially computer programs that attempt to emulate the performance of human experts. As an example of what can be achieved, an expert system for ready-to-eat meals has been described (Adair and Briggs, 1993). The system contained databases on product design, manufacturing and microorganisms, and several predictive bacterial growth models. In response to user inputs, a rule base was applied and the output comprised the required assembly and packaging conditions, the minimum thermal process and the maximum shelf-life to ensure a microbiologically safe product.

1.3.2 Microbiological testing in HACCP

Successful implementation of a fully validated HACCP study means that the supposed reliance on microbiological testing, with all its sampling limitations, is relinquished, resulting in a significant reduction in the volume of testing. Some in the food industry went so far as to surmise that microbiological testing would become obsolete (Struijk, 1996). In reality, however, microbiology testing has continued, albeit with a shift in application and emphasis and accompanying changes in the role of the microbiologist (Kvenberg and Schwalm, 2000).

It is recognised that microbiological testing can serve several functions within HACCP (de Boer and Beumer, 1999) and that the extent and scope of microbial testing is likely to vary with differences in facilities and equipment, the scales of processes and the types of products involved (Brown *et al.*, 2000).

Hazard analysis

The first step of the HACCP process is to conduct a hazard analysis. This is the process of collecting and evaluating information on hazards and conditions leading to their presence in order to decide which are significant for food safety and therefore should be addressed in the HACCP plan (CAC, 1997). For each raw material and process step, this includes consideration of the likely occurrence of the hazards, qualitative/quantitative evaluation of the hazards, survival or multiplication of the hazardous agents and identification of appropriate preventative measures.

Published sources of microbiological data, including epidemiological and surveillance data, together with knowledge gained through commercial experience can provide the HACCP team with relevant information on the likely hazards associated with the product and process. However, when existing data are lacking, microbiological testing is often needed (Kvenberg and Schwalm, 2000). This may involve determining the incidence of pathogens or indicator organisms in raw materials, the presence of pathogens (e.g. *Listeria monocytogenes*) in the environment and microbial loads in foods and on equipment (Stier, 1993). Here the links with PRPs are important.

Validation of the technical accuracy of the hazard analysis and effectiveness of the preventative measures are important before the HACCP study is finalised and implemented. Examples where microbiological methods may be used for validation include pre-operation checks of cleaning and sanitising, screening of sensitive raw materials, challenge testing and monitoring of critical sites for microbiological build-up during processing (Hall, 1994).

Predictive models can be used to help assess the risk and determine the consequence of a microbiological hazard in food during the different process steps (Elliott, 1996). The advantage of using predictive models is that the effect of adjusting the product formulation and/or processing parameters can be assessed rapidly. To obtain predictions from most models, a starting concentration of microorganisms is required and here information from microbial testing can be of value. Where predictive models are appropriate, this may allow a reduction in, or negate the need for, challenge testing, which might otherwise be required to provide this information. Predictive models have a particularly important role to play in obtaining information about microorganisms that require specialist facilities for data generation. For example, models for *Clostridium botulinum* have been used to predict the safe refrigerated shelf-life of sous vide-type food products (Baker and Genigeorgis, 1993) and vacuum-packed fish (Hyytiä *et al.*, 1999). Even though the FSIS (2002) has stated that 'generally, a microbial pathogen computer model (MPCM) would not be the only documentation relied upon to support an element of a HACCP plan' it is conceded that 'in certain circumstances, a microbiologist . . . may determine the MPCM program is the most appropriate (and sole) source of data . . .', and *Cl. botulinum* is used as an example.

The use of molecular characterisation techniques has further increased the microbiologist's armoury, and epidemiological tracking of strains can provide a more in-depth knowledge of the food process. This may enable the determination of sites of cross-contamination, or sites where strains appear and disappear, thus pin-pointing the positions contributing to the final flora of the product, permitting more precise identification of CCPs (Dodd, 1994).

Critical Control Points (CCPs)
The second HACCP principle involves determining CCPs. A CCP is a step at which control can be applied and is essential to prevent or eliminate a food safety hazard or reduce it to an acceptable level (CAC, 1997). The third and fourth principles involve setting critical limits, which are criteria that separate acceptability from unacceptability (CAC, 1997), and establishing a CCP monitoring system, which involves conducting a planned sequence of observations or measurements of control parameters to assess whether a CCP is under control (CAC, 1997).

Challenge testing and predictive models can provide useful information for both the determination of CCPs and the setting of critical limits (Baker, 1995; Elliott, 1996; Griffiths, 1997; Fujikawa and Kokubo, 2001; IFT, 2001; FSIS, 2002). This information is often required to set maximum times and temperatures for storage conditions and minimum times and temperatures for heat processes. Microbial testing can play a major role in the validation of CCPs to demonstrate their effectiveness (van Schothorst, 1998; Blackburn, 2000; Kvenberg and Schwalm, 2000). For safe product design, a defined reduction in target organisms may be required, either in one CCP or over a series of process steps. Quantitative data may be required to demonstrate that the process can deliver the defined level of microbial kill or that the end product meets the specification for safety and/or stability. This is particularly true if unconventional or unique control measures and/or critical limits are used.

Predictive microbiology models can be used for 'what if' scenarios to provide an indication of the severity of problems caused by process deviations or the complete breakdown of any of the CCPs (FSIS, 2002). They can also be used to provide useful information on the assessment of equivalence of HACCP plans (Fujikawa and Kokubo, 2001). In many cases it is still necessary to conduct challenge tests to validate CCPs, since current models will not be appropriate for all situations that may be encountered in food production.

The HACCP process requires the establishment of systems to monitor all identified CCPs. In most cases, it is not feasible to use microbial testing to monitor CCPs due to the long analytical time, low method sensitivity and heterogeneous nature of most microbial contamination. However, there are some notable exceptions. The receipt of raw materials within defined microbiological specifications is often identified as a CCP. As a consequence, preventative measures are likely to focus on the supplier's own microbiology assurance procedures and may include a Certificate of Analysis for selected contaminants, with the use of 'in-house' laboratory testing to confirm acceptability and when selecting new suppliers.

The development of rapid methods based on microbial adenosine triphosphate (ATP) bioluminescence provided techniques that could be used in a CCP monitoring context. Such a test was shown to be an adequate means of assaying the microbial load of poultry carcasses, and it was stated that the assay had potential for monitoring the microbial load of carcasses at poultry-processing CCPs (Siragusa *et al.*, 1996). Although limited in their availability, other 'real-time' methods such as flow cytometry have been proposed for CCP monitoring (Griffiths, 1997).

Rather than detecting microorganisms *per se*, ATP bioluminescence kits are widely used for checking the sanitation of equipment, as judged by quantifying the presence of non-microbial ATP. Since results from these methods can be obtained in only a few minutes, there is sufficient time for equipment to be re-sanitised before production begins, thus preventing contamination. Consequently, sanitation of equipment and monitoring with ATP bioluminescence may be identified as a CCP, although this is frequently covered as part of a GHP programme. Although care in the application of these methods is required to avoid being lulled into a false sense of security (Stier, 1993), the methodology can have a beneficial impact in demonstrating to cleaning staff the importance of their role.

Within the poultry industry, bacteriological and serological monitoring of flocks, accompanied by high standards of animal management, are being implemented to prevent the introduction and spread of *Salmonella* in breeder flocks (Hafez, 1999). It has been suggested that, as a pre-harvest control measure, infected flocks should be destroyed (Hafez, 1999).

Verification and review

The sixth HACCP principle involves the establishment of verification procedures to confirm that the HACCP system, once operational, is working effectively. Finished product testing can be one of the means by which its successful implementation is verified. Although there is a view that verification does not need to include microbial testing, because it is largely accomplished by reviewing HACCP monitoring records (Kvenberg and Schwalm, 2000), it is clear that microbial testing is often used in verification programmes for both incoming ingredients and finished products. In theory, a well-functioning HACCP plan should only require occasional testing as part of the verification process. However, sometimes local legislation, customer requirements or the companies' own standards demand a higher level of testing (Stier, 1993).

Microbial testing for verification purposes may involve pathogen testing, although quantitative indicators can provide a much more effective tool for verifying that HACCP is properly implemented (Swanson and Anderson, 2000). The choice of appropriate indicators is product- and process-specific. For example, testing for coliforms provides an effective verification technique for pasteurised milk and water potability. However, in certain applications, finished product testing for even indicator organisms provides no meaningful data, e.g. in the case of canned products.

In the meat and poultry industries, the importance of testing carcasses periodically to verify that HACCP is working has been highlighted (Quinn and Marriott, 2002). As part of an external audit of three Greek meat plants, microbiological examination of samples related to general hygiene control points and the CCPs of the system was conducted (Metaxopoulos *et al.*, 2003). The data helped to show that the GHP and HACCP systems were adequate, but that more effort was required on the microbiological quality of incoming materials and processing. During a review of the role of microbiological testing in systems for assuring the safety of beef, it was concluded that the main purpose should be to implement and maintain effective HACCP systems (Brown *et al.*, 2000). It was also concluded that testing for indicator organisms was necessary for that purpose, with the microbiological performance of the processes involved being assessed against appropriate FSOs (Brown *et al.*, 2000). This would include investigating the microbiological effects of the operations in or affecting a process, to validate the procedures and to verify the maintenance of control over the microbiological condition of the product (Brown *et al.*, 2000).

In a cost–benefit analysis of HACCP implementation in the Mexican meat industry, the main benefit reported was a reduction in microbial counts (Maldonado *et al.*, 2005). In the USA, to verify that PRP/HACCP systems are effective in controlling pathogen contamination of raw meat and poultry products, product-specific *Salmonella* performance standards must be met by slaughter establishments and establishments producing raw ground products (Rosc *et al.*, 2002). Test results showed that *Salmonella* prevalence in most product categories was lower after the implementation of PRP/HACCP than in pre-PRP/HACCP baseline studies.

Microbiological data can provide valuable sources of information for trend analysis and statistical process control and, for this purpose, they are generally under-utilised. Here, quantitative tests are more informative to a processor than negative pathogen tests, since trends can be examined and early warnings of problems or loss of control can be obtained. Loss of operational control may give rise to dramatic changes in microbiological test results; however, it may manifest itself in much more subtle and gradual changes in microbial counts, only detectable via trend analysis. If microbiological data are examined proactively it is then conceivable for microbiological problems to be prevented, making the exercise compatible with the QA approach to food safety. The goal should be for data to be directed towards process improvement, and microbiological analyses should not be done solely for the sake of generating data. By applying statistical techniques, it has been possible to demonstrate the usefulness of particular microbiological tests for verifying HACCP in the final phases of poultry meat production (González-Miret *et al.*, 2001).

Many organisations have test results and baseline data for indicator organisms that were collected over many years. New criteria that replace historic baselines must be reviewed carefully to ensure that the processor retains a solid understanding of the microbial profiles of their processes and products. For example, the implementation of a new test method can make previously developed baseline data worthless, if the new test protocol does not provide equivalent results.

Although not defined as a separate stage, validation of the HACCP study is a vital part of verification and is concerned with obtaining evidence that the elements of the HACCP plan are effective (CAC, 1997), i.e. 'doing the right things'. This contrasts with verification, which is the application of methods, procedures, tests and other evaluations, in addition to monitoring, to determine compliance with the HACCP plan (CAC, 1997), i.e. 'doing things right'. Validation of the technical accuracy of the hazard analysis and effectiveness of the preventative measures is particularly important before the HACCP study is finalised and implemented. There are a number of approaches to validation, including the use of scientific publications, historical knowledge, regulatory documents, experimental trials, scientific models, operational data and survey results and, generally, a combination of these approaches will be used (Scott, 2005).

Examples where microbiological methods may be used for validation include pre-operational checks on cleaning and sanitising, screening of sensitive raw materials, challenge testing to assess product design, and monitoring of critical sites for microbial build-up during processing (Hall, 1994). For safe product design, a defined reduction (e.g. 5 or 6 $\log_{10}$ units) in target organisms may be required, delivered either at one CCP or over a series of process steps. Quantitative data may be required to demonstrate that the process can deliver the specified level of microbial kill or that the end product meets the requirements for safety and/or stability. Experimental trials to document the adequacy of a control measure may involve laboratory challenge testing or in-plant challenge testing, using surrogate microorganisms (Scott, 2005).

In the egg industry *Salmonella* testing has been used to validate the identified CCPs of culling at various stages of handling, washing and marketing, following macroscopic classification of the eggs as cracked and/or dirty (Poppe *et al.*, 1998). Challenge testing using *Salmonella* and *Campylobacter* has provided validation data for irradiation as an intervention measure (and possible CCP) in the production of pathogen-free eggs (Cabo Verde *et al.*, 2004).

Quantitative microbial testing (total viable counts and counts of *E. coli*, Enterobacteriaceae and coliforms) has been used to validate a reduction in the incidence of faecal contamination in a beef slaughter plant following the implementation of a novel information technology-based online monitoring system as part of dehiding and evisceration CCPs (Tergney and Bolton, 2006). The use of microbiological testing to enumerate *Campylobacter* has demonstrated the reduction (1–2 $\log_{10}$ cfu/ml of carcass rinse) achievable using a variety of antimicrobial processes applied during the processing of broiler chickens (Oyarzabal, 2005).

Microbial methods, particularly those involving molecular characterisation, can be useful in answering questions that may arise as part of the

HACCP validation exercise. For example, if a hazardous organism appears in a product at a point in the production line beyond the CCP designed to control it, it could mean failure of the CCP, or it could indicate post-process contamination (Dodd, 1994).

HACCP is a 'living' system and therefore review of the HACCP plan is an important aspect to ensure that it remains fully valid and implemented. A formal review should be triggered if there is a change to the product or process but, if this is not the case, then it should be reviewed at regular intervals, e.g. annually. In these reviews, it may be decided that microbiological data are required to assess the significance of a new hazard or to ensure that the CCPs can still control the existing hazards in the light of any proposed changes to the product or process.

1.3.3 Microbiological testing and pre-requisite programmes (PRPs)

Microbiological testing has been shown to have a role to play within the PRPs targeted at primary production. For example, *Salmonella* testing has been used to show that much pre-harvest *Salmonella enterica* infection in pigs occurs immediately before slaughter, during lairage in the contaminated abattoir holding pens (Rostagno *et al.*, 2005). Subsequently, a potential intervention strategy to reduce the prevalence of *S. enterica*-positive pigs at slaughter, which consisted of resting pigs prior to slaughter in their transport vehicle instead of in the abattoir holding pen, was evaluated. In addition, programmes for monitoring *Salmonella* in the pork production chain have begun in several European countries, and this has included the use of serological assays for the presence of antibodies against *Salmonella*. The aim is to give farms a particular status, reflecting a certain level of prevalence of *Salmonella*, in order to make decisions on the implementation of certain hygiene measures and slaughtering logistics (Achterberg *et al.*, 2005).

GMP/GHP systems have been found to be effective provided that they are: well documented with standard operating procedures (SOPs); fully implemented; and include monitoring records and verification procedures (Kvenberg and Schwalm, 2000). From a manufacturing perspective, there are several key sources of microbial contamination of a product that require control: raw materials, equipment, process/production environment and people. The extent to which microbial testing plays a role, and the degree of sampling required, should reflect the category of risk associated with the particular raw material, area or operation. For example, a 'high-risk' raw material that is added to a product post-pasteurisation may require more testing to verify compliance with a specification than one added before pasteurisation. In addition, the food contact surfaces and air quality in a 'high-care hygiene' area may also require a higher level of sampling.

Determining the sources and significance of hazards

Whether GMP/GHP or HACCP eventually controls the hazards, hazard identification is an important first step to ensure safe food products. Microbiological testing can play an important role in identifying potential hazards, as well as linking them to a source, assessing their significance for the final product and verifying that controls are effective and implemented successfully. For example, in a meat processing plant, microbial testing demonstrated that the most important factor contributing to contamination of ground beef and retail cuts was from incoming raw materials obtained from different suppliers (Eisel *et al.*, 1997). Environmental sources of contamination were shown not to be a significant source of overall microbial contamination, although it was demonstrated that cleaning and sanitation programmes and safe handling were still important.

In a similar way, predictive microbiology models have helped determine the significance of different microbial hazards in establishing the shelf-life of pasteurised milk (Griffiths and Phillips, 1988). This exercise highlighted the importance of good hygienic processing to reduce post-pasteurisation contamination.

The use of molecular characterisation techniques has further increased the microbiologist's armoury, and epidemiological tracking of strains can provide a more in-depth knowledge of the food process. This may enable sites of contamination to be determined, indicating where controls are required, whether via GHP or CCPs (Dodd, 1994).

Raw materials

The quality of raw materials can affect the overall quality of the finished product. Microbiological testing may often be required to verify that raw materials are delivered to the agreed specification and as a means of monitoring or selecting suppliers. Although frequently covered as part of a HACCP study, raw material specifications may not be identified as a CCP, in which case they are usually covered by GMPs/GHPs. Testing may then involve confirming the absence of specific pathogens or that indicator organisms are within defined limits.

Raw materials may also be the means of introducing contamination into the food-processing environment. This is particularly important from the point of view of controlling contamination in animal husbandry. Although control is particularly difficult in this environment, preventing feed contaminated with pathogens being introduced into, e.g. broiler flocks is an important control point. Here, microbial testing can be a means of verifying that pathogens are absent and that the batch conforms to agreed microbial criteria.

Equipment

Food contact surfaces are particularly important as a potential source of contamination, and sanitation (cleaning and disinfection) is the major day-to-day control measure. When undertaken correctly, sanitation programmes have been shown to be cost-effective and easy to manage and, if diligently applied, can significantly reduce the risk of microbial contamination (Holah and Thorpe, 2002). In this regard, microbial testing is useful in the validating standard sanitary operating procedures (SSOPs) and verifying that they have been carried out effectively.

Although a microbial surface may not be a source of contamination after sanitation, food residue on that surface during production can provide the opportunity for microbial growth, which could then be a source of re-contamination for the product. With production pressures to keep lines running as long as possible between SSOPs, microbial testing can provide valuable information to maximise line efficiency without compromising the microbial safety or quality of the product.

Environment

The food production/processing environment can be a source of general contamination. Many surfaces not directly in contact with food may harbour microorganisms, e.g. non-food-contact equipment surfaces, walls, floors, drains, overhead structures. These microorganisms can then be transferred to the food in the air via water droplets and dust. Sampling of this environment can provide information on the likely presence and incidence of pathogens, their distribution in relation to processing lines and thus the risk of product contamination (Cordier, 2002). This allows preventative measures to be established in the framework of GHP, such as layout of processing lines and zoning within the factory.

Sampling the cleaning equipment is a very useful index of what is actually present in a production environment, because cleaning 'collects' dirt and bacteria from all parts of the factory, e.g. floor mops, brushes, vacuum cleaners (Fraser, 2002). In a similar way, sampling of drains also gives a better chance of determining whether a particular pathogen is present in the production environment, e.g. *L. monocytogenes*. This can often be a better approach than sampling finished products. In addition, other wet areas such as sinks, taps, cleaning cloths and brushes, and boot-washing baths should be checked routinely. Aerosols can be created from such areas and contaminants can find their way into products on the manufacturing line. Testing for indicator organisms generally gives the most useful information on environmental hygiene, an exception to this being the testing for *L. monocytogenes* in high-hygiene environments.

Air quality can be a good index of the overall sanitary condition of a production environment. Air can contain microbes from both external and internal sources, depending on the set-up of the factory. For example, if filters and air-conditioning units are not properly maintained, microorganisms can enter the plant from outside. Internal contamination can occur from skin particles shed by factory personnel, dust particles from packaging materials and aerosols created during either production or on-going cooling or cleaning processes. The records from routine monitoring of air quality can build a picture of the general standard of air hygiene in a plant (and identify the areas/sources of highest contamination). Generally, the methods for measuring air quality are either settle plates or the use of a portable, battery-operated, air sampler. Because airborne counts can fluctuate widely depending on activities around the area, it is important to note what is happening, e.g. cleaning, shift-change, in order to correlate data with events (Fraser, 2002).

The temperature of the production environment can obviously affect microbial proliferation and here predictive microbiology can play a role in GMP/GHP. For example, a dynamic Temperature Function Integration (TFI) model was used jointly by regulators and processors to develop justifiable criteria for the management of refrigeration during the production of hot- and warm-boned meat, the post-slaughter handling of ovine carcasses and the handling of offals (Armitage, 1997). Similarly, predicted lag times and growth rates of coliform bacteria have been used to support a proposal to alter the temperature of cutting rooms for chilled meat carcasses, as stipulated by public health authorities in several countries (Baker, 1995).

People/training

Food production staff and food handlers are a potential source of contamination for food products. For this reason, it is important that adequate training is given, and that proper supervision ensures adherence to all hygiene measures, particularly hand washing.

The use of microbiological testing should not be under-estimated as a part of hygiene training. The impact of seeing agar plates covered in colonies that have been isolated from swabs taken from hands pre-washing or surfaces pre-cleaning, and the reduction achieved following washing or sanitation, can be significant. The rapid results achievable by ATP bioluminescence can be particularly useful for the motivation and training of sanitation and production staff by providing a means for them to judge their own performance and by demonstrating the importance of their work. Regular swabbing of hands can also help to reinforce hygiene procedures.

1.3.4 Trouble-shooting and 'forensic' investigation

It has been pointed out that, in spite of meticulous adherence to HACCP-based good practices, occasional human, instrumental or operational hiatuses can and will occur (Struijk, 1996). Microbiological testing may be required for trouble-shooting and 'forensic' investigation in order to identify the cause of the problem and rectify it. Usually, the first action required is to identify and control the affected product. Microbiological testing may be appropriate to determine, or confirm, whether there is a microbiological problem and, if so, whether it is a safety or spoilage/quality incident. In combination with a review of the process records, particularly at CCPs, and any historical microbiological test data, it may be necessary to instigate a sampling and testing plan to determine how much product is affected. As speed is often critical, rapid test methods can play an important role (Stier, 1993). In addition, predictions from microbiological models may help to determine the extent of the problem (Fujikawa and Kokubo, 2001). Once this information has been obtained, decisions can be made regarding segregation, blocking, recall and salvaging of affected batches and the status of further production.

Microbiological analysis is often required to determine the cause or source of the problem, and the type and extent of testing required will vary enormously, depending on the situation. Rapid techniques like ATP bioluminescence can be useful trouble-shooting tools to quickly identify problem areas. Tests ranging from indicator organisms, through specific pathogen detection methods, to genetic finger-printing of strains may also be appropriate.

Following this immediate action, an assessment of the integrity of the HACCP plan is required. It has to be determined whether the HACCP has failed due to its validity or its implementation. Here again, microbiological analysis may have a role to play in any subsequent review and re-validation.

1.3.5 Microbiological testing and risk assessment

Risk assessment relies on the use of microbiological data derived from a variety of sources. For example, in order to determine how much of a risk cracked eggs were to human health, Agriculture and Agri-Food Canada requested that a risk assessment be conducted (Todd, 1996). On the basis of outbreak data, the main hazard in these eggs was identified as *Salmonella* and it was found that cracked eggs were 3–93 times more likely than uncracked eggs to cause outbreaks. Subsequent to this risk assessment, *Salmonella* test results were used to validate identified CCPs that had been proposed as part of a risk management plan (Poppe *et al.*, 1998).

Risk assessment approaches can identify gaps in knowledge that are crucial to providing better estimates of risk, and this may, in fact, lead to an increase in the level of microbiological testing (de Boer and Beumer, 1999). Assessing the risk posed by a 'new' or 'emerging' organism may also highlight deficiencies in current methodology, requiring the need for method development.

1.4 Applying microbiological testing

1.4.1 To test or not to test?

There are many reasons for performing microbiological tests, as described in the preceding text. Testing to meet microbiological criteria may be enforced by legislation (e.g. microbiological standards) or customers (e.g.

microbiological specifications); or identified as part of HACCP or PRPs (e.g. microbiological guidelines). However, one of the most useful questions to ask before deciding which type of test method is most suitable is 'what is the purpose of the test and how will the result be used?' If there is no clear justification for the test or a use for the result, with identified actions depending on the outcome of the test, then the testing is likely to be of no value. Asking this question will help to target testing more effectively and lead to the implementation of the most appropriate methods and more effective use of test results.

1.4.2 Choice of microbiological test

The numbers of pathogenic microorganisms in most raw materials and food products are usually low and so pathogen tests may provide little information of use for the implementation and maintenance of GMP and HACCP systems. Instead, the enumeration of so-called 'indicator organisms' has an important role. Indicator organisms are single species or groups that are indicative of the possible presence of pathogens. Although there is not necessarily a relationship between indicator and pathogen numbers, it can be generally assumed that the likely numbers of a pathogen will be less than those of the organisms indicative of its presence. It can also be assumed that reducing the numbers of indicator organisms will produce a similar reduction in the numbers of any corresponding pathogen (Brown *et al.*, 2000). For the same reasons indicator organisms can also provide a measure of post-pasteurisation contamination that might lead to pathogen contamination.

Since different indicator organisms imply the possible presence of different pathogens, there are several groups of tests that may be appropriate, e.g. total aerobic counts, coliforms, Enterobacteriaceae, *E. coli*, faecal streptococci and aeromonads (Brown *et al.*, 2000).

Microbiological methods can differ widely in their comparative advantages and disadvantages. These relative benefits and limitations may influence the choice of method for a particular task (Table 1.4 and Chapter 10). For example, for products with a short shelf-life, the rapidity with which a test result becomes available may be an important factor. However, when maximising the volume of sampled material is crucial, sample throughput and low cost/test may be higher on the priority list. In recent years, a plethora of rapid test kits have become available that, to a greater or lesser extent, have helped to expedite, simplify, miniaturise and automate methodology. The drive for standardisation, validation and international acceptance of methods, with regard to good laboratory practice and accreditation, means that this is often a constraint on method selection.

The role of a particular test will help to determine the relative importance of possible selection criteria (Table 1.4). For example, tests used for validation and verification may not need to be particularly rapid, but

Table 1.4 Factors that may influence the choice of microbiological method

Factor	Considerations
Performance	Sensitivity, specificity, accuracy, precision, reproducibility, repeatability
Time	Total test time (presumptive/confirmed results), 'hands-on' time, time constraints
Ease of use	Complexity, automation, robustness, training requirement, sample throughput, result interpretation
Standardisation	Validation, accreditation, international acceptance
Cost	Cost per test, capital outlay, equipment running cost, labour costs

accuracy may be more critical. Alternatively, speed is more likely to be important for methods used to monitor CCPs.

1.4.3 Requirements for testing facilities

The requirements for laboratory facilities will depend on the role of the microbiological tests being employed and the suitability of available methods. For example, some methods can be used in the factory (e.g. hygiene monitors) and do not require laboratory facilities. However, these methods are the exceptions and most require the use of a microbiological laboratory. It is important that, if such a laboratory is located at a production site, it should be isolated from production areas. Also, careful attention to laboratory design and layout will facilitate safety, assist good microbiological practice, ensure separation of activities and aid process flow (CCFRA, 1994). Attention needs to be paid to laboratory management, the equipment used, media preparation and the microbiological procedures themselves (CCFRA, 2004b). This is equally important whether testing is an 'in-house' capability or is out-sourced to a contract laboratory. The implementation of QA in the laboratory is covered in detail elsewhere (see Chapter 13).

1.4.4 Interpretation of test results

The importance of reliable laboratory management and test methodology is highlighted by the fact that 'false-positive', 'false-negative' and erroneous results can occur for a number of reasons, including: insufficiently experienced personnel; laboratory contamination or cross-contamination from controls; contamination from packaging (e.g. yeast/mould in aseptically packed beverage); the use of an unvalidated method; insufficient controls; variation in methodology between different laboratories; identification errors and misinterpretation (e.g. product residues/particles appearing similar to microbial colonies).

Inaccurate results can lead to poor management decisions, ranging from unnecessary blocking, disposal or recall of product to failure to identify a problem leading to a food poisoning outbreak. Thus, microbiological testing should be used to complement a preventative approach to food safety and quality, rather than being relied upon as the only approach to managing microbiological hazards.

1.5 Future trends

The increasing application of HACCP and HACCP-based approaches across the entire food supply chain is likely to continue, with particular attention to primary production. With increasing demands for minimally processed foods, the control of microbiological hazards will need to focus on reducing their incidence and levels at the primary production stage.

Although microbiological testing has a different role in complementing the preventative, QA approach to managing microbial safety and quality, there is still a need for method development. The 'holy grail' for microbial methodology would be the ability to analyse a batch of food non-destructively, on-line and with the required accuracy, sensitivity and specificity. Although current technical capabilities fall well short of this, the diversity of applications and user requirements and the shift from QC to QA mean that new and improved methods still have the potential to bring benefits. Methods that are faster, cheaper, easier to use, more accurate and/or more sensitive are likely to find welcome recipients. There is also a push to standardise methods and demonstrate equivalence to address the increasingly global market for food.

The rapid monitoring of hygiene using ATP bioluminescence is probably the best example of a 'microbiological' test applied in a GHP/HACCP environment. It is likely that the range of other compounds that could be used to monitor hygiene will extend and their application increase. Increased use of genetic finger-printing methods to better understand the microbial ecology of the factory, manufacturing line and production process may also bring benefits in permitting better control of the hazards. Biosensor development for very rapid pathogen detection and indicator organism enumeration could also be of benefit for application within GHP and HACCP (Fung, 2002).

The new '-omics' technologies are likely to provide opportunities to achieve goals that were previously beyond reach. In particular, DNA microarray technology has the potential to generate data that could be used to improve the safety of our food supply (Al-Khaldi *et al.*, 2002). By simultaneously examining the presence and expression of all genes of a specific microorganism at a given point in time, it is possible to study microbial ecology related to food safety, e.g. the relationships between virulence genes and fitness genes (Al-Khaldi *et al.*, 2002). By understanding growth potential and survival characteristics, it is conceivable that the technology could be applied in process setting, based on the microbial characteristics of the raw material contaminants, and safe shelf-life setting, based on knowledge of the degree of injury of any survivors (Brul *et al.*, 2002). The technology is also being used to understand how pathogenic microorganisms overcome lag phase, with a view to being able to extend safe shelf-life. By identifying unknown cellular pathways, and hence fresh antimicrobial targets, existing preservation systems may be optimised or new preservatives identified. By hybridising genomic DNA to microarrays, a system known as genomotyping (Lucchini *et al.*, 2001), large sets of bacterial strains can be characterised and compared, thus making the technique applicable to epidemiology and in tracing the sources of bacterial contamination (van der Vossen *et al.*, 2005).

The use of predictive models, particularly those based on probabilities, for GMP and HACCP has yet to be fully realised. It has been stated that their utility will be further enhanced when predictive microbiology is recognised as a rapid method (McMeekin *et al.*, 2002). This will require an increased availability and applicability of models, and improvements in the accuracy of predictions, as well as greater understanding by the user of the benefits and limitations. The deviations that can occur between predictions from current models and observed data from foods are often due to a factor not included in the model (e.g. a preservative) or differences in the factors used (e.g. type of acid or humectant). The physiological state of microorganisms in food, particularly if injured or pre-conditioned, can have a dramatic effect on their fate and growth or survival kinetics (Blackburn and Davies, 1994; McMeekin *et al.*, 2002). Combining knowledge of microbial kinetics in food with an understanding of the underlying physiological processes offers great benefits for the management of food safety in the future. Ultimately, the combining of predictive models with rule bases in expert systems offers the potential for greater assurance of food safety, while still providing scope for innovation by food developers and producers.

Intelligent packaging (IP) is an emerging technology that uses the communication function of the package to facilitate decision making and achieve the benefits of enhanced food quality and safety (Yam *et al.*, 2005). The terms IP and Smart Packaging are often used interchangeably, but Yam *et al.* (2005) have proposed a definition of IP as a packaging system that is capable of carrying out intelligent functions (such as detecting, sensing, recording, tracing, communicating and applying scientific logic), thus facilitating decision making to extend shelf-life, enhance safety, improve quality, provide information and warn about possible problems. Potential IP applications include: leak/pack integrity indicators; 'freshness' indicators; cooking instructions; and information on the health value of food (Ahvenainen, 2002). Pathogen indicators have been developed and include: Toxin Guard™ (Toxin Alert Inc., Ontario, Canada), which involves a polyethylene-based packaging material that is claimed to be able to detect the presence of pathogenic bacteria (*Salmonella*, *Campylobacter*, *Escherichia coli* O157 and *Listeria*) with the aid of immobilised antibodies (Smolander, 2003); and Food Sentinel System™ (SIRA Technologies, Pasadena, CA, USA), based on an immunochemical reaction taking place in a bar code that is then converted to being unreadable (Smolander, 2003). Perhaps the IP application with the most potential is a time–temperature integrator/indicator (TTI): a simple, inexpensive device that can show an easily measurable, time–temperature-dependent change that reflects the full or partial temperature history of a food product to which it is attached (Taoukis and Labuza, 2003). As well as the potential for monitoring shelf-life during distribution, the use of TTIs in a risk assessment-based context could allow the development of a suitable management system to assure both safety and quality in the food chill chain (Taoukis and Labuza, 2003).

In the future, risk assessment, which is a tool that sits within the decision-making framework of risk analysis, is likely to become more widely adopted and used by governments to set FSOs that are distinct levels of foodborne hazards, which cannot be exceeded at the point of consumption (Gorris, 2005). The food industry will need to ensure that their food safety management systems achieve these goals and, in certain circumstances, testing against microbiological criteria that are based on performance objectives (POs: distinct levels of foodborne hazards earlier in the food chain) can be an effective means of verifying the POs (ICMSF, 2005). The application of risk assessment approaches by the food industry will further help to tackle certain of its more complex food safety issues. A risk assessment approach is compatible with several factors that are likely to be important in food safety assurance in the future, including: the incorporation of probabilistic data and models; and the move from purely hazard control to, where appropriate, risk control. Integration of this approach across the whole food supply chain potentially provides information to target the most effective measures in order to provide step-change improvements. To be successful, such approaches will still need to be linked with existing management systems, such as HACCP and appropriate pre-requisite programmes.

1.6 Sources of further information and advice

BLACKBURN C DE W and McCLURE P J (2002), *Foodborne Pathogens: Hazards, Risk Analysis and Control*, Cambridge, UK, Woodhead Publishing.

FSA, CookSafe, http://www.food.gov.uk/foodindustry/hygiene/cooksafe/

FSA, Safer food, better business, http://www.food.gov.uk/foodindustry/hygiene/sfbb/

ICMSF (2002), *Microorganisms in Food 7 – Microbiological testing in food safety management*, New York, USA, Kluwer Academic/Plenum Publishers.

MAYES T and MORTIMORE S (2001), *Making the most of HACCP*, Cambridge, UK, Woodhead Publishing.

McMEEKIN T A (2003), *Detecting Pathogens in Food*, Cambridge, UK, Woodhead Publishing.

1.7 References

ACHTERBERG R, MANESCHIJN-BONSING J, BLOEMRAAD R, SWANENBURG M and MAASSEN K (2005), 'Detecting *Salmonella* antibodies in pork', *New Food*, **4**, 56–58.

ADAIR C and BRIGGS P A (1993), 'The concept and application of expert-systems in the field of microbiological safety', *Journal of Industrial Microbiology*, **12**, 263–267.

AHVENAINEN R (2002), 'Minimal processing in the future: integration across the supply chain', in *Minimal Processing Technologies in the Food Industry*, eds Ohlsson T and Bengtsson N, Cambridge, UK, Woodhead Publishing, 267–281.

Al-KHALDI S F, MARTIN S A, RASOOLY A and EVANS J D (2002), 'DNA microarray technology used for studying foodborne pathogens and microbial habitats: minireview', *Journal of AOAC International*, **85**, 906–910.

ARMITAGE N H (1997), 'Use of predictive microbiology in meat hygiene regulatory activity', *International Journal of Food Microbiology*, **36**, 103–109.

ARMSTRONG G L, HOLLINGSWORTH J and MORRIS J G (1996), 'Emerging foodborne pathogens: *Escherichia coli* O157:H7 as a model of entry of a new pathogen into the food supply of the developed world', *Epidemiologic Reviews*, **18**, 29–51.

BAKER D A (1995), 'Application of modelling in HACCP plan development', *International Journal of Food Microbiology*, **25**, 251–261.

BAKER D A and GENIGEORGIS C (1993), 'Predictive modeling', in Clostridium botulinum: *Ecology and Control in Foods*, eds Hauschild A H W and Dodds K L, New York, USA, Marcel Dekker, 343–406.

BLACKBURN C DE W (2000), 'Modelling shelf-life', in *The Stability and Shelf-Life of Food*, eds Kilcast D and Subramaniam P, Cambridge, UK, Woodhead Publishing, 55–78.

BLACKBURN C DE W (2003), 'Microbiological analysis and food safety management: GMP and HACCP systems', in *Detecting Pathogens in Food*, ed. McMeekin T A, Cambridge, UK, Woodhead Publishing, 3–19.

BLACKBURN C DE W (2006), 'Managing microbial food spoilage: an overview', in *Food Spoilage Microorganisms*, ed. Blackburn C de W, Cambridge, UK, Woodhead Publishing, 147–170.

BLACKBURN C DE W and DAVIES A R (1994), 'Effects of preconditioning on pathogenic bacteria in foods', *Food Technology International Europe* ed. Turner A, London, UK, Sterling Publications, 37–40.

BROWN M (2002), 'Safe process design and operation', in *Foodborne Pathogens – Hazards, Risk Analysis and Control*, eds Blackburn C de W and McClure P J, Cambridge, UK, Woodhead Publishing, 197–228.

BROWN M H, GILL C O, HOLLINGSWORTH J, NICKELSON II R, SEWARD S, SHERIDAN J J, STEVENSON T, SUMNER J L, THENO D M, USBORNE W R and ZINK D, (2000), 'The role of microbiological testing in systems for assuring the safety of beef', *International Journal of Food Microbiology*, **62**, 7–16.

BRUL S, KLIS F M, OOMES S J C M, MONTIJN R C, SCHUREN F H J, COOTE P and HELLINGWERF K J (2002), 'Detailed process design based on genomics of survivors of food preservation processes', *Trends in Food Science and Technology*, **13**, 325–333.

CABO VERDE S, TENREIRO R and BOTELHO M L (2004), 'Sanitation of chicken eggs by ionizing radiation: HACCP and inactivation studies', *Radiation Physics and Chemistry*, **71**, 29–33.
CAC (1997), *Hazard analysis and critical control point (HACCP) system and guidelines for its application*, Annex to CAC/RCP 1-1969, Rev. 3 (1997), Rome, Italy, Codex Alimentarius Commission.
CAST (2004), 'Intervention strategies for the microbiological safety of foods of animal origin', Issue Paper, Number 25, January 2004, Council for Agricultural Science and Technology, http://www.cast-science.org/cast/src/cast_top.htm
CATES S C, ANDERSON D W, KARNS S A and BROWN P A (2001), 'Traditional versus hazard analysis and critical control point-based inspection: results from a poultry slaughter project', *Journal of Food Protection*, **64**, 826–832.
CCFRA (1994), 'Guidelines for the design and safety of food microbiology laboratories', Technical Manual No. 42, Chipping Campden, UK, Campden and Chorleywood Food Research Association.
CCFRA (2004a), 'Evaluation of product shelf-life for chilled foods', Guideline No. 46, Chipping Campden, UK, Campden and Chorleywood Food Research Association.
CCFRA (2004b), 'A code of practice for microbiology laboratories handling food, drink and associated samples (third edition)', Guideline No. 9, Chipping Campden, UK, Campden and Chorleywood Food Research Association.
CORDIER J-L (2002) Sampling and testing for pathogens essential in safe food manufacture, *New Food*, **2**, 37–40.
DE BOER E and BEUMER R R (1999), 'Methodology for detection and typing of foodborne microorganisms', *International Journal of Food Microbiology*, **50**, 119–130.
DODD C E R (1994), 'The application of molecular typing techniques to HACCP', *Trends in Food Science and Technology*, **5**, 160–164.
EC (2002), 'Regulation (EC) No 178/2002 of the European Parliament and of the Council of 28 January 2002 laying down the general principles of food law, establishing the European Food Safety Authority and laying down procedures in matters of food safety', *Official Journal of the European Communities*, **L 31**, 1–24.
EISEL W G, LINTON R H and MURIANA P M (1997), 'A survey of microbial levels for incoming raw beef, environmental sources, and ground beef in a red meat processing plant', *Food Microbiology*, **14**, 273–282.
ELLIOTT P H (1996), 'Predictive microbiology and HACCP', *Journal of Food Protection*, Suppl. S, 48–53.
FAO/WHO (1995), 'Application of risk analysis to food standards issues. Report of the Joint FAO/WHO Expert Consultation', Geneva, Switzerland, World Health Organisation.
FDA (1998), 'Guide to minimize microbial food safety hazards for fresh fruits and vegetables', Food and Drug Administration, http://www.cfsan.fda.gov/~mow/arprodgu.html
FRASER E (2002), 'Environmental monitoring in the food industry', *New Food*, **5**, 9–14.
FSIS (2002), 'Use of microbial pathogen computer modelling in HACCP plans', USDA Food Safety and Inspection Service, Washington, DC, FSIS Notice 55-02.
FUJIKAWA H and KOKUBO Y (2001), 'Practical applications of predictive microbiology software programs to HACCP plans', *Journal of the Food Hygienic Society of Japan*, **42**, 252–256.
FUNG D Y C (2002), 'Predictions for rapid methods and automation in food microbiology, *Journal of AOAC International*, **85**, 1000–1002.

GALE P (1996), 'Developments in microbiological risk assessment models for drinking water – a short review', *Journal of Applied Bacteriology*, **81**, 403–410.

GAZE J (2005), 'Microbiological aspects of thermally processed foods', *Journal of Applied Microbiology*, **98**, 1381–1386.

GOFTI L, ZMIROU D, MURANDI F S, HARTEMANN P and POTELON J L (1999), 'Waterborne microbiological risk assessment: a state of the art and perspectives', *Revue d' epidemiologie et de Sante Publique*, **47**, 61–73.

GONZÁLEZ-MIRET M L, COELLO M T, ALONSO S and HEREDIA F J (2001), 'Validation of parameters in HACCP verification using univariate and multivariate statistics. Application to the final phases of poultry meat production', *Food Control*, **12**, 261–268.

GORRIS L G M (2005), 'Food safety objective: an integral part of food chain management', *Food Control*, **16**, 801–809.

GRIFFITH C (2002), 'Good practices for food handlers and consumers', in *Foodborne Pathogens – Hazards, Risk Analysis and Control*, eds Blackburn C de W and McClure P J, Cambridge, UK, Woodhead Publishing, 257–276.

GRIFFITHS M W (1997), 'Rapid microbiological methods with hazard analysis critical control point', *Journal of AOAC International*, **80**, 1143–1150.

GRIFFITHS M W and PHILLIPS J D (1988), 'Modeling the relation between bacterial growth and storage temperature in pasteurised milks of varying hygienic quality', *Journal of the Society of Dairy Technology*, **41**, 96–102.

HAFEZ H M (1999), 'Poultry meat and food safety: pre- and post-harvest approaches to reduce foodborne pathogens', *World's Poultry Science Journal*, **55**, 269 280.

HALL P A (1994), 'Scope of rapid microbiological methods in modern food production', in *Rapid Analysis Techniques in Food Microbiology*, ed. Patel P D, Glasgow, UK, Blackie Academic, 255–267.

HOLAH J and THORPE R (2002), 'Hygienic plant design and sanitation', in *Foodborne Pathogens – Hazards, Risk Analysis and Control*, eds Blackburn C de W and McClure P J, Cambridge, UK, Woodhead Publishing, 151–195.

HYYTIÄ E, HIELM S, MOKKILA M, KINNUNEN A and KORKEALA H (1999), 'Predicted and observed growth and toxigenesis by *Clostridium botulinum* type E in vacuum-packaged fishery product challenge tests', *International Journal of Food Microbiology*, **47**, 161–169.

ICMSF (1986), *Microorganisms in Foods 2. Sampling for Microbiological Analysis; Principles and Specific Applications*, 2nd edition, eds Roberts T A, Bryan F L, Christian J H B, Kilsby D, Olson J C Jr and Silliker J H, International Commission on Microbiological Specifications for Foods, Toronto, Canada, University of Toronto Press.

ICMSF (2005), 'A simplified guide to understanding and using Food Safety Objectives and Performance Objectives', International Commission on Microbiological Specifications for Foods, http://www.icmsf.iit.edu/pdf/Simplified%20 FSO9nov05.pdf (accessed 13th February 2006).

IFT (2001), 'Microbiological challenge testing', in *Evaluation and Definition of Potentially Hazardous Foods*, A Report of the Institute of Food Technologists for the Food and Drug Administration of the US Department of Health and Human Services, IFT/FDA Contract No. 223-98-2333.

JOHNSTON M (2002), 'HACCP in farm production', in *Foodborne Pathogens – Hazards, Risk Analysis and Control*, eds Blackburn C de W and McClure P J, Cambridge, UK, Woodhead Publishing, 127–150.

KÄFERSTEIN F and ABDUSSALAM M (1999), 'Food safety in the 21st century', *Bulletin of the World Health Organisation*, **77**, 347–351.

KELLY L A, HARTNETT E, GETTINBY G, FAZIL A, SNARY E and WOOLDRIDGE M (2003), 'Microbiological safety of poultry meat: risk assessment as a way forward', *World Poultry Science Journal*, **59**, 495–508.
KILSBY D (2001), Personal communication.
KVENBERG J E and SCHWALM D J (2000), 'Use of microbial data for hazard analysis and critical control point verification – Food and Drug Administration perspective', *Journal of Food Protection*, **63**, 810–814.
LAWRENCE E (1997), 'Advisory Committee on Dangerous Pathogens (ACDP) seminar on microbiological risk assessment 28 January 1997', *Journal of Applied Microbiology*, **83**, 659–664.
LUCCHINI S, THOMPSON A and HINTON J C D (2001), 'Microarrays for microbiologists', *Microbiology*, **147**, 1403–1414.
MALDONADO E S, HENSON S J, CASWELL J A, LEOS L A, MARTINEZ P A, ARANDA G and CADENA J A (2005), 'Cost–benefit analysis of HACCP implementation in the Mexican meat industry', *Food Control*, **16**, 375–381.
MAUNSELL B and BOLTON D J (2004), 'Guidelines for food safety management on farms', Teagasc – The National Food Centre, Dublin, Republic of Ireland.
MAYES T and MORTIMORE S eds (2001), *Making the Most of HACCP*, Cambridge, UK, Woodhead Publishing.
MCLAUCHLIN J, MITCHELL R T, SMERDON W J and JEWELL K (2004), '*Listeria monocytogenes* and listeriosis: a review of hazard characterisation for use in microbiological risk assessment of foods', *International Journal of Food Microbiology*, **92**, 15–33.
MCMEEKIN T A, OLLEY J, RATKOWSKY D A and ROSS T, (2002), 'Predictive microbiology: towards the interface and beyond', *International Journal of Food Microbiology*, **73**, 395–407.
MEMBRÉ J-M, JOHNSTON M D, BASSETT J, NAAKTGEBOREN G, BLACKBURN C DE W and GORRIS L G M (2005), 'Application of modelling techniques in the food industry: Determination of shelf-life for chilled foods', in Hertog M L A T M and Nicolaï B M, *III International Symposium on Applications of Modelling as an Innovative Technology in the Agri-Food Chain; MODEL-IT*, Leuven, Belgium, ISHS Acta Horticulture **674**, 407–414.
MEMBRÉ J-M, AMÉZQUITA A, BASSETT J, GIAVEDONI P, BLACKBURN C DE W and GORRIS L G M (2006), 'A probabilistic modeling approach in thermal inactivation: estimation of post-process *Bacillus cereus* spore prevalence and concentration', *Journal of Food Protection*, **69**, 118–129.
METAXOPOULOS J, KRITIKOS D and DROSINOS E H (2003), 'Examination of microbiological parameters relevant to the implementation of GHP and HACCP system in Greek meat industry in the production of cooked sausages and cooked cured meat products', *Food Control*, **14**, 323–332.
MORTIMORE S and MAYES T (2001), 'Conclusions', in *Making the Most of HACCP*, eds Mayes T and Mortimore S, Cambridge, UK, Woodhead Publishing, 235–264.
MORTIMORE S and MAYES T (2002), 'The effective implementation of HACCP systems in food processing', in *Foodborne Pathogens – Hazards, Risk Analysis and Control*, eds Blackburn C de W and McClure P J, Cambridge, UK, Woodhead Publishing, 229–256.
OYARZABAL O (2005), 'Reduction of *Campylobacter* spp. by commercial antimicrobials applied during the processing of broiler chickens: a review from the United States perspective, *Journal of Food Protection*, **68**, 1752–1760.
PETERS R E (1998), 'The broader application of HACCP concepts to food quality in Australia', *Food Control*, **9**, 83–89.

POPPE C, DUNCAN C L and MAZZOCCO A (1998), '*Salmonella* contamination of hatching and table eggs: A comparison', *Canadian Journal of Veterinary Research*, **62**, 191–198.

QUINN B P and MARRIOTT N G (2002), 'HACCP plan development and assessment: a review', *Journal of Muscle Foods*, **13**, 313–330.

ROSE B E, HILL W E, UMHOLTZ R, RANSOM G M and JAMES W O (2002), 'Testing of *Salmonella* in raw meat and poultry products collected at federally inspected establishments in the United States, 1998 through 2000', *Journal of Food Protection*, **65**, 937–947.

ROSS T and McMEEKIN T (2002), 'Risk assessment ad pathogen management', in *Foodborne Pathogens – Hazards, Risk Analysis and Control*, eds Blackburn C de W and McClure P J, Cambridge, UK, Woodhead Publishing, 97–126.

ROSTAGNO M H, HURD H S and McKEAN J D (2005), 'Resting pigs on transport trailers as an intervention strategy to reduce *Salmonella enterica* prevalence at slaughter', *Journal of Food Protection*, **68**, 1720–1723.

SCHLUNDT J (2000), 'Comparison of microbiological risk assessment studies published', *International Journal of Food Microbiology*, **58**, 197–202.

SCOTT V N (2005), 'How does industry validate elements of HACCP plans?', *Food Control*, **16**, 497–503.

SIRAGUSA G R, DORSA W J, CUTTER C N, PERINO L J and KOOHMARAIE M (1996), 'Use of a newly developed rapid microbial ATP bioluminescence assay to detect microbial contamination on poultry carcasses', *Journal of Bioluminescence and Chemiluminescence*, **11**, 297–301.

SMOLANDER M (2003), 'The use of freshness indicators in packaging', in *Novel Food Packaging Techniques*, ed Ahvenainen R, Cambridge, UK, Woodhead Publishing, 127–143.

SRIKAEO K and HOURIGAN J A (2002), 'The use of statistical process control (SPC) to enhance the validation of critical control points (CCPs) in shell egg washing', *Food Control*, **13**, 263–273.

STIER R F (1993), 'Development and confirmation of CCP's using rapid microbiological tests', *Journal of Rapid Methods and Automation in Microbiology*, **2**, 17–26.

STRUIJK C B (1996), 'The Hamlet option in food microbiology: to analyze or not to analyze food specimens as marketed once HACCP implemented', *Acta Alimentaria*, **25**, 57–72.

SUWANRANGSI S (2001), 'HACCP implementation in the Thai fisheries industry', in *Making the most of HACCP*, eds Mayes T and Mortimore S, Cambridge, UK, Woodhead Publishing, 183–201.

SWANSON K M J and ANDERSON J E (2000), 'Industry perspectives on the use of microbial data for hazard analysis and critical control point validation and verification', *Journal of Food Protection*, **63**, 815–818.

SYPOSS Z, REICHART O and MÉSZÁROS L (2005), 'Microbiological risk assessment in the beverage industry', *Food Control*, **16**, 515–521.

TAOUKIS P S and LABUZA T P (2003), 'Time-temperature indicators (TTIs)', in *Novel Food Packaging Techniques*, ed. Ahvenainen R, Cambridge, UK, Woodhouse Publishing, 103–126.

TAYLOR L H, LATHAM S M and WOOLHOUSE M E J (2001), 'Risk factors for human disease emergence', *Philosophical Transactions of the Royal Society of London Series B – Biological Sciences*, **356**, 983–989.

TERGNEY A and BOLTON D J (2006), 'Validation studies on an online monitoring system for reducing faecal and microbial contamination on beef carcasses', *Food Control*, **17**, 378–382.

TODD E C D (1996), 'Risk assessment of use of cracked eggs in Canada', *International Journal of Food Microbiology*, **30**, 125–143.
VAN DER VOSSEN J, SCHUREN F and MONTIJN R (2005), 'Genomics in microbial food quality and safety', *New Food*, **8**, 74–77.
VAN SCHOTHORST M (1998), 'Principles for the establishment of microbiology food safety objectives and related control measures', *Food Control*, **9**, 379–384.
VAN SCHOTHORST M and JONGENEEL S (1994), 'Line monitoring, HACCP and food safety', *Food Control*, **5**, 107–110.
VENUGOPAL R, TOLLEFSON L, HYMAN F N, TIMBO B, JOYCE R E and KLONTZ K C (1996), 'Recalls of foods and cosmetics by the U.S. Food and Drug Administration', *Journal of Food Protection*, **59**, 876–880.
VESTERGAARD E M (2001), 'Building product confidence with challenge studies', *Dairy, Food and Environmental Sanitation*, **21**, 206–209.
YAM K L, TAKHISTOV P T and MILTZ J (2005), 'Intelligent packaging: concepts and applications', *Journal of Food Science*, **70**, R1–R10.

2

Legislation for red meat, poultry and eggs: which way forward?

M. Fogden, Consultant, UK

2.1 Introduction

This chapter shows how microbiological standards for red meat, poultry and eggs, and their products, are being developed under Codex Alimentarius codes and European Union (EU) hygiene legislation. A series of scares has reduced consumer confidence in food safety, and microbes cause many cases of foodborne illness every year. Thus, it is important to reassure consumers and restore their confidence. This requires elimination of the basis for consumer concern by the industry promising and providing safe food through the application of effective quality management systems, backed by regulatory measures.

Within the EU, microbiological standards have been adopted and refined, and it is reasonable to expect that these will be developed further. Existing measures will be extended, as food control mechanisms, increasingly based on risk management, but retaining an underlying foundation of prescriptive rules, seek to ensure food safety for the benefit of consumers. As a consequence, commercial understanding of microbiological issues and the capability of businesses for analysis and control will need to be improved in many cases, in order to comply with the regulations. This will add extra costs to the production and distribution of food, and such costs will, of course, fall on consumers, as well as operators in the agri-food supply chain. Therefore, there will be a tension between the understandable desire of politicians, legislators and consumers to secure a very high level of food safety and the consumer expectation that food should be available at the lowest possible price. While each regulatory advance will be met by conscientious businesses, the food industry will face adverse market pressures that will favour any players who avoid complying with costly microbiological controls until they encounter stringent enforcement. Historically, such enforcement has not always been in place and therefore the position of non-compliers is unlikely to weaken, at least in the short to medium term. Fortunately, failure to carry out microbiological tests does not equate with unsafe food, as is shown by the great volume of untested, but safe, food on the market, although microbiological testing is just one aspect of food-safety control. Consumers able to do so may be willing to pay more for food that has been tested microbiologically, but it is essential for legislators to ensure that the measures they introduce and enforce are proportionate, so poorer members of society are not compelled to purchase cheaper food of less certain safety, and reputable traders are not put at financial risk.

As is always the case for an overview of regulatory rules, this chapter cannot provide full details of the legal requirements, which, anyway, will change with time. Therefore, readers must refer in each case to the original regulatory text and seek such further advice as is necessary and appropriate to their own circumstances (see Section 2.7).

2.2 Codex Alimentarius (food code)

The Codex Alimentarius Commission (CAC) was established in 1962 as a subsidiary body of the Food and Agriculture Organisation (FAO) of the United Nations and the World Health Organisation (WHO) to implement the Joint FAO/WHO Food Standards Programme. It is inter-governmental, covers over 97 % of the world's population and is administered by the FAO in Rome. Its objectives are the global protection of consumers' health and economic interests, and assurance of fair food-trading practices. The CAC is intended to facilitate trade, not to restrict it or to interpose unnecessary or artificial barriers. Its standards allow governments to know that compliant products will not jeopardise the health or interests of consumers and, similarly, the agri-food industry can trade in compliant products, confident in the knowledge that they are accepted internationally as safe.

Commonly known in government and trading circles simply as 'Codex', the Codex Alimentarius is a systematic collection of internationally-adopted food standards that are presented in a uniform manner. They cover provisions relating to the hygienic quality of food, including microbiological standards, when considered both necessary and feasible, as well as other aspects, such as methods of sampling and analysis. Codex standards do not have regulatory authority and they have no direct role in areas such as environmental protection or animal welfare, unless such issues affect food quality or safety directly.

Sanitary and phytosanitary (SPS) measures were adopted as part of the 1994 (Uruguay Round) General Agreement on Tariffs and Trade (GATT) to provide internationally-recognised hygiene rules in the context of world trade. The SPS agreement (WTO, 1994) has impacted on CAC activities, with increasing numbers of Codex standards incorporating microbiological criteria, although few specify micobial limit values prescriptively. For example, the CAC code on hygienic practice for poultry processing (CAC, 1976) requires food production to be carried out using hygienic equipment and techniques, in premises designed with sanitation in mind. The code is not written in great detail, but does indicate the principles that should be followed, including recommendations for protection against contamination, infestation, development of a public health hazard and product deterioration, together with specific requirements for temperature control. However, paragraph 4.6 on laboratory control procedures only states:

> In addition to any control by the official agency having jurisdiction, it is desirable that each plant in its own interest should have access to laboratory control of the sanitary quality of the products processed. The amount and type of such control will vary with the food product as well as the needs of management. Such control should reject all foods that are unfit for human consumption. Analytical procedures used should follow recognized or standard methods in order that the results may be readily interpreted.

It is understandable that CAC codes provide only basic rules, based on principles rather than detail, since they need to be met by establishments that vary widely in competence and are situated in undeveloped, developing and developed countries. On the other hand, more emphasis could have been placed on the responsibility of the operator to ensure the safety of the food being produced, rather than relying on the competent authority in the country concerned. Also, since poultry meat is susceptible to microbial contamination during processing and readily supports microbial growth, mere access to laboratory control facilities is by no means an onerous recommendation. It may be implied, but could have been stated beneficially, that each plant should undertake appropriate laboratory controls, rather than just having access to the facilities.

2.2.1 Codex microbiological criteria

As with other codes, there are no specific microbiological criteria given in the code of practice for mechanically separated meat (CAC, 1983) nor, perhaps more surprisingly, in the guide to the microbiological quality of herbs and spices used in processed meat products (CAC, 1991). The latter aims principally to minimise microbial contamination, deterioration and the development of microorganisms that will impair the suitability of the product for its intended use. Thus, means of achieving these aims are required, and there is a recommendation that a Quality Control programme, developed in accordance with Hazard Analysis Critical Control Point (HACCP) principles, should be in place. Furthermore, the microbiological specifications for treated herbs or spices should be compatible with the preservation parameters of the products in which they are used. However, there are no quantitative estimates of the quality that should be achieved.

The code for processed meat and poultry products (CAC, 1985) is more helpful. Again, while specifying access to a laboratory, it does recommend specifically that laboratory facilities should be available to monitor hygiene. It recognises that microbiological criteria may be necessary, depending on the nature of the product, and that the application of HACCP principles is more effective than intensive end-product testing. The code states that products should be free from pathogenic microorganisms in numbers representing a hazard to health, and should not contain substances of microbial origin in amounts that may represent such a hazard. It defines procedures for the examination of meat products in hermetically-sealed containers in the context of detecting *Clostridium botulinum* toxin, while acknowledging that such testing is unlikely to be carried out on shelf-stable products. The code provides various microbiological criteria, including detention criteria based on aerobic plate counts, and tabulates the probability of identifying a defective product according to the number of sample units tested per batch. Other relevant Codex standards contain similar guidance.

2.2.2 Equivalency and HACCP principles

Historically, many Codex standards have concentrated on end-point inspection and tended to be restrictive about ways of meeting their objectives. Nowadays, desirable objectives are set, but scope is allowed for different approaches to achieving them. The focus is on 'equivalency', which means that the measures taken need not be identical, rather that the results attained are both identical and verifiable. A key feature now in the development of a Codex text is the use of risk analysis. This has two elements, the assessment of risk and development of approaches to managing that risk, and these are seen in more recent standards. Quality Assurance systems are also a focal point, and the CAC has adopted guidelines (CAC, 2003a) for application of the HACCP system, which is mentioned throughout this book. Application of the principles themselves has been described (CCFRA, 1997; CAC, 2003b) and recent legislative changes in the EU have strengthened the application of Principles 6 and 7, which were insufficiently provided for in the preceding legislation:

- *Principle 6*: Establish procedures for verification to confirm that HACCP is working effectively (which may include appropriate supplementary tests, together with a review); and
- *Principle 7*: Establish documentation concerning all procedures and records appropriate to these principles and their application.

The particular value of the HACCP system is that it can be applied throughout the food chain, from the primary producer to the final consumer, and that, in addition to enhancing food safety, it allows a better use of resources and a more timely response to any problems. Control based on risk analysis and management is, however, less easy to enforce than that based on prescriptive measures, but both types of control should be measurable. Control based on microbiological testing of food and related materials should relate to objective criteria set in regulations, codes or guidance documents, or a risk management system, such as HACCP. A certain degree of flexibility may be necessary in applying such criteria, because of the inherent variation in levels of microorganisms in and on foods. This requires operators and enforcement authorities to have sufficient understanding of the realities behind microbiologically-based controls, in order to exercise their powers transparently, responsibly and proportionately. In practice, deviation from a normal pattern can be as important in identifying a problem as exceeding a recommended or regulatory limit. Before implementing the HACCP system, however, food businesses should always ensure that they are operating according to food safety legislation and Codex standards, including the General Principles of Food Hygiene (CAC, 2003b).

2.2.3 Codex standards

Table 2.1 shows Codex standards that are relevant to the hygienic production of red meat, poultry, eggs and their products (as at June 2005). Many of these standards and codes are quite old, considering the rapid rate at which food hygiene and food-safety control measures have developed. This may reflect the difficulty of negotiating such texts internationally or suggest a general comfort with the level of guidance in existing documents.

2.2.4 Codex procedures

Codex standards, recommended codes of practice and guidelines have been produced through consultative procedures, essentially as follows. The CAC Secretariat decides on the need for a standard (step 1) and arranges, with input from Members of the Commission and interested international organisations, for the preparation of a 'proposed draft standard' that is based on scientific principles and incorporates the principles of risk analysis and process control (steps 2–4). This is then circulated to governments and developed further, in the light of their comments, into a 'draft standard' (step 5). If CAC adopts the draft, it is passed to governments for further comment (step 6), is modified accordingly (step 7) and then returns to the Commission (step 8), which may finally adopt it as a 'Codex standard'. An accelerated procedure is available for appropriate cases.

The Commission or its subsidiary working body may return a draft at any step in the procedure and the CAC may also decide that the draft

Table 2.1 Codex standards

Reference	Title	Adoption or revision
STAN 88	Corned beef	1991
STAN 89	Luncheon meat	1991
STAN 96	Cooked cured ham	1991
STAN 97	Cooked cured pork shoulder	1991
STAN 98	Cooked cured chopped meat	1991
STAN 106	General standard for irradiated foods	2003
STAN 193	General standard for contaminants and toxins in foods	2004
STAN 228	General methods of analysis for contaminants	2001
STAN 231	General Codex methods for the detection of irradiated foods	2003
STAN 233	Sampling plans for prepackaged foods	1969
STAN 234	Recommended methods of analysis and sampling	1999
CAC/RCP 1	General principles of food hygiene	2003
CAC/RCP 8	Processing and handling of quick frozen foods	1983
CAC/RCP 11	Fresh meat	1993
CAC/RCP 13	Processed meat and poultry products	1985
CAC/RCP 14	Poultry processing	1976
CAC/RCP 15	Egg products	1985
CAC/RCP 19	International code of practice for radiation processing of food	2003
CAC/RCP 23	Low-acid and acidified low-acid canned foods	1993
CAC/RCP 29	Game	1993
CAC/RCP 30	Processing of frog legs	1983
CAC/RCP 32	Production, storage and composition of mechanically separated meat and poultry meat intended for further processing	1983
CAC/RCP 39	Precooked and cooked foods in mass catering	1993
CAC/RCP 40	Aseptically processed and packaged low-acid foods	1993
CAC/RCP 41	Ante-mortem and post-mortem inspection of slaughter animals and for anti-mortem and post-mortem judgement of slaughter animals and meat	1993
CAC/RCP 46	Refrigerated packaged foods with extended shelf-life	1999
CAC/RCP 47	Transport of food in bulk and semi-packed food	2001
CAC/GL 14	Guide for the microbiological quality of spices and herbs used in processed meat and poultry products	1991
CAC/GL 21	Principles for the establishment and application of microbiological criteria for foods	1997
CAC/GL 30	Principles and guidelines for the conduct of microbiological risk assessment	1999
CAC/GL 52	General principles of meat hygiene	2003
CAC/GL 53	Judgement of equivalence of sanitary measures associated with food inspection and certification systems	2003

Source: http://www.codexalimentarius.net.

shall be held at step 8. Revision of an existing standard follows a similar procedure, with discretionary omission of steps being possible, notably where proposed amendments would have insubstantial effects.

Codex standards are published and sent to governments for acceptance, as well as to international bodies that have the necessary authority to do so on behalf of their member states. Governments may choose to adopt the standard in full, with deviations, or not at all, the last two cases applying particularly where such adoption would adversely affect a national economic interest.

2.2.5 Codex and the EU

What is the relationship between Codex and hygiene regulation in the EU? The SPS agreement, created under GATT and now operating through the World Trade Organisation (WTO), obliges contracting parties to ensure that SPS measures do not unjustifiably discriminate between countries, that they have a scientific basis, do not create direct or indirect trade restrictions, are not more restrictive on trade than is necessary to secure the appropriate level of consumer protection, and are established and maintained in an open and transparent manner. Where a state applies Codex standards, codes of practice and guidelines, there is a presumption that it is complying with WTO/SPS requirements. Indeed, for harmonisation purposes, the WTO text allows individual countries to base their hygiene measures on international standards, guidelines or recommendations, where these exist, and gives them the opportunity to play a full part in the activities of the CAC.

Since the EU itself and all EU Member States are contracting parties, the EU can be expected to regulate on food hygiene in sympathy with the SPS rules. Nevertheless, since it has set itself the task of achieving a high level of health protection, the EU frequently regulates at a higher level than the globally appropriate SPS standard. This, if scientifically justifiable and justified, lies within the WTO rules.

The EU has regulated to control the hygiene of meats, eggs and their products, and the rules thus instituted and implemented do generally pursue control mechanisms that are consistent with the relevant Codex standards. Of course, it would be somewhat surprising if that were not the case, since controlling the hygiene of these foods in global trade has a long, international history. Nevertheless, it is only in the last iterative development of EU hygiene controls that the institutions and Member States of the Community have begun to rely substantially on the application of HACCP. principles in relation to products of animal origin, rather than demanding compliance with numerous prescriptive requirements. This development must be in the interests of relevant food businesses, since it provides a mechanism based on equivalence, rather than on controls that were sometimes disproportionate to the risk.

2.3 Regulation in the EU

The EU regulatory process is based on confluence through compromise and consensus between its institutions, notably the Council, the European Parliament and the European Commission. In general terms, the first two represent respectively the Member State governments and the citizenry, while the Commission is the Community's administrative bureaucracy, which is charged with ensuring compliance with the principles of the Treaties.

Regulatory measures take several forms. Regulations apply directly in every Member State, and need no separate, national implementation. However, as in the UK, a national process may be needed to provide for enforcement and create penalties for non-compliance. Decisions have a similar, direct effect, but only on the Member States or parties named in them. Directives, on the other hand, define objectives that the Member States have agreed to pursue and these do require implementing measures at national level. In addition, the EU institutions, notably the Commission, issue Recommendations and other documents, which lack determinative legal status, but are extremely influential in indicating the correct interpretation of legal requirements. Ultimately, residual differences of opinion on appropriate interpretation are settled in and by the European Court of Justice (Van der Meulen and Van der Velde, 2004).

One source of confusion can be the relationship between EU regulatory requirements and the national measures of a Member State. Whether the latter pre-date the EU rules or are created subsequently, it is a fundamental rule that they must always be interpreted so as to be consistent with EU law. This may demand a strained interpretation of the words used in the national text.

In terms of international trade, as indicated above, similar principles apply to the relationship between EU requirements and WTO rules, especially the SPS agreement, insofar as food hygiene is concerned. As also mentioned, contracting parties to WTO have a right to stipulate a higher level of protection than that arising from the specific SPS provisions, if this is scientifically justified. The EU and its Member States frequently do so, although global competitors claim that subjective consumer pressures can play a significant role in EU decisions on the level of protection to be achieved and the mechanism for doing so.

2.3.1 Harmonisation at an appropriate level

One area of consumer concern, following from the series of scares in various Member States, is food safety. In relation to food hygiene, and particularly that relating to products of animal origin, the desired level of protection for human health (and animal health, where appropriate) can be expressed in terms of microbiological criteria that are relevant to the product concerned, thus providing an objective, scientific control system. Both the evaluation of products and the enforcement of compliance with the desired level of protection can be achieved through microbiological monitoring. This may be effected as part of a HACCP-based risk analysis and management system, where compliance with properly defined microbiological limits may be used to determine whether the processing of a food is within specification. Alternatively or additionally, the end-product can be monitored, although this is less efficient and less effective for control purposes.

From the early days of the development of the EU, its Member States have moved towards harmonised food-hygiene control through Community laws. Attention was given initially to meats crossing Community frontiers in substantial amounts, e.g. fresh meat (EEC, 1964) and poultry meat (EEC, 1971). Within the Common Agriculture Policy, the relevant institutions gradually developed legislation for products of animal origin in a number of (vertical) directives, each covering a restricted range of foodstuffs, usually in considerable detail and often including non-sanitary matters. They contained numerous inconsistencies, often for no obvious technical reason (Fogden, 1991, 1994a, b, c, d, 1995a, b, 1996), and there was no general (horizontal) measure covering basic requirements for all foods.

Hygiene rules cannot be considered satisfactory unless those concerned in their application and enforcement can interpret them effectively and consistently (Lugt, 1999). They must be capable of ensuring the protection of public health, but should be sufficiently flexible to satisfy diverse, essential needs. In many cases, this requirement was met, but improvements were (and remain) possible.

Hygiene requirements were reconsidered in tortuous negotiations between 1986 and 1993. With the elimination of border controls in a single market, there was concern to simplify and harmonise the rules, so the existing directives were updated although, undoubtedly, the objectives of simplification and harmonisation were not fully achieved. Additionally, a horizontal directive that provided general hygiene rules for matters and foods not covered by the vertical legislation was added (EC, 1993). This was enforced under national food control systems, often with a Ministry of Health as the competent authority, while the vertical rules were under veterinary control, usually through a Ministry of Agriculture.

2.3.2 New millennium, new rules

The above directives led to a comprehensive review, which began in 1996, and this proved to be another cumbersome exercise, with national political interests interfacing and interfering with scientific objectivity, as is unfortunately common in EU regulation, despite the best efforts of the European Commission. Where food hygiene and safety are concerned, and particularly in circumstances where simplification and harmonisation are to be achieved, politicians fear repercussions from the electorate, if any measure is withdrawn, even if it is replaced by one having an equivalent, but perhaps less specific, effect. For this reason, if no other, it proved extremely difficult to simplify, harmonise and consolidate the previous legislation, since achieving those objectives demanded the replacement of specific, prescriptive requirements with the implementation of operator-driven HACCP controls based on the risks present in the circumstances of individual businesses. Nevertheless, this was largely achieved by 2004, despite the fact that many prescriptive requirements remain in place. However, some of these will probably fall away following the next regulatory review. The numerous measures, generally directives, of the 1990s, have now been replaced by two directly applicable regulations. Regulation 852/2004 ('Hygiene 1'; EU, 2004a) controls general food hygiene and Regulation 853/2004 ('Hygiene 2'; EU, 2004b) provides rules that are specific to certain types of food of animal origin. Related regulatory texts provide controls on animal health and animal feed (EU, 2003, 2005a) and controls for the production and supply of food and feed (EU, 2004c, e), thus causing the revocation of the previous laws (EU, 2004d). All of these relate directly or indirectly to control of the microbiological quality of food (Fogden and Van der Meulen, 2005).

2.3.3 Self-regulation versus prescription

The global trend is towards self-regulation, and the new legislative system allows this for businesses that can demonstrate their competence and effectiveness in this respect. It makes food businesses solely responsible for the safety of the food they supply, rather than sharing this responsibility with the competent authorities, as previously. Therefore, businesses can now profit from introducing flexible and effective systems that meet their needs which, in appropriate cases, will include monitoring the levels of microbial contamination, bearing in mind that microorganisms occur naturally in products of animal origin. The question is: how are they to learn to do so? The GB Meat and Livestock Commission (MLC) has unique experience in introducing HACCP into butchers' shops, following the fatal outbreak of *Escherichia coli* O157 in Scotland (Pennington, 1997). The MLC introduced HACCP-based systems into 7500 of these establishments within 18 months, using most of the expertise available in the UK, including the preparation of new trainers to use a purpose-designed system. Projection of these numbers into a broad scenario of introducing the HACCP system into every food business throughout the EU, including the foodservice sector, and allowing for the transient nature of small food businesses, shows that the necessary resources simply do not exist to obtain compliance within only a few years. This presents a dilemma, since the 'honeymoon' period before the regulations are fully implemented should be as short as possible although, in reality, high levels of compliance may never be achieved.

It is easier to enforce detailed rules than to assess individual control systems, so inspectorates need to be trained thoroughly to ensure that they are able to satisfy themselves that food hygiene standards are being met appropriately and adequately. HACCP-based systems are currently weak at best in many premises, due to limited understanding, competence, commitment or active implementation, so there is still a considerable need for education and encouragement, preferably before resorting to strong enforcement (except in dangerous situations). 'Give a man a fish and you feed him for a day . . . show him how to fish and you feed him for life' (P Bache, personal communication).

2.4 EU regulation on microbiological criteria

Regulations 852/2004 and 853/2004 provide a framework for the adoption and application of specific microbiological criteria through a Commission Regulation (2073/2005; EU, 2005b) that details these and requires compliance with them.

The requirements are scientifically based on an opinion of the EU Scientific Committee on Veterinary Measures relating to Public Health (SCVMPH) concerning the evaluation of microbiological criteria (EC, 1999), and on the opinions of SCVMPH, the EU Scientific Committee on Food and the European Food Safety Authority on various, specific microbiological risks. The requirements also took account of discussions within the EU Standing Committee on the Food Chain and Animal Health. They include revised versions of the microbiological criteria contained in previous regulatory directives and decisions. They also follow the Codex 'Principles for the establishment and application of microbiological criteria for foods' (CAC, 1997) and other standards.

The requirements apply the principle that food should not contain microorganisms or their toxins or metabolites in quantities that present an unacceptable risk to human health. They recognise that microbiological criteria give guidance on the acceptability of food and its manufacturing, handling and distribution processes, and that the use of such criteria should form an integral part of the implementation of HACCP-based procedures and other hygiene control measures.

Article 4 of Regulation (EC) 852/2004 requires food business operators to comply with microbiological criteria, which should include testing against the set values by taking samples, analysing them and implementing any necessary corrective action, all in accordance with food law and any instructions given by the competent authority.

2.4.1 Articles of Regulation 2073/2005

The preambles (recitals) indicate the underlying basis and the required objectives.

Article 1 requires competent authorities to verify, in accordance with Regulation 882/2004, that food businesses comply with this Regulation's rules, without prejudice to any other enforcement powers or regulatory requirements.

Article 2 provides definitions, including 'microorganisms', 'microbiological criterion', 'food safety criterion', 'process hygiene criterion', 'sample' and 'compliance with microbiological criteria'. A microbiological criterion defines the acceptability of food or a process, based on the absence, presence or quantity of microorganisms (or their toxins/metabolites). Food safety criteria and process hygiene criteria indicate respectively the acceptability of food placed on the market and the functioning of the production process, the latter setting an indicative contamination value, above which corrective actions are required to maintain process hygiene. Compliance with microbiological criteria means obtaining satisfactory or acceptable results, as laid down in Annex I, applying appropriate methodology.

Article 3 requires food business operators to ensure that the food they supply complies with the microbiological criteria in Annex I. They must also ensure that they implement good hygienic practice and apply procedures based on HACCP principles throughout the food production and supply chain, so that the relevant food safety criteria can be met under reasonably foreseeable conditions of distribution, storage and use throughout the shelf-life of each product. Where necessary, and particularly in relation to ready-to-eat foods, where there is a public health risk from *Listeria monocytogenes*, operators must undertake investigations to determine compliance with the criteria throughout the product shelf-life.

Article 4 requires operators to test as appropriate against the Annex I microbiological criteria, when they are validating or verifying the correct functioning of their HACCP-based procedures. It is for them to decide the sampling frequency although, at least, this must match the frequencies prescribed in Annex I. Frequencies may be adapted to the nature and size of the business, provided that food safety will not be endangered.

Article 5 states that the sampling plans, sampling methods and analytical methods in Annex I are the reference methods. Where necessary, sampling of processing areas and production equipment is to be undertaken, with ISO 18593 being the reference standard (ISO, 2004). Specific provisions apply to testing for *L. monocytogenes* in the manufacture of ready-to-eat foods. Flexibility is built into the application of the Annex I sampling plans, subject to the availability of documentary evidence that the HACCP-based procedures being used are effective. However, those sampling plans are the minimum requirement, if the specific aim of testing is to assess the acceptability of a particular batch of food or a process. Nevertheless, operators may use other sampling and testing methods and procedures, provided they can demonstrate to the satisfaction of the competent authority that these are equally effective and that the analytical methods have been validated according to internationally accepted protocols. Such procedures may include trend analyses.

Article 6 states that, where *Salmonella* requirements in Annex I relating to minced meat, meat preparations and meat products intended to be eaten cooked are applied, the relevant foods must, when placed on the market, be labelled clearly to indicate the need for thorough cooking prior to consumption. Such labelling will no longer be required after 2009.

Article 7 requires operators to take certain measures, if unsatisfactory test results are obtained, as well as other corrective actions, whether defined in their own HACCP-based procedures or otherwise necessary to protect consumer health. They must also determine the cause of the unsatisfactory results to prevent recurrence of the unacceptable microbiological contamination, which may require them to modify their procedures or control methods. If testing has provided unsatisfactory results, as per Annex I, Chapter 1, the product or batch of food must be withdrawn or recalled in accordance with Article 19 of the 'General Food Law Regulation' (EC, 2002). However, products that have not yet reached retail level may be further processed by operators other than retailers, using a treatment that eliminates the relevant hazard. Alternatively, the batch may be used for purposes other than those originally intended, provided this (i) does not pose a risk to public or animal health, (ii) is provided for in the HACCP-based procedures, (iii) is in accordance with good hygienic practice, and (iv) is authorised by the competent authority. Any batch of mechanically separated meat that is classified as unsatisfactory in relation to *Salmonella* may only be used in the food chain to manufacture heat-treated meat products in appropriately approved establishments. Finally, unsatisfactory results relating to process hygiene criteria must be dealt with in accordance with Annex I, Chapter 2.

Article 8 allows Member States to apply a transitional, discretionary derogation until the end of 2009 in relation to compliance with the Annex I criteria for *Salmonella* in minced meat, meat preparations and meat products intended to be eaten cooked, where these are to be marketed only in that country; they must not enter intra-Community trade. Under the derogation, the product must be appropriately labelled, using a special mark, and indicating clearly that it must be cooked thoroughly prior to consumption. In relation to derogated foods, no more than one out of five sample units shall be found positive, when tested for *Salmonella*.

Article 9 requires operators to analyse trends in the test results. They must also take any appropriate action without undue delay, if they observe a trend towards unsatisfactory results, to prevent the occurrence of microbiological risks.

Article 10 requires this regulation to be reviewed 'taking into account progress in science, technology and methodology, emerging pathogenic microorganisms in foodstuffs, and information from risk assessments'. In so doing, particular attention is to be given to criteria and conditions relating to the presence of *Salmonella* on the carcasses of cattle, sheep, goats, horses, pigs and poultry in the light of any changes observed in *Salmonella* prevalence. No review date is specified, but this is unlikely to be later than 2009.

Article 11 repeals an earlier measure and Article 12 makes the regulation apply from 1 January 2006. As is usual in Community regulations, there is a statement that its provisions are binding in their entirety and are directly applicable in all Member States.

2.5 Annex I: criteria

This Annex is divided into three chapters: the first deals with food safety criteria and the second with process hygiene criteria. Chapter 3 covers sampling and the preparation of test samples, providing detailed requirements for sampling procedures and the frequencies at which sampling should take place.

2.5.1 Annex I, Chapter 1: food safety criteria

This chapter tabulates food safety sampling-plan requirements, reference analytical methods and microbial test limits applicable at specified points in the supply chain for various categories of food, principally in relation to *L. monocytogenes* and *Salmonella*. Microbial thresholds are expressed as values for m and M although, in most cases in this chapter, $m = M$. Sampling plans include n (the number of units comprising a sample) and, where appropriate, c (the number of sample units allowed to give test results in excess of m, but not greater than M).

Sections 1.2 and 1.3 deal with ready-to-eat foods that are respectively able and unable to support the growth of *L. monocytogenes*, but are not intended for infants or for special medical purposes. In each case, $n = 5$ and $c = 0$, so all sample units must satisfy the microbial limit. For those foods that are able to support the growth of this bacterium, it must be absent from 25 g of the product, when the food leaves the immediate control of the manufacturer. Also, there must be no more than 100 colony forming units (cfu) per gram throughout the shelf-life of the product, once it has been placed on the market. For ready-to-eat foods that cannot support the growth of *L. monocytogenes*, the same 100 cfu/g limit applies throughout shelf-life, but there is no pre-marketing requirement.

Limits are set for *Salmonella* in meat-based foods: in each case, $n = 5$ and $c = 0$. *Salmonella* must be absent from 25 g of minced meat and any meat preparation that is intended to be eaten raw throughout its shelf-life on the market. For such foods made from poultry meat and intended to be eaten cooked, including meat products, a lower threshold (absence from 10 g) applies until 1 January 2010. That lower threshold is without such a time limit for minced meat and meat preparations made from other animal species and intended to be eaten cooked, and for mechanically separated meat. 'Meat preparations' essentially comprise raw meat with added ingredients that have not been further processed to provide a significant preservative action. These must be distinguished in EU food hygiene law from 'meat products', which are basically meat with or without added ingredients that has been processed to change the nature of the meat, often resulting in a degree of preservation against microbial deterioration (see Regulation 853/2004, Annex I, paragraphs 1.15 and 7.1). For meat products intended to be eaten raw, excluding products where the manufacturing process or product composition will eliminate the risk from *Salmonella*, the microbial limit is absence from 25 g; this also applies to gelatine and collagen.

For *Salmonella* in egg products and ready-to-eat foods containing raw egg, $n = 5$, $c = 0$ and the organism must be absent from 25 g of the product throughout its shelf-life on the market. Products for which the manufacturing process or product composition eliminates the *Salmonella* risk are excluded from this requirement.

The chapter concludes with the manner in which the test results must be interpreted. The limits given for each category refer to each sample unit tested. The test results demonstrate the microbiological quality of the batch tested; they may also be used to demonstrate the effectiveness of the HACCP system or the hygiene of the process.

In relation to *L. monocytogenes*, the result is satisfactory if none of the values obtained exceeds the 100 cfu/g limit or the organism appears to be absent, as appropriate. Otherwise the result is unsatisfactory. For *Salmonella*, the result is satisfactory if the organism is not detected and unsatisfactory if it is found in any of the sample units.

2.5.2 Annex I, Chapter 2: process hygiene criteria

This chapter deals with process hygiene criteria and is divided into five sub-chapters. Meat and meat products are covered in 2.1 and egg products in 2.3. The requirements are more complex than in Chapter 1, and only indicative examples are provided here.

Chapter 2, structured much like the previous one, tabulates process hygiene sampling-plan requirements, reference analytical methods and microbial limits for various categories of food. However, the sampling points in 2.1 and 2.3 are carcasses after dressing, but before chilling, or the end of the manufacturing process. The microbiological focus is now on the aerobic colony count (ACC) and the presence or levels of Enterobacteriaceae (including *E. coli* and *Salmonella*). Microbial limits are again expressed as values for m and M although, in many cases, measurements relate to surface area rather than being per gram. In most cases in this chapter, m is about a tenth of M. Sampling plans again include n (the number of units comprising a sample) and c (the number of sample units allowed to give test results in excess of *m* and between *m* and *M*). An additional and important column is included in the table: 'Action in case of unsatisfactory results'.

2.5.3 Process hygiene criteria for meat and meat products

As a measure of process hygiene, requirements are established for red meat carcasses in terms of ACC and Enterobacteriaceae, in each case giving values for *m* and *M* as log cfu/cm^2 on a daily mean log basis (the daily mean log is calculated by first taking the $\log_{10}$ value of each individual test result and then calculating the mean of these log values). As an example, the daily mean log ACC for pig carcasses (excision sampling) is set at 4.0 log cfu/cm^2 for *m*, with *M* at 5.0 cfu/cm^2. Similarly, the limits for Enterobacteriaceae are $m = 2.0$ log cfu/cm^2, with $M = 3.0$ cfu/cm^2.

Salmonella criteria apply to carcasses of both red-meat species and certain poultry. Sampling rules are given in Annex I, 3.2; they require samples to be obtained from five red meat carcasses on each occasion, using an abrasive sponge method. For poultry, 15 carcasses must be sampled and examined as five pooled samples, taking 10 g of neck skin from each of three carcasses and testing 25 g. The criteria include $n = 50$ (samples from the last 10 consecutive sampling occasions of five units each); $c = 2$ (for cattle, sheep, goats and horses) or $c = 5$ (for pigs) or $c = 7$ (for broilers and turkeys), or less in each case, depending upon particular local circumstances. The required result is absence of *Salmonella* from the area tested per carcass, or from 25 g of pooled neck skin in the case of poultry.

Enterobacteriaceae and ACC results are considered satisfactory if the daily mean is $\leq m$; acceptable if the mean is between *m* and *M*; and unsatisfactory if the mean exceeds *M*. *Salmonella* results are considered satisfactory if the organism is detected in a maximum of c/n samples, and unsatisfactory if the organism is detected in a larger proportion of samples. Where there are unsatisfactory results for ACCs or levels of Enterobacteriaceae, the corrective action to be taken is described as 'improvements in slaughter hygiene and review of process controls'. The same action applies in the case of unsatisfactory *Salmonella* results, together with a review of the origin of the animals and biosecurity measures on the relevant farms.

Comminuted meat

Microbiological testing of minced meat, mechanically separated meat and meat preparations takes place at the end of the manufacturing process. In each case, $n = 5$ and $c = 2$, with testing for ACC and/or *E. coli*, which is used as an indicator of faecal contamination.

For minced meat, other than minced meat produced at retail level with a shelf-life of less than 24 hours, the ACC values are $m = 5 \times 10^5$ cfu/g and $M = 5 \times 10^6$ cfu/g. The *E. coli* requirement is $m = 50$ cfu/g and $M = 500$ cfu/g. There are no specific criteria for verotoxigenic *E. coli*. The same criteria apply to mechanically separated meat, but only to that produced using the techniques referred to in Annex III, Section V, Chapter III, paragraph 3 of Regulation (EC) No 853/2004. These criteria do not apply to other meat, whether produced by a mechanical separation process or otherwise. For meat preparations, there is no requirement for an ACC, and the relevant values for *E. coli* are $m = 500$ cfu/g (or per cm^2) and $M = 5000$ cfu/g (or per cm^2).

ACC and *E. coli* results for minced meat and meat preparations (and, presumably, also for mechanically separated meat) are considered satisfactory if all the values observed are $\leq m$; acceptable if a maximum of c/n values are between m and M and the remainder are $\leq m$ (i.e. none exceed M); and unsatisfactory if one or more of the values obtained exceeds M or more than c/n values are between m and M. In the case of unsatisfactory results, the action to be taken is 'improvements in production hygiene and improvements in selection and/or origin of raw materials'.

2.5.4 Process hygiene criteria for egg products

The criteria applicable to egg products for Enterobacteriaceae are $n = 5$, $c = 2$, $m = 10$ cfu/g (or /ml) and $M = 100$ cfu/g (or /ml). Testing takes place at the end of the manufacturing process. Results are considered satisfactory if all the values observed are $\leq m$; acceptable if a maximum of c/n values are between m and M and the remainder are $\leq m$; and unsatisfactory if one or more of the values exceeds M or if more than c/n values are between m and M. The action required in the event of results being unsatisfactory is 'checks on the efficiency of the heat treatment and prevention of re-contamination'.

2.6 Future trends

The on-going initiatives to reduce risks to food safety and to assure and reassure consumers that food is safe to eat undoubtedly seem to conflict with any possible relaxation of the hygiene rules that apply to the food industry. Also, there is pressure on competent authorities at national level to increase their control over the supply chain by improving their enforcement of regulatory requirements. However, there is equally a growing trend towards placing responsibility directly and unequivocally on farmers, processors and others in the supply chain to ensure the products that they produce, handle, store and supply are, and will remain, safe throughout their anticipated shelf-life. In the light of this trend, regulatory authorities are increasingly able and willing to reduce prescriptive and stringent control measures in favour of more flexible, systematic arrangements that can be implemented and controlled by the operator concerned, with the burden of demonstrating that these are maintaining food safety falling on that operator. This leaves the authorities with the functions of verifying and enforcing operator control, rather than controlling food safety themselves.

It is highly likely that the belief and enthusiasm of the CAC in the application of HACCP principles as a risk management tool will continue and that future standards, guidelines and codes of practice will incorporate such a system. It is also likely that objective, quantitative criteria will be used, where appropriate, to monitor food hygiene and there can be a reasonable expectation that microbiological criteria will be built into Codex specifications in the future. While it is sometimes reasonable to compromise and lower compositional standards for food produced under less sophisticated systems in developing countries, this is not the case in relation to food safety. Therefore, it seems appropriate that the control and monitoring of microbial contaminants, whether themselves hazardous or as indicators of potential hazards, should be harmonised on a global basis, because of the ubiquitous nature of international food trade. This can best be achieved through Codex working with individual countries and with trading blocs, such as the EU.

The EU, now with 25 Member States and likely to expand further, not only has a significant impact on the microbiological quality of food within those states and in countries that are candidates for membership, but also elsewhere in the world, because it rightly applies the same criteria to imports as it does to its own production, for the safety of its citizens. EU regulatory requirements, such as those discussed in this chapter, therefore have global significance and will help to improve microbiological standards in countries where the necessary level has yet to be reached. EU regulations have been created by taking into account Codex standards and the SPS Agreement, and will undoubtedly develop further in order to meet internal and external pressures, introducing additional requirements of other trading blocs, which EU industries must meet in order to export into those markets. For example, it would be possible or even necessary to increase the range of microbiological testing by applying such tests to other foods, by introducing new test requirements or increasing the stringency of existing criteria. In any of these cases, there would be a consequential burden on the food industry, which can only be warranted if the additional testing is practicable, scientifically based and justifiable. It would also be necessary to apply the principle of proportionality, which is fundamental to regulation within the Union. Notwithstanding these constraints, it must be borne in mind that consumer, media and political pressures will have a strong influence on the EU institutions, especially in an area such as food safety, with claims of anti-competitive international effects being pursued by other countries through WTO processes.

Equally important to bear in mind is that compliance with EU microbiological criteria in any Member State may not be sufficient from the political viewpoint. Individual Member States may legislate independently to provide higher levels of public health protection for their citizens, provided this does not introduce an unjustifiable barrier to intra-Community trade. However, with the recent introduction of EU legislation in the form of regulations rather than directives, disparities between national interpretations, resulting (often deliberately) from variations in transposing the directives into national laws, have been eliminated, providing a greater level of harmonisation between countries. It is appropriate to acknowledge the considerable and rapid efforts made by those that have recently joined the Union to reach or exceed the standards achieved in existing Member States over many decades. Many operators in those new Member States have benefited from investments in modern equipment and technology, which will provide them with a temporary advantage over businesses in other countries, where facilities may have been in place for some years, making it more difficult to comply with ever-greater demands for raising standards.

In practice, implementation of flexible risk management systems is inherently more difficult to enforce than detailed, prescriptive requirements. There is a need to employ officials that are both competent and thorough, with a good understanding of hygiene as it applies to the particular businesses they are required to inspect. Even then, problems arise, because detailed aspects of good hygiene practice are often open to interpretation and faults are difficult to challenge objectively, so as to satisfy a court of law. However, those able to demonstrate a history of good attitude, understanding and control capability should be permitted to manage their hygiene in a business-efficient manner, without undue interference from enforcement officials. The way forward appears promising for both businesses and consumers.

2.7 Sources of further information and advice

For further or more detailed information on legislated microbiological standards and other food law matters, readers are invited to contact the author of this chapter (Medvék Consultancy Limited; tel: +44 (0) 7770 320486; e-mail: medve@medvek.com).

Other sources include:

- Codex Alimentarius Commission: http://www.codexalimentarius.net
- European Commission: http://europa.eu.int/comm/index_en.htm
- European Food Safety Authority: http://www.efsa.eu.int
- First in Brussels Limited (agri-food law): tel: +44 (0) 7770 320487; http://www.FirstinBrussels.com
- Food Law Enforcement Practitioners' Forum: www.flep.org.
- Food Safety Authority of Ireland: www.fsai.ie.
- Food Standards Agency (UK): http://www.food.gov.uk

Table 2.2 Trade associations

Sector	Contact (telephone; website; email)
Agriculture (COPA-COGECA)	+32 (0) 2 287 27 11; www.copa-cogeca.be; mail@copa-cogeca.be
Agri-food trade (CELCAA)	+32 (0) 2 230 0370; www.schuman9.com/celcaa.htm; celcaa@schuman9.com
Meat (EMA)	+32 (0) 2 502 6423; e-euroconsulting.srl@skynet.be
Animal feed (FEFAC)	+32 (0) 2 285 0050; www.fefac.org; fefac@fefac.org
Butchers (CIBC)	+32 (0) 2 230 38 76; members-only site; info@cibc.be
Minced meat (EMMA)	+44 (0) 1908 665431; peterm@mckey.co.uk
Poultry (AVEC)	+32 (0) 2 238 1082; www.avec.dk; cv@poultry.dk
Poultry, egg and game (EUPG)	+49 (0) 228 959600; www.epega.org; info@epega.org
Fat processors and renderers (EFPRA)	+32 (0) 2 203 51 41; devries@skypro.be
Food and drink (CIAA)	+32 (0) 2 514 11 11; www.ciaa.be; ciaa@ciaa.be
Livestock and meat (UECBV)	+32 (0) 2 2304603; www.uecbv.be; uecbv@scarlet.be
Meat processors (CLITRAVI)	+32 (0) 2 203 51 41; devries@skypro.be
Retail (EUROCOMMERCE)	+32 (0) 2 737 0598; www.eurocommerce.be; vervondel@eurocommerce.be

- Meat and Livestock Commission (GB): http://www.mlc.org.uk.
- Trade associations: the organisations in Table 2.2 perform representative functions for national associations and individual companies at European level. See also the European Public Affairs Directory.
- Internet access to principal documents:
  - Codex standards: http://www.codexalimentarius.net/web/standard_list.do?lang=en.
  - EU hygiene legislation: http://europa.eu.int/eur-lex/lex/en/index.htm.

2.8 References

CAC (1976), *Recommended Code of Hygienic Practice for Poultry Processing* (CAC/RCP 14-1976), Rome, Italy, Codex Alimentarius Commission.

CAC (1983), *Recommended International Code of Practice for the Production, Storage and Composition of Mechanically Separated Meat and Poultry Meat Intended for Further Processing* (CAC/RCP 32-1983), Rome, Italy, Codex Alimentarius Commission.

CAC (1985), *Recommended International Code of Hygienic Practice for Processed Meat and Poultry Products* (CAC/RCP 13-1976, Rev. 1-1985), Rome, Italy, Codex Alimentarius Commission.

CAC (1991), *Guide for the Microbiological Quality of Spices and Herbs Used in Processed Meat and Poultry Products* (CAC/GL 14-1991), Rome, Italy, Codex Alimentarius Commission.

CAC (1997), *Principles for the Establishment and Application of Microbiological Criteria for Foods* (CAC/GL 21-1997), Rome, Italy, Codex Alimentarius Commission.

CAC (2003a), *Guidelines for the Application of the HACCP System* (Annex to CAC (2003b)), Rome, Italy, Codex Alimentarius Commission.

CAC (2003b), *Recommended International Code of Practice – General Principles of Food Hygiene* (CAC/RCP 1-1969, Rev. 4-2003), Rome, Italy, Codex Alimentarius Commission.

CCFRA (1997), *HACCP: a Practical Guide*, Chipping Campden, UK, Campden and Chorleywood Food Research Association.

EC (1993), 'Council Directive 93/43/EEC of 14 June 1993 on the hygiene of foodstuffs', *Official Journal of the European Communities*, **L175**, 1–11.

EC (1999), *Scientific Opinion on the Evaluation of Microbiological Criteria and Temperatures for Storage and Transport of Products of Animal Origin Intended for Human Consumption*, Brussels, Belgium, European Commission.

EC (2002), 'Regulation (EC) No. 178/2002 of the European Parliament and of the Council of 28 January 2002 laying down the general principles and requirements of food law, establishing the European Food Safety Authority and laying down procedures in matters of food safety', *Official Journal of the European Communities*, **L31**, 1–24.

EEC (1964), 'Council Directive 64/433/EEC of 26 June 1964 on health problems affecting intra-Community trade in fresh meat', *Official Journal of the European Communities*, **L121**, 2012–2032.

EEC (1971), 'Council Directive 71/118/EEC of 15 February 1971 on health problems affecting trade in fresh poultrymeat', *Official Journal of the European Communities*, **L55**, 23–39.

EU (2003), 'Council Directive 2002/99/EC of 16 December 2002 laying down the animal health rules governing the production, processing, distribution and introduction of products of animal origin for human consumption', *Official Journal of the European Communities*, **L18**, 11–20.

EU (2004a), 'Regulation (EC) No. 852/2004 of the European Parliament and of the Council of 29 April 2004 on the hygiene of foodstuffs', *Official Journal of the European Union*, **L139**, 1–54; Corrigendum: *Official Journal of the European Union*, **L226**, 3–21.

EU (2004b), 'Regulation (EC) No. 853/2004 of the European Parliament and of the Council of 29 April 2004 laying down specific hygiene rules for food of animal origin', *Official Journal of the European Union*, **L139**, 55–205; Corrigendum: *Official Journal of the European Union*, **L226**, 22–82.

EU (2004c), 'Regulation (EC) No. 854/2004 of the European Parliament and of the Council of 29 April 2004 laying down specific rules for the organisation of official controls on products of animal origin intended for human consumption', *Official Journal of the European Union*, **L139**, 206–319; Corrigendum: *Official Journal of the European Union*, **L226**, 83–127.

EU (2004d), 'Directive 2004/41/EC of the European Parliament and of the Council of 21 April 2004 repealing certain Directives concerning food hygiene and health conditions for the production and placing on the market of certain products of animal origin intended for human consumption and amending Council Directives 89/662/EEC and 92/118/EEC and Council Decision 95/408/EC', *Official Journal of the European Union*, **L157**, 33–44; Corrigendum: *Official Journal of the European Union*, **L195**, 12–15.

EU (2004e), 'Regulation (EC) No. 882/2004 of the European Parliament and of the Council of 29 April 2004 on official controls performed to ensure the verification of compliance with feed and food law, animal health and animal welfare rules', *Official Journal of the European Union*, **L165**, 1–141; Corrigendum: *Official Journal of the European Union*, **L191**, 1–52.

EU (2005a), 'Regulation (EC) No. 183/2005 of the European Parliament and of the Council of 12 January 2005 laying down requirements for feed hygiene', *Official Journal of the European Union*, **L35**, 1–22.

EU (2005b), 'Commission Regulation (EC) No. 2073/2005 of 15 November 2005 on microbiological criteria for foodstuffs', *Official Journal of the European Union*, **L338**, 1–26.

FOGDEN M (1991), 'European Community minced meat legislation', *European Food Law Review*, **2/3**, 150–173.

FOGDEN M (1994a), 'European Community food hygiene legislation. Part I. Introduction and preambles – the principles', *European Food Law Review*, **5**, 65–79.

FOGDEN M (1994b), 'European Community food hygiene legislation. Part 2. Areas of control', *European Food Law Review*, **5**, 181–194.

FOGDEN M (1994c), 'European Community food hygiene legislation. Part 3. Structural requirements', *European Food Law Review*, **5**, 267–283.

FOGDEN M (1994d), 'European Community food hygiene legislation. Part 4. Controls on raw materials', *European Food Law Review*, **5**, 367–390.

FOGDEN M (1995a), 'European Community food hygiene legislation. Part 5. Processing controls. Part 6. Control on finished products', *European Food Law Review*, **6**, 21–45.

FOGDEN M (1995b), 'European Community food hygiene legislation. Part 7. Composition, labelling and microbiology', *European Food Law Review*, **6**, 379–398.

FOGDEN M (1996), 'European Community food hygiene legislation. Part 8. Administration and personnel', *European Food Law Review*, **7**, 15–36.

FOGDEN M and VAN DER MEULEN B (2006), *The EU Food Hygiene Regulations*, Wageningen, Netherlands, Wageningen Academic Publishers.

ISO (2004), ISO 18593: 2004, 'Microbiology of food and animal feeding stuff – Horizontal methods for sampling techniques from surfaces using contact plates and swabs', Geneva, Switzerland, International Organization for Standardization.

LUGT M (1999), *Enforcing European and National Food Law in the Netherlands and England*, Lelystad, Netherlands, Koninklijke Vermande BV.

PENNINGTON H (1997), *Report on the Circumstances Leading to the 1996 Outbreak of Infection with E. coli O157 in Central Scotland, the Implications for Food Safety and the Lessons to be Learned*, Edinburgh, UK, The Scottish Office (of the UK Government), http://www.scotland.gov.uk/library/documents-w4/pgr-00.htm.

VAN DER MEULEN B and VAN DER VELDE M (2004), *Food Safety Law in the European Union*, Wageningen, Netherlands, Wageningen Academic Publishers.

WTO (1994): 'Sanitary and phytosanitary measures agreement', in *Agreement Establishing the World Trade Organisation*, 69–83, Geneva, Switzerland, World Trade Organisation.

3

Escherichia coli O157 and other VTEC in the meat industry

I. D. Ogden, University of Aberdeen, UK

3.1 Introduction

The first known report outlining the pathogenic nature of *E. coli* O157 was made by Riley *et al.* (1983), after they had investigated an 'unusual gastro-intestinal illness characterised by severe crampy abdominal pain, initially watery diarrhoea followed by grossly bloody diarrhoea, and little or no fever'. In subsequent decades, these symptoms were to become all too familiar to public health officials throughout the world, as O157 became, arguably, the most notorious foodborne pathogen of that time. The illness described by Riley *et al.* (1983) was linked to dining in fast-food restaurants in Oregon and Michigan, USA, and involved under-cooked beef in burgers, which no doubt contributed to the name 'burgerbug' used by certain members of the popular press for this hitherto rare *E. coli* serotype. The term is, perhaps, inappropriate, at least in some countries, as demonstrated later in this chapter. However, there is no doubt that the organism is inexorably linked to ruminants, and especially to cattle. In the context of the numbers of food poisoning cases reported in Europe and N. America, *E. coli* O157 thankfully remains low in comparison with *Campylobacter* and *Salmonella*. Figures for the UK suggest that, on average, there are 7.6 times more cases of human campylobacteriosis annually (Cumberland *et al.*, 2003). Although the extent of under-reporting of O157 is unclear, it is most probably less than the figure for *Campylobacter*, due to the severity of symptoms and need for medical attention. Similarly, infections from *Salmonella* (probably the best known food poisoning 'bug' to the lay person) are themselves approximately four-fold greater than O157 infections. Rates for the latter vary (Figs 3.1 and 3.2) but can reach almost 10 per 100 000 of

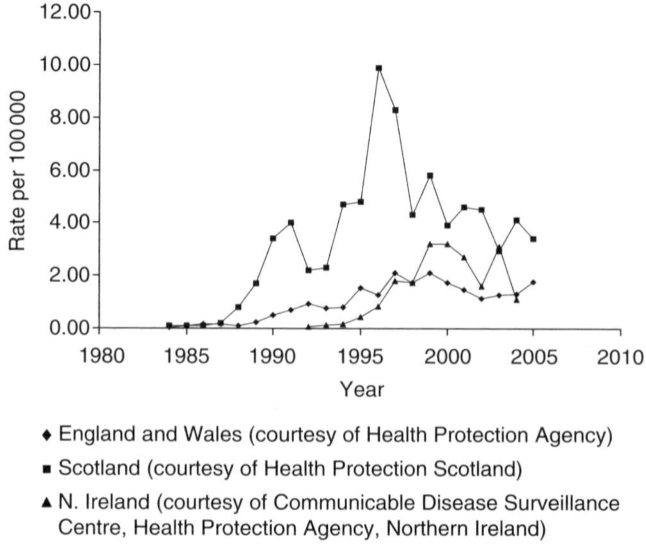

Fig. 3.1 Laboratory isolation rates for *E. coli* O157 in the UK since 1984.

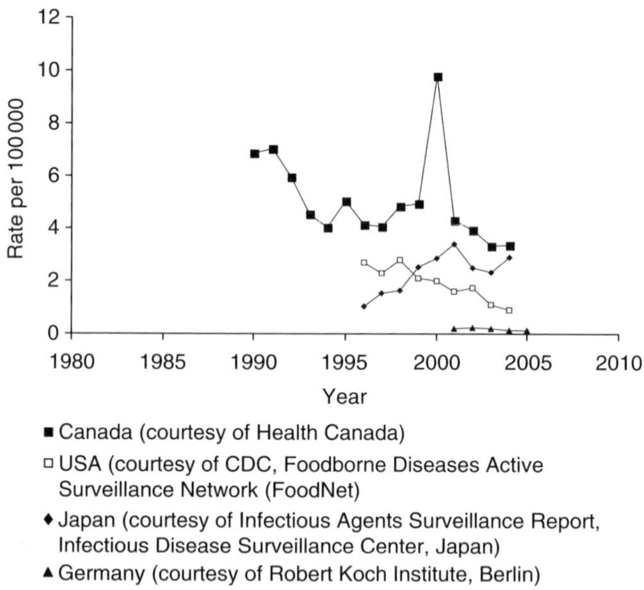

Fig. 3.2 Laboratory isolation rates for *E. coli* O157 in countries outside the UK.

the population in Scotland and as high as 15 per 100 000 in NE Scotland (Grampian), where infection is, historically, highest in the UK.

The definitions relating to *E. coli* O157 are somewhat confusing and require clarification at this point. The organism belongs to the verocytotoxigenic group of *E. coli* (VTEC), due to possession of a gene or genes associated with the production of verotoxin. Verotoxin itself is also widely known as 'shigatoxin', because of its similarity to the toxin produced by *Shigella*, the dysentery pathogen. Hence, O157 strains are also known as shigatoxin *E. coli* or STEC. The possession of VT (or ST) genes alone does not necessarily imply pathogenicity for humans. In addition, VTEC strains require a means to attach to the gut wall, where they can proliferate, produce toxin and cause illness. The disease-causing strains are known as enterohaemorrhagic *E. coli* (EHEC) and, as the name suggests, it is these that produce severe symptoms, such as bloody diarrhoea, in humans. There are many different strains of VTEC (and EHEC), although serotype O157 is currently the most commonly reported in the UK, N. America and Japan, while other serotypes may dominate elsewhere, and this has prompted the World Health Organisation (WHO, 1998) to include O26, O103, O111 and O145 as serotypes of particular concern. A typical feature of VTEC infections in continental Europe is the variety of serotypes involved. Huppertz *et al.* (1996) identified 20 different O-groups, of which those listed by WHO were the most common. The total number of VTEC serotypes linked to human disease is probably >100 and, to some extent, this reflects the ease with which virulence factors are transferred between *E. coli* strains. Table 3.1 presents the distribution of serotypes from human infections in Europe

Table 3.1 Quarterly data (July–September) for human VTEC infections reported to Enternet over a two-year period

Serogroup	Cases %			
	2004	%	2005	%
O157	252	45.2	188	38.6
O26	46	8.3	56	11.5
O103	20	3.6	42	8.6
O91	28	5.0	21	4.3
O145	16	2.9	20	4.1
O111	8	1.4	16	3.3
O128ab	1	0.2	7	1.4
O128	4	0.7	6	1.2
O5	3	0.5	5	1.0
O55	8	1.4	5	1.0
NT	94	16.9	73	15.0
Other	77	13.8	48	9.9
Total	557		487	

Details refer to eleven EU countries that have supplied data electronically for 2004 and 2005. The total number of reports in the database shows a decrease of 12.6 % over the same period, see: http://www.hpa.org.uk/hpa/inter/enter-net/vtec05q3.rtf.

that have been reported to Enternet (for details of Enternet, see section on further information below), during the peak months of infection (July–September) in 2004 and 2005, and shows that serotype O26 was the predominant non-O157 type of pathogenic *E. coli*. Brooks *et al.* (2005) reported serotype O26 to be the most common in the USA and gave the following figures: O26–22 %, O111–16 %, O103–12 %, O121–8 %, O45–7 % and O145–5 %. Whilst serotypes are usually described according to their O or somatic antigens, the H or flagellar antigen composition is often included. Thus, *E. coli* O157 : H7 is frequently associated with disease, although pathogenic, but non-motile (H0) strains do occur. There are also asymptomatic cases of EHEC carriage in humans that may be due to a specific immune response.

Abdominal pain, followed by bloody diarrhoea, as described by Riley *et al.* (1983), are, indeed, routinely associated with O157 infections, but other symptoms are also reported. Fever occurs in <30 % of cases and vomiting in 30–60 %. Most, but not all, diarrhoea is bloody (haemorrhagic colitis) and a significant number of cases give rise to additional complications, such as haemolytic uraemic syndrome (HUS) or the closely related thrombotic thrombocytopaenic purpura (TTP). These complications are present in 2–7 % of sporadic cases and up to 20 % of cases in outbreaks (Todd *et al.*, 2001). The mortality of children suffering from HUS and TTP is 5 % and the figure is even higher in the elderly. Treatment usually involves correction of the electrolyte and fluid balance, but the use of antibiotics is thought to be counter-productive, because they can cause an increase in release of the toxin.

3.2 Pathogenic potential

3.2.1 Pathogenicity factors

The key pathogenicity factors of O157 and other EHEC are the ability to attach to the human gut wall and, from there, produce the potent verotoxins, briefly described above. These factors are investigated routinely by molecular microbiologists, immediately after the isolation of a potential EHEC from food, an environmental source or a patient. The presence of certain genes other than those relating to the above properties may also be important in the aetiology of infection from this pathogen. The following list describes the most important virulence factors associated with clinical disease from EHEC infections:

1. The primary trait of EHECs is their ability to produce one or more verotoxins, usually labelled VT1 and VT2. Variants of both exist. Some genetic variants of *vt*1 (*vt*1ox3/*vt*1c) and *vt*2 (*vt*2e and *vt*2d-ount) are not present in classical EHEC strains, but are frequently found in intimin-negative VTEC strains (see below) from patients with uncomplicated diarrhoea or asymptomatic infections. Other variants, such as

*vt*2e, are rarely associated with STEC from humans and yet others, e.g. *vt*2f, not at all. The virulence of VTEC for humans may also be related to the VT type produced. The presence of the *vt*2 gene in the infecting strain was reported to correlate with severe disease in humans (Boerlin *et al.*, 1999), and the administration of purified VT2, but not VT1, was shown to cause HUS in experimentally treated primates (Siegler *et al.*, 2003).

2. The *eae* gene mediates attachment of the organism to the gut wall and encodes the protein intimin. Several types of intimin are known, of which γ, ε and θ are often associated with EHEC strains. Other colonisation factors, such as adhesins and pili, have also been identified in *eae*-negative strains that cause infection (Paton *et al.*, 2001; Srimanote *et al.*, 2002). The intimin gene is found in a pathogenicity island referred to as the locus of enterocyte effacement (McDaniel *et al.*, 1995), which harbours other key virulence-factor genes, such as *tir* (the intimin receptor), and a number of proteins required for the type III secretion system, as well as those involved in pore formation, through which effector proteins can transfer to the host cell (Roe *et al.*, 2003).
3. Haemolysin is found in the majority of O157 strains and many non-O157 VTEC (Nataro and Kaper, 1998), although its role in pathogenesis is still unclear. Haemolysin is distinguished by the appearance of a zone of lysis on blood agar containing washed erythrocytes. The frequent association of enterohaemolysin with verotoxin production makes it useful as an epidemiological marker for VTEC detection. Beutin *et al.*, (1989) reported that 89 % of strains tested showed such an association.

3.2.2 Infective dose of *E. coli* O157

The infective dose of a pathogen is not a precise number of bacterial cells, guaranteed to cause disease in the host. Resistance or immune response to any pathogen varies from person to person, depending on, for example, the current health status of the individual or their antibody level to the pathogen in question. The infective dose was estimated traditionally from feeding studies on volunteers, perhaps acceptable in the past, but less so in the second millennium and, presumably, there would be few volunteers to consume O157, knowing the dire consequences. Therefore, the estimated infective dose for this pathogen was based upon other criteria.

Some early reports of O157 outbreaks, where the implicated food was available for testing, indicated that few viable organisms were present, thus suggesting that the infective dose was low (Willshaw *et al.*, 1994). In other studies, surrogate models were used, such as the response of rabbits inoculated orally with different concentrations of O157 (Haas *et al.*, 2000) or the previously formulated human response data to *Shigella* (Crockett *et al.*, 1996). This was based on the fact that *Shigella* and *E. coli* O157 produce similar toxins and therefore have similar dose–responses. Powell *et al.*, (2000) proposed a dose–response envelope for O157, with bounding values defined by two separate, β-Poisson dose–response curves fitted to human clinical trial data for two surrogate pathogens (*Shigella dysenteriae* and enteropathogenic *E. coli*: EPEC). Strachan *et al.* (2005b) used a similar approach to Willshaw *et al.* (1994), but combined worldwide outbreak data, where not only the likely number of ingested organisms was known, but also the attack rate. Unfortunately, few outbreaks provide reliable data for modelling purposes and only eight such outbreaks were used to create a dose–response model. Figure 3.3 illustrates this model, alongside the *Shigella* and rabbit surrogates, and suggests the former is more plausible. Interestingly, the two Japanese outbreaks used in the model of Strachan *et al.* (2005b) demonstrate the forward-thinking and enlightened approach of the Japanese authorities. Following the 1996 radish-sprout outbreak in Japan that affected >6000 people (see account below), it is now necessary for retail portions of institutionally prepared food to be stored for subsequent analysis, if required.

What can be gained from a reliable dose–response model? In practice, it would allow accurate predictions to be made of the risk of infection, following exposure to a pathogen from either food or any other route of infection. The Meat Industry might be particularly interested in a model for *E. coli* O157, because red meat is most often linked to cases of foodborne human infection from this organism. Bearing in mind that the

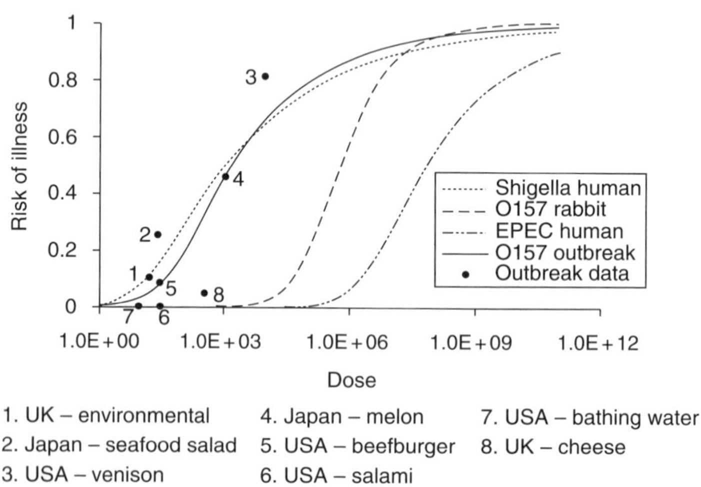

Fig. 3.3 Beta-Poisson dose–response model fits for animal/human feeding studies and outbreak data (Strachan *et al.*, 2005b).

majority of human cases are sporadic, i.e. the causal link is unknown (Slutsker *et al.*, 1998), an increasing number of cases appear to be due to environmental contact (Locking *et al.*, 2001). This should not be surprising, knowing that ruminants are the main reservoir of O157, but many of the major outbreaks (e.g. Washington, 1992; Central Scotland, 1996) have been associated with meat. Following the work done on the outbreak dose–response model for O157, Strachan *et al.* (2006) used the data to estimate the relative risk of eating a 'fast-food' beefburger and environmental exposure from camping in a field recently grazed by cattle. No doubt, the Meat Industry will be relieved that the camping episode carried a 100-fold greater risk, although this should be qualified by stating that the calculation is based on data from Scotland, where it is probable that burgers are seldom eaten rare, unlike N. America. The dose–response model could also be used to predict the size of an outbreak, if the number of O157 cells in a recently distributed food were known or, on the other hand, the extent of unhygienic food production, if only the attack rate of an outbreak were available.

3.3 Key outbreaks of *E. coli* O157 and other VTEC

Key outbreaks are defined not only by their size and severity, but also by the route of infection and the serotype involved. When compiling a list, the inclusion or omission of certain outbreaks is to some extent subjective and therefore no excuse is offered for placing the 1996 Central Scotland outbreak at the top. One particular reason for this position is that the outbreak was responsible for more deaths than any other, although the main reason it is foremost in the author's mind is more to do with its timing. During 1996, the author relocated to the University of Aberdeen at Foresterhill Hospital which, at that time, housed the Scottish *E. coli* O157 Reference Laboratory and thus was at the centre of all laboratory testing for the organism. The Head of Department was Professor Hugh Pennington, who went on to chair the public enquiry concerning the outbreak so, although the epicentre occurred some 15 miles SW of Glasgow, Aberdeen was the hub of analysis and investigation.

3.3.1 Central Scotland outbreak, 1996 – contaminated red meat

The following text is taken entirely from the abstract of Cowden *et al.* (2001), which neatly summarises the key clinical points.

> On Friday, 22 November 1996, the microbiologist at a hospital in Lanarkshire, Scotland, UK, identified presumptive *E. coli* O157 in faecal specimens submitted by three patients with bloody diarrhoea, and confirmed its presence in one. Over the next 6 h, 12 more potential cases were identified. Investigations first indicated then confirmed a single food premises as the source of infection. Effective control measures were applied promptly. The outbreak was declared

> over on 20 January 1997, by which time 512 cases had been identified, and infection with the outbreak strain confirmed in 279. Twenty deaths occurred in cases during the outbreak and there were two more in cases during 1997. Seventeen of these deaths resulted from the outbreak.

The butcher whose meat products were contaminated with O157 turned out to be more than a retail outlet, since he supplied a number of other establishments across central Scotland which, no doubt, contributed to the size of the outbreak. The deaths occurred after an old people's outing, where steak pie was consumed by the majority. The enquiry following the outbreak led to the licensing of retail butchers in Scotland and the separation of cooked and uncooked products presented for retail sale. However, the butcher involved in the outbreak is no longer operating.

3.3.2 Washington hamburger outbreak, 1992 – contaminated red meat

Between November 1992 and February 1993, an outbreak of *E. coli* O157 involving more than 700 cases occurred in the western USA and was associated with eating undercooked ground beef patties at restaurants of a major fast-food chain. Bell *et al.* (1994) designated 398 primary cases in Washington itself. Isolates of the organism obtained from re-called ground beef patties that were epidemiologically associated with the outbreak were indistinguishable by pulsed field gel electrophoresis (PFGE) from those isolated from patients. In considering this outbreak, the USDA risk assessment for O157 in ground beef (National Academy of Sciences, 2002) estimated the number of contaminated patties to be 5634, taking into account under-reporting.

3.3.3 Osaka outbreak, 1996 – contaminated radish sprouts

During July 1996, an outbreak of *E. coli* O157:H7 affected >6000 schoolchildren in Sakai City, Osaka, Japan, and three died (Michino *et al.*, 1999). In a food-consumption study, school lunches were associated with the infection. White radish sprouts from a single farm were the only uncooked food that was common to the most strongly implicated meal that included sweet and sour chicken with lettuce and chilled Japanese noodles. Thus, it was concluded that the cause of the outbreak was the radish sprouts, shipped from the farm on 7–9 July.

3.3.4 Walkerton outbreak, 2000 – contaminated drinking water

An estimated 2300 people became seriously ill and seven died after drinking contaminated water in the town of Walkerton, Ontario, Canada (Hrudey *et al.*, 2003). The severity of this outbreak resulted in the Government of Ontario calling a public inquiry to address the cause of the problem, the role (if any) of government policy and, ultimately, the implications of the experience for the safety of drinking water across the Province of Ontario.

3.3.5 New Deer outbreak, 2000 – ingested mud contaminated with sheep faeces

A scout camp was held at the New Deer agricultural showground, UK (Howie *et al.*, 2003), where 20 people (of 228 attendees) aged between 8 and 20 were later confirmed with *E. coli* O157 infection and dates of onset suggestive of a point-source outbreak. Investigations revealed that the field had been grazed by sheep prior to the camp, and subsequent testing of 28 animals showed that 17 were shedding O157. Samples taken from the field for microbiological analysis showed that O157 was present in soil, sheep faeces, standing water and on a climbing frame. Drinking water at the site and remaining food from the camp showed no presence of the pathogen. Isolates of O157 from animal, environmental and human sources were indistinguishable by PFGE. Heavy rainfall during the camping period caused localised flooding, resulting in mud and faecal material being widespread. The likely route of pathogen transmission was *via* hands contaminated with mud.

3.3.6 Australian outbreak of serotype O111, 1995 – contaminated red meat

During January and February 1995, the South Australian Communicable Disease Control Unit received reports of 23 cases of HUS among children aged less than 16 years (CDC, 1995; Paton *et al.* 1996). A preliminary investigation suggested that HUS occurred as a complication of an infection associated with the consumption of uncooked, semi-dry, fermented sausage. Sixteen patients required dialysis and one four-year-old girl died. During the illness, stools from all 23 patients were screened by a polymerase chain reaction (PCR) method for the genes encoding VT toxins; 20 were positive for both VT1 and VT2, one was positive for VT2 only, and two were negative. *E. coli* O111:NM (non-motile) was isolated from stool specimens obtained from 16 of the patients. In addition to the 23 cases of HUS, 30 others with bloody diarrhoea were investigated. Stool samples from eight were PCR-positive for VT genes, but O111:NM was isolated from only one. From 105 reports of patients with gastro-intestinal illness other than bloody diarrhoea, 32 were linked to consumption of the implicated sausage. Stools from 20 of these were VT-positive. VTEC were isolated from all 20 of these PCR-positive specimens, and isolates from two patients were identified as O111:NM. Of 10 sausage samples taken from the homes of patients, eight were positive for VT by PCR, and serotype O111:NM was isolated from four. Eighteen additional sausage samples produced by the same manufacturer and taken from homes where diarrhoeal illness without HUS had occurred, as well as from retail stores, were PCR-positive; three yielded O111:NM (CDC, 1995).

3.4 Occurrence in red meat and poultry

Not surprisingly, the epidemiology of O157 infections was linked to beef products in the early days of VTEC research, due in part to the Washington outbreak of 1992. The association was strengthened, when the consumption of contaminated steak pie in the Central Scotland outbreak led to the death of several elderly people at a church social event. Field studies were carried out at this time, mainly in the UK and USA, to determine the prevalence of O157 in cattle. Initial estimates in Scotland were low (approx. 1 %), but the values rose as isolation techniques improved with the advent of immunomagnetic separation, in conjunction with optimisation of enrichment conditions and new selective agars. A comprehensive survey of Scottish beef cattle in 1998–2000 (Table 3.2) indicated that 8.6 % of cattle were carrying the pathogen and a quarter of the herds tested contained at least one positive animal (Synge and Paiba, 2000).

Such studies are still being undertaken and values vary widely, depending on animal origin, season, diet and isolation method. Furthermore, Greenquist *et al.* (2005) showed that the prevalence of O157 in feedlot

Table 3.2 Prevalence of VTEC in cattle in different countries

Country	Sampling period	Serotype	Prevalence (%)	Cattle type	Reference
England and Wales	1999	O157	4.2	All types	Paiba *et al.* (2003)
Ireland	1998–1999	O157	3.0	Beef cattle at abattoir	McEvoy *et al.* (2003)
Scotland	Summer 2002	O157	7.5	Beef cattle at abattoir	Omisakin *et al.* (2003)
Scotland	Winter 2003	O157	11.2	Beef cattle at abattoir	Ogden *et al.* (2004)
Scotland	1998–2000	O157	8.6	Beef cattle	Synge and Paiba (2000)
USA	2001–2002	O157	6.5	Dairy cattle	Dunn *et al.* (2004)
USA	1997	O157	7.0	Beef calves	Laegreid *et al.* (1999)
Canada	1995–1996	O157	8.0	Beef cattle	Van Donkersgoed *et al.* (1999)
Japan	1999–2001	NA	5.1	Beef cattle on farms	Kijima-Tanaka *et al.* (2005)

NA: not available; reported as VTEC.

cattle approximately doubled (9.5 % compared to 4.7 %), when recto-anal swab sampling was used, instead of the more routine faecal pat testing. The author's laboratory has carried out studies in beef abattoirs and found the prevalence to be in line with that of Synge and Paiba (2000) but, perhaps of more importance, was the finding that 7.5 % of positive cattle contained $>10^4$ / g in their faeces. Such high-shedding animals accounted for >96 % of the total O157 found in the study (Omisakin *et al.*, 2003). Mathews *et al.* (2006) have also reported high-shedding animals and suggest that they transmit infection to other animals, accounting for the high prevalence levels in O157-positive herds.

Table 3.2 presents some of the more recently reported prevalence values for O157 and VTEC in general in cattle worldwide. The majority of countries listed are amongst those with the highest reported human infection rates and it is interesting to note the uniformity of prevalence values.

Other ruminants have also been implicated in human infections with O157, including those of the New Deer outbreak described above. In that case, not only were approximately 50 % of 28 sheep tested found to be shedding O157, but there were high-shedding animals amongst them. Unfortunately, few surveys of sheep have been reported from countries with high rates of human infection. Paiba *et al.* (2002) showed that approximately 2 % of sheep in Great Britain contained O157 at slaughter. However, Ogden *et al.* (2005) estimated that 6.5 % of sheep grazing on pasture in Grampian, Scotland carried the pathogen, of which 34 % were high shedders, i.e. with concentrations $>10^4$ / g, which may partly explain why the Grampian area has such a high rate of human infection. By contrast, Blanco *et al.* (2003) reported a prevalence of only 0.4 % in 93 Spanish flocks. Chapman *et al.* (2001) found that O157 was more prevalent in cattle and their carcasses than on sheep, but observed that twice as many food products derived from lamb were contaminated than those of beef origin. Kudva *et al.* (1997) found multiple strains of O157 in sheep, in addition to other *E. coli* strains containing VT genes.

Knowing the prevalence and concentrations of O157, together with the volumes of faecal material produced by both cattle and sheep, Strachan *et al.* (2005a) suggested that as many as 10^{13} O157 are shed daily by each animal species in N. E. Scotland. It is somewhat surprising, however, that epidemiological data showing cattle as the source of human infection far outnumber those implicating sheep. To the author's knowledge, eating contaminated lamb has yet to be linked clearly to human illness from O157. In contrast, other ruminants, including deer and goats, have been implicated in foodborne disease from this organism. In 1995, venison jerky caused an outbreak involving 11 people in Oregon, USA (Keene and Sazie, 1997) and, in 1999, cheese made from goat's milk in Scotland (Grampian Health Board, 2000) caused a larger outbreak, in which 27 of 28 children consuming the cheese became ill.

Similar to the situation with sheep, no O157 infections appear to have been linked directly to pigs or pork, and studies in Norway (Johnsen *et al.*,

2001), Sweden (Eriksson *et al.*, 2003), the Netherlands (Schouten *et al.*, 2005) and Japan (Kijima-Tanaka *et al.*, 2005) have shown a prevalence of <2 %. In Chile, however, Borie *et al.* (1997) found a prevalence of 10.8 % compared to cattle, which had a 3 % infection rate! Cornick and Helgerson (2004) reported that potentially pathogenic strains of O157 were capable of colonising pigs. Recent studies on stored animal waste in the UK (Hutchison *et al.*, 2004) showed that O157 was present in 11.9 % (fresh) and 15.5 % (stored) of waste from pigs. Extensive data on the virulence status of porcine isolates has yet to be obtained. If pigs are shown unequivocally to harbour pathogenic strains of VTEC, then the lack of epidemiological links to human illness may be due to farm management practices or carcass processing procedures, which differ to some extent from those of cattle and sheep.

Although poultry is implicated in many human gastro-intestinal illnesses, epidemiological and surveillance data suggest that VTEC infections are not among them. Chapman *et al.* (1997) obtained completely negative results from 1000 chickens reared in the UK and Kijima-Tanaka *et al.* (2005) also failed to find *E. coli* O157 in any chickens they tested.

3.5 Regulatory measures

New food hygiene regulations are in force across the EU from 11 January 2006, their primary objective being to optimise public health protection. They will affect all food businesses, including caterers, primary producers (e.g. farmers), manufacturers, distributors and retailers, and one of the key pathogens these laws are designed to control is *E. coli* O157. For the UK, and particularly for Scotland, the regulations replace the licensing laws for butchers, implemented after the Central Scotland outbreak.

Food handlers must receive training in food hygiene to minimise the risk of transferring pathogens to food products. Personnel responsible for company food safety procedures (based on Hazard Analysis Critical Control Point (HACCP) principles) must be trained to carry out such procedures. However, there is no legal requirement to attend a formal training course or obtain a qualification.

The Microbiological Criteria for Foodstuffs Regulation (EU, 2005) came into effect in January 2006 and complements other EU food hygiene legislation. Microbiological criteria in current EU legislation have been revised as part of a risk-based approach to food safety. The criteria can be used by food businesses to validate and verify their food safety management procedures and in assessing the acceptability of foodstuffs or their manufacturing, handling and distribution processes. The legislation is structured so that it can be applied flexibly in all food businesses, regardless of their type or size. Microbiological criteria should be applied within the framework of procedures based on HACCP principles.

In the USA, the Food Safety Inspection Service of the US Department of Agriculture has declared a zero tolerance for faecal material and *E. coli* O157:H7 in meat and poultry produced within a HACCP framework. Recalls of meat (often ground (minced) beef) are legendary in the USA, involving the voluntary removal of products from trade and consumer access. Canada has a similar policy of zero tolerance for O157 in certain foods. Because of the irregular distribution and low numbers of the pathogen in contaminated meat products, presence or absence of the organism is often difficult to determine with certainty.

3.6 Laboratory methods of isolation and identification

Much of the credit for developing sound methods of isolating *E. coli* O157 must go to Dr Peter Chapman of the UK Health Protection Agency (HPA), and he and his colleagues were the first to suggest adding cefixime and tellurite as additional selective agents to sorbitol-MacConkey agar (CT-SMAC), as described by Zadik *et al.* (1993). The UK national standard method for isolating and identifying *E. coli* O157 can be found on the website of the HPA: http://www.hpa-standardmethods.org.uk/documents/bsopid/pdf/bsopid22.pdf. This is aimed at isolating the organism from clinical specimens. Briefly, the method entails isolation on cefixime–tellurite–sorbitol–MacConkey (CT-SMAC) agar incubated in air at 35–37 °C for 16–24 hours. Enrichment culture in modified tryptone–soya broth (MTSB) may be required in investigating outbreaks.

The standard EN/ISO 16654 for food and animal feeding stuffs (ISO, 2001) details enrichment of samples in MTSB at 41.5 °C for six hours, followed by immuno-magnetic separation (IMS) and plating onto selective agars. The University of Aberdeen has worked extensively on optimising media to isolate O157 from a range of foods and prefers a slightly different approach. As with most non-clinical testing, the use of IMS is considered advisable to increase sensitivity and reduce non-target microflora. Also, prior to IMS, sample enrichment at 42 °C in buffered peptone water + vancomycin (BPW-V) for six hours has been found to improve selectivity further, while allowing a better recovery of physiologically stressed VTEC (Ogden *et al.*, 2001). Stressed cells are not usually an issue in clinical analyses, but are relevant in food microbiology, where products may be frozen (e.g. minced beef) or subject to high salt and/or low pH (e.g. salami and certain other sausages). The preference for BPW-V over MTSB is based on its being less inhibitory than TSB, which contains bile salts that are known to hinder the recovery of damaged organisms. Interestingly, the same enrichment medium was also optimal for VTEC O26 and O111, when isolated from a range of foods (Drysdale *et al.*, 2004).

Furthermore, it is advisable to use two selective agars for incubating the IMS beads. The use of CT-SMAC involves recognition of sorbitol-negative colonies and is a good choice for the majority of O157 strains encountered in the UK, N. America and Japan; in many European countries, however, sorbitol-positive strains are encountered frequently (Bielaszewska *et al.*, 2000). Plating one half of the beads onto a chromogenic VTEC agar allows the recognition of sorbitol-fermenting strains and reduces the chances of reporting false-negative results.

Confirmation or typing of VTEC is achieved by identifying virulence characteristics, such as VT type and the presence of intimin and/or enterohaemolysin. While this may differentiate some strains of the same serotype, many strains of clinical significance from the same country tend to have a similar profile. For example, in the UK, the majority of O157 strains are VT1-negative, VT2-positive and intimin-positive, and hence there is a need for further discrimination. Phage typing (Ahmed *et al.*, 1987) has been used extensively, although certain types have tended to predominate over specific periods of time, reducing the value of the technique in this case. The 'gold standard' typing technique is PFGE (Willshaw *et al.*, 1997) and is used by the majority of Reference Laboratories. The main problem with this approach is that the results are not easily transferable, making interlaboratory comparisons somewhat difficult, and the technique is both labour-intensive and time-consuming. An alternative DNA-based method, multiple-locus variable-number tandem repeat analysis (MLVA), has recently found favour for typing *E. coli* O157, and this targets areas of the genome that evolve rapidly and show polymorphism in the number of tandem repeats. The advantage of the technique is in assigning allele numbers to the loci (usually seven), as the basis for strain identification. Thus, each unique isolate has a seven-digit profile that is easily transferable for comparative purposes. MLVA shows a high degree of co-clustering with PFGE (Lindstedt *et al.*, 2004) and, in some cases, has proved to be more discriminatory, with a better correlation to epidemiological data than PFGE (Noller *et al.*, 2003).

3.7 Industry measures

It should be understood at the outset that the vast majority of VTEC are shed asymptomatically by farm ruminants. Because there is no obvious illness in the animal, and the presence of VTEC cannot be determined quickly prior to slaughter, there is no easy way of identifying and removing carrier animals from the food chain. However, a series of hurdles can be put in place to prevent pathogens from reaching the final food product. These can be summarised thus:

- on-farm practices;
- control in the abattoir – *ante-mortem*;
- HACCP along the processing line – *post-mortem*;
- correct cooking procedures.

3.7.1 On-farm practices

The most effective way to prevent food poisoning bacteria entering the food chain is to have animals free from pathogens (like VTEC) and, to this end, a range of techniques has been promoted. Much work has been done in N. America on the effect of diet, bacteriophage, vaccines, probiotics and orally-administered chemicals to reduce or eradicate *E. coli* O157. None are in regular use (in 2006) and, somewhat like human medicines, their approval is a lengthy procedure in the hands of the regulatory authorities. Brief descriptions of each follow.

Diet

The use of different feeding regimes to reduce carriage of O157 in cattle in the days prior to slaughter has proved a contentious issue. Feeding barley to cattle prior to slaughter is routine in the USA and has been reported to increase the shedding of O157 (Dargatz *et al.*, 1997). Diez-Gonzalez *et al.* (1998) showed that changing to a 100 % hay diet led to a decline in faecal *E. coli* populations. Similarly, Keen *et al.* (1999) studied cattle that were naturally infected with O157 and compared one half fed on barley with the remainder that were given hay. Of the grain-fed cattle, 52 % remained O157-positive, compared to only 18 % of the hay-fed animals. However, Hovde *et al.* (1999) and Buchko *et al.* (2000) reported results indicating the contrary. Further investigation is clearly required.

Competitive exclusion and probiotics

'Competitive exclusion' has been widely used in the poultry industry in Finland and Sweden to reduce *Salmonella* colonisation of broilers (Nurmi *et al.*, 1992). Initially, probiotics were thought to be unsuitable for those animals with a complex gut microflora already established and a long production life (e.g. cattle), but recent work has shown more promise (Zhao *et al.*, 2003). Commensal strains of *E. coli* produce protein inhibitors, termed colicins, that can displace O157 carried by cattle. The well-known human probiotic, *Lactobacillus acidophilus*, has been shown to reduce the carriage of O157 by 50 %, when fed to cattle pre-slaughter (Brashears *et al.*, 2003) although, preferably, such interventions should have a greater effect, bearing in mind that a 99.9 % reduction in an animal carrying 10^6/g *E. coli* O157 in faeces still allows 1000/g to enter the abattoir. Nevertheless, any reduction is a step in the right direction and the probiotic approach is likely to be approved by regulatory bodies.

Vaccination

Vaccinations against a wide range of microorganisms are used routinely by farmers, and experiments have been performed to identify a suitable vaccine for inhibiting colonisation by O157. Dean-Nystrom *et al.* (2002) used passive immunisation of neonatal piglets as a model to determine whether antibodies against intimin (the attaching and effacing protein) from EHEC O157 could inhibit further colonisation by this pathogen. The piglets were fed by vaccinated or placebo-treated mothers, before being inoculated with atoxigenic (for humane reasons) *E. coli* O157. Piglets that ingested colostrum-containing antibodies from vaccinated dams, but not those from placebo-treated pigs, were protected from colonisation by O157 and intestinal damage. The results suggest that this is a viable vaccine against the pathogen.

Chemicals

Incorporating chlorate into feed or drinking water before slaughter was shown to effectively reduce numbers of *E. coli* and *Salmonella* Typhimurium in the intestines of cattle, sheep, pigs and poultry (Anderson *et al.*, 2005). Most Enterobacteriaceae (including VTEC) can metabolise chlorate, which is reduced to the toxic chlorite and eventually to harmless chloride. The beneficial gut bacteria lack respiratory nitrate reductase activity and therefore remain uninhibited, allowing the basic metabolism of the animal to function normally. It was also reported that certain nitro-compounds can act as selective bacterial inhibitors (Anderson *et al.*, 2005). Furthermore, these compounds are most effective alongside chlorate, where they have the additional benefit of decreasing ruminal methanogenesis, involving a costly waste product from feed digestion in cattle and sheep. As yet, neither type of compound is licensed for commercial use.

Antibiotics

The role of antibiotics is made difficult, because colonisation by VTEC does not usually cause illness in ruminants. Neomycin sulphate has been shown to decrease significantly carriage of O157 in cattle (Elder *et al.*, 2002; Ransom *et al.*, 2003) and is now undergoing further trials.

Treatment of waste

Effective treatment of animal waste contaminated with a range of potential human pathogens, including *E. coli* O157, is sound farming practice to prevent environmental spread and re-infection of farm ruminants. Environmental control minimises transfer of the organism to watercourses, including private well-water for human consumption. Approved treatments include a ban on the spreading of slurry onto frozen ground and regular turning of manure heaps to ensure thorough composting (and therefore sterilisation), before application to farm land.

3.7.2 Ante-mortem meat inspection

In the UK, the Meat Hygiene Service, an executive body of the Food Standards Agency (FSA), controls the hygiene and welfare aspects of animal slaughter at the abattoir. Initially, livestock are inspected by veterinarians for signs of disease, injury, fatigue, stress or mishandling, and only clean, dry animals are allowed to progress to slaughter, thus reducing the risk of contaminating the final product. The requirement for clean animals has caused much controversy, since 'clipping', the term used to describe shearing of the hide to remove excessive faecal material, has resulted in injury to animal handlers. Compared to numbers of O157 in the intestine, the numbers on external surfaces of the animal are probably small.

3.7.3 Post-mortem inspection in the abattoir

As described in the section on regulatory measures, use of the HACCP system is the main means of minimising the transfer of pathogens to the meat. The HACCP system does not involve routine microbiological analysis, but many scientific studies have reported on the microbiological condition of raw meat products at the end of processing. Testing for VTEC on the final product would necessitate much sampling, because of the relatively low incidence and irregular distribution, and commensal *E. coli* is often used as a surrogate. How effective is the HACCP system in this context? The following illustrates the difficulty in making such a judgement.

Two abattoir studies were carried out in the author's laboratory to quantify O157 in beef cattle (Ogden *et al.*, 2004). In each case, several animals were found with faecal material containing numbers in excess of 10^4/g. However, in the time period following slaughter, during which the meat almost certainly would have been sold and consumed, there were no reports of O157 infection in humans. Thus, it is possible that the meat may have been sent to locations where illnesses would not have come to the attention of the laboratory, or all of it may have been cooked properly, thus destroying any vegetative pathogens present. Nevertheless, the study showed that 10 % of the animals tested were positive for O157 and almost 1 % were shedding $>10^4$/g. Inevitably, very large numbers of O157 would have been passing through the abattoir and, if effective HACCP procedures, including meat hygiene inspection, had not been implemented, the burden of human disease could have been considerable. Also, in a properly regulated system, meat products are chilled throughout the supply chain, i.e. during storage, transport and retail sale, to prevent any significant increase in pathogen numbers.

3.7.4 Correct cooking practices

Fast-food premises attempt to reduce microbial contamination by physically separating raw meat from the cooked product and implementing strict time-temperature cooking regimes to ensure destruction of pathogens. However, cooking in the home and in restaurants is more difficult to regulate. Nevertheless, there are basic rules of which every cook should be aware. The obvious one is to cook the meat thoroughly until the juices run clear, but a more accurate test is to use a thermometer inserted into deep tissue, where temperatures should reach a minimum of 62 °C for meats. Correct thawing of frozen foods to ensure that the middle part is completely thawed prior to cooking is equally important. Perhaps the commonest error seen in many domestic kitchens is the washing of meat before cooking. Washing might remove some bacteria, including VTEC, but certainly fails to remove them all. The most likely result is that any pathogens present are spread around the cooking area in water droplets. These may contaminate ready-to-eat foods, such as salads, or cause cross-contamination of surfaces and dishcloths, where subsequent transfer to food or the human mouth may cause illness. Therefore, the rule is never wash meat – put it straight into the roasting tray and dispose of the wrapper immediately. Then, wash hands and utensils in a pre-prepared basin of hot, soapy water.

3.8 Consumer responses

Not surprisingly, outbreaks of *E. coli* O157 receive widespread coverage in the press, because serious illness and death are emotive issues, especially when they affect young children. One of the first outbreaks in the UK involved contamination of milk and resulted in the death of a child. This was quickly followed by the Central Scotland outbreak, which, at the time of writing, has been responsible for more deaths than any other VTEC outbreak in the world. The Washington outbreak in 1992 provoked the same response in the USA and, like the Central Scotland outbreak, resulted in hundreds of cases. Both were eclipsed in terms of numbers affected by the events in Japan and Canada (see section above), where >6000 and >2000 respectively became ill. That all these cases occurred in countries of supposedly advanced health-care status came as a shock to people that were unaccustomed to foodborne illnesses on a large scale.

In Scotland, the 1996 outbreak resulted in a public enquiry that made 32 separate recommendations, summarised under the following headings.

1. **Farms and livestock** – education of farm workers on the nature of *E. coli* O157 and the spread of infection; the need to treat animal waste correctly; the requirement to deliver clean animals for slaughter, supported by official inspection.
2. **Abattoirs** – official enforcement of rigorous hygiene; introduction of HACCP; consideration of end-process treatment, e.g. steam treatment of carcasses. Meat production premises and butchers' shops – introduction of HACCP and licensing; physical separation of cooked and raw products.
3. **Point of consumption** – implementation of hygiene training in schools and for food handlers; adoption of HACCP, where appropriate.
4. **Enforcement** – government to lead on the need for enforcement of food safety measures and accelerated implementation of HACCP; local

authorities to ensure that there are sufficient environmental health officers to address enforcement.

5. **Surveillance** – government to ensure testing for O157 in stools; consultants in public health to inform Health Protection Scotland of all results, including full reports on outbreaks.
6. **Research** – proposals for research relating to O157 should be subject to normal processes of peer review and consideration for funding, but with appropriate weight given to the threat the organism represents to public health.
7. **Handling and control of outbreaks** – authorities to make available appropriate personnel, together with laboratory facilities, to meet their disease-control obligations.

A similar public enquiry has been set up following a recent outbreak of *E. coli* O157 in Wales (linked to contaminated meat), and it will be interesting to see if the recommendations listed above are implemented in this case. The chairman of the enquiry holds the view that such lessons are seldom learnt (Prof. H Pennington, personal communication)!

A number of charities, organisations and legal aid bodies have been formed as a direct result of O157 outbreaks. Haemolytic Uraemic Syndrome Help (HUSH) is a charitable organisation set up in 1997, following the Central Scotland outbreak, and has sought to bring together families from throughout the UK whose lives were affected by this pathogen. HUSH has the following aims:

1. To create a means of communication between families affected by O157, in order to prevent a feeling of isolation and to put recent sufferers in contact with those who have suffered this illness previously.
2. To increase public awareness of O157 and disseminate information on the steps that can be taken to avoid infection.
3. To increase the knowledge/awareness of GPs and Accident and Emergency Departments in order to speed up diagnosis.
4. To encourage the Government to legislate effectively in order to reduce the occurrence of O157 poisoning.

HUSH has its own website (http://www.ecoli-uk.com/news/php?pageNum.News=2) and publishes electronic newsletters. It also has links to other organisations, personal injury lawyers and government bodies.

3.9 Future trends

Have more than 20 years of research had any effect on the number of cases of *E. coli* O157? Some data in Figs 3.1 and 3.2 indicate no apparent reduction in infections over the last ten years, although such figures are often distorted by large outbreaks (see Scotland and Canada in particular). Breaking down these data into known routes of infection reveals some

Table 3.3 General outbreaks of *E. coli* O157 reported to Health Protection Scotland, 1994–2003, and related to meat/dairy products or contaminated water/environments

Period	Foodborne	Environmental
1994–1999	12 (71 %)	4 (20 %)
2000–2003	5 (29 %)	16 (80 %)

Source: Strachan *et al.* (2006).

surprising trends. Table 3.3 lists outbreaks related to meat/dairy products and contaminated water/environments where a dramatic reversal is demonstrated in the two time-frames, which suggests that food is not as great a risk post-2000 as before that date. This may reflect improved hygiene in food processing, since food companies have implemented recommendations made from the Central Scotland outbreak investigation (Pennington, 1998).

However, this would not explain the increase in environmental exposure, which may be partly due to an increased awareness that this route exists (previously health officials may have concentrated their efforts on the food-borne route). There is no evidence that increased numbers of O157 are being shed by farm ruminants (and hence a greater chance of infection). Furthermore, the author is not aware of any increase in the proportion of shedding animals or in the incidence of visits made by people to the countryside. In fact, the very opposite occurred in some areas of the UK under special restrictions during the foot and mouth disease outbreak in 2001. Also, the Task Force on *E. coli* (2001), which was formed to review the health of the public in Scotland and prevent human infection with O157, specifically cautioned against camping on pasture immediately following ruminant grazing. Despite this, the number of outbreaks linked to consumption of contaminated meat has declined, while environmental outbreaks have increased. Nevertheless, UK data suggests that the largest outbreaks are mainly foodborne, although it should be remembered that this is not always true: the Walkerton outbreak (2300 cases) was waterborne. Further investigation is needed to explain these anomalies. Whilst the implementation of research findings is aimed at reducing the burden of disease, it is also possible that methods of detection have improved and some older disease data were misleadingly low. This was certainly the case in the O157 prevalence figures for cattle discussed above.

An increasing number of O157 outbreaks, particularly in the USA, are being linked to contaminated, ready-to-eat salads, which may be partly due to their increased popularity with consumers. Law suits relating to illness from these foods are currently underway. Cabbage, alfalfa sprouts, celery, coriander, cress, sprouts and lettuce have all been implicated (Brackett,

1999; Hilborn *et al.*, 1999) and the contamination route is thought to be *via* dirty irrigation water, although inadequate composting of manure used as fertiliser and direct splashing of growing produce by rainwater contaminated from the soil may also have contributed. Sproston *et al.* (2006) suggested slugs as vectors of O157, transferring the pathogen either on their outer surfaces or in the intestines, after feeding on sheep faeces containing the organism, prior to infesting salad crops.

Can microbial infections be forecast with any accuracy? Whilst this is always difficult, some predictions can be made in the light of recent evidence, e.g. a study of non-O157 VTEC in Scottish cattle by Pearce *et al.* (2006), which showed that the proportions of positive farms were 23 % for O26, 22 % for O103 and 10 % for O145. This compares with approximately 15 % for O157. The prevalence in faecal pats was 4.6 % for O26, 2.7 % for O103, and 0.7 % for O145, while no O111 were detected. The presence of VTEC was rare in serotypes O103 and O145, and therefore they are not likely to be a threat to human health, but 49 % of O26 strains were VTEC and thus potentially pathogenic. Anecdotal evidence suggests that some cattle shed high numbers of O26, so it is puzzling that there are so few cases of human infection in the UK from this particular serotype. One reason may be that the majority of these VTEC O26 were VT1-positive, but VT2-negative, a combination that Brooks *et al.* (2005) suggest is less likely to cause severe symptoms than one including VT2. The majority of UK VTEC O157 is VT2-positive, which is why the disease burden from this pathogen is high. It may be that serotype O26 could acquire VT2 from the farming environment and so become as virulent as O157. Further research is currently underway to address this issue.

3.10 Sources of further information and advice

Specific information on VTEC can be found at the following websites:

1. UK Food Standards Agency: http://www.eatwell.gov.uk/healthissues/foodpoisoning/abugslife/#cat236133

   Guidance on the 2006 food hygiene regulations can be seen at the Agency website: http://www.food.gov.uk/foodindustry/guidancenotes/foodguid/fhlguidance/

   There is information on the following topics, which is designed to assist those working in the meat industry:

   - requirements of food hygiene legislation and the Summary Guides;
   - the requirements of the Microbiological Criteria Regulation;
   - revised draft code of practice on food law and practice guidance for England;

- draft guide to food hygiene and other regulations affecting the meat industry;
- guidance produced by the Dairy Hygiene Inspectorate and the Egg Marketing Inspectorate;
- guidance from the European Commission on various aspects of the legislation.

A link to further information from the FSA on meat and meat hygiene can be seen at: http://www.food.gov.uk/foodindustry/meat/.

FSA advice on clean beef at slaughter: http://www.foodstandards.gov.uk/search?p=Q&mainresults=mt_mainresults_yes&ts=custom&w=clean+beef

2. The Health Protection Agency: http://www.hpa.org.uk/infections/topics_az/ecoli/O157/menu.htm
3. Enternet: international surveillance network for the enteric infections, Salmonella and VTEC O157: http://www.hpa.org.uk/hpa/inter/enter-net_menu.htm.
4. FoodNet: The Foodborne Diseases Active Surveillance Network (FoodNet) is a surveillance system for identifying and characterising culture-confirmed infections that may be foodborne, including those due to *E. coli* O157:H7. Annual FoodNet reports that include data on this organism can be found at http://www.cdc.gov/foodnet/reports.htm.
5. PulseNet: The National Molecular Subtyping Network for Foodborne Diseases Surveillance is a national network of public health laboratories that perform PFGE on certain foodborne bacteria, including *E. coli* O157:H7. More information on PulseNet can be found at http://www.cdc.gov/pulsenet.
6. The Centers for Disease Control and Prevention (CDC) in Atlanta, USA. There is a series of fact sheets that cover both *E. coli* O157 and EHEC. These give background information, prevention measures, methods of treatment and long-term consequences: http://www.cdc.gov/ncidod/dbmd/diseaseinfo/escherichiacoli_g.htm; http://www.cdc.gov/ncidod/dbmd/diseaseinfo/enterohemecoli_t.htm

3.11 References

AHMED R, BOPP C, BORCZYK A and KASATIYA S (1987), 'Phage-typing scheme for *Escherichia coli* O157:H7', *Journal of Infectious Diseases*, **155**, 806–809.

ANDERSON R C, HARVEY R B, BYRD J A, CALLAWAY T R, GENOVESE K J, EDRINGTON T S, JUNG Y S, McREYNOLDS J L and NISBET D J (2005), 'Novel preharvest strategies involving the use of experimental chlorate preparations and nitro-based compounds to prevent colonization of food-producing animals by foodborne pathogens', *Poultry Science*, **84**, 649–654.

BELL B P, GOLDOFT M, GRIFFIN P M, *et al.* (1994), 'A multistate outbreak of *E. coli* O157:H7 – associated bloody diarrhea and hemolytic uremic syndrome from hamburgers', *Journal of the American Medical Association*, **272**, 1349–1353.

BEUTIN L, MONTENEGRO M A, ORSKOV I, ORSKOV F, PRADA J, ZIMMERMANN S and STEPHAN R (1989), 'Close association of verotoxin (Shiga-like toxin) production with enterohemolysin production in strains of *Escherichia coli*', *Journal of Clinical Microbiology*, **27**, 2559–2564.

BIELASZEWSKA M, SCHMIDT H, LIESEGANG A, PRAGER R, RABSCH W, TSCHÄPE H, CÍZEK A, JANDA J, BLÁHOVÁ K and KARCH H (2000), 'Cattle can be a reservoir of sorbitol-fermenting shiga toxin-producing *Escherichia coli* O157:H^- strains and a source of human diseases', *Journal of Clinical Microbiology*, **38**, 3470–3473.

BLANCO M, BLANCO J E, MORA A, REY J, ALONSO J M, HERMOSO M, HERMOSO J, ALONSO M P, DAHBI G, GONZALEZ E A, BERNARDEZ M I and BLANCO J (2003), 'Serotypes, virulence genes, and intimin types of Shiga toxin (verotoxin)-producing *Escherichia coli* isolates from healthy sheep in Spain', *Journal of Clinical Microbiology*, **41**, 1351–1356.

BOERLIN P, McEWEN S A, BOERLIN-PETZOLD F, WILSON J B, JOHNSON R P and GYLES C L (1999), 'Associations between virulence factors of Shiga toxin-producing *Escherichia coli* and disease in humans', *Journal of Clinical Microbiology*, **37**, 497–503.

BORIE C, MONREAL Z, GUERRERO P, SANCHEZ M L, MARTINEZ J, ARELLANO C and PRADO V (1997), 'Prevalencia y caracterizacion de *Escherichia coli* enterohemorragica aisladas de bovinos y cerdos sanos faenados en Santiago, Chile', *Archives Medical Veterinarian*, **29**, 205–212.

BRACKETT R E (1999), 'Incidence, contributing factors, and control of bacterial pathogens in produce', *Postharvest Biology and Technology*, **15**, 305–311.

BRASHEARS M M, JARONI D and TRIMBLE J (2003), 'Isolation, selection, and characterization of lactic acid bacteria for a competitive exclusion product to reduce shedding of *Escherichia coli* O157 : H7 in cattle', *Journal of Food Protection*, **66**, 355–363.

BROOKS J T, SOWERS E G, WELLS J G, GREENE K D, GRIFFIN P M, HOEKSTRA R M and STROCKBINE N A (2005), 'Non-O157 Shiga toxin-producing *Escherichia coli* infections in the United States, 1983–2002', *Journal of Infectious Diseases*, **192**, 1422–1429.

BUCHKO S J, HOLLEY R A, OLSON W O, GANNON V P J and VEIRA D M, (2000), 'The effect of fasting and diet on fecal shedding of *Escherichia coli* O157:H7 by cattle', *Canadian Journal of Animal Science*, **80**, 741–744.

CDC (1995), 'Community outbreak of hemolytic uremic syndrome attributable to *Escherichia coli* O111:NM-South Australia, 1950', *Morbidity Mortality Weekly Report*, **44**, 550–551.

CHAPMAN P A, SIDDONS C A, GAERDAN MALO A T and HARKIN M A (1997), 'A 1-year study of *Escherichia coli* O157 in cattle, sheep, pigs and poultry', *Epidemiology and Infection*, **119**, 245–250.

CHAPMAN P A, MALO A T C, ELLIN M, ASHTON R and HARKIN M A (2001), '*Escherichia coli* O157 in cattle and sheep at slaughter, on beef and lamb carcasses and in raw beef and lamb products in South Yorkshire, UK', *International Journal of Food Microbiology*, **64**, 139–150.

CORNICK N A and HELGERSON A F (2004), 'Transmission and infectious dose of *Escherichia coli* O157:H7 in swine', *Applied and Environmental Microbiology*, **70**, 5331–5335.

COWDEN J M, AHMED S, DONAGHY M and RILEY A (2001), 'Epidemiological investigation of the central Scotland outbreak of *Escherichia coli* O157 infection, November to December 1996', *Epidemiology and Infection*, **126**, 335–341.

CROCKETT C S, HAAS C N, FAZIL A, ROSE J B and GERBA C P (1996), 'Prevalence of shigellosis in the U.S.: consistency with dose–response information', *International Journal of Food Microbiology*, **30**, 87–99.

CUMBERLAND P D, SETHI P J, RODERICK J G, WHEELER J M, COWDEN J A, ROBERTS L C, RODRIGUES M J, HUDSON D S and TOMPKIN D S; IID STUDY EXECUTIVE (2003), 'The infectious intestinal disease study of England: a prospective evaluation of symptoms and health care use after an acute episode', *Epidemiology and Infection*, **130**, 453–460.

DARGATZ D A, WELLS S J, THOMAS L A, HANCOCK D D and GARBER L P (1997), 'Factors associated with the presence of *Escherichia coli* O157 in feces of feedlot cattle', *Journal of Food Protection*, **60**, 466–470.

DEAN-NYSTROM E A, GANSHEROFF L J, MILLS M, MOON H W and O'BRIEN A D (2002), 'Vaccination of pregnant dams with intiminO157 protects suckling piglets from *Escherichia coli* O157:H7 infection', *Infection and Immunity*, **70**, 2414–2418.

DIEZ-GONZALEZ F, CALLAWAY T R, KIZOULIS M G and RUSSELL J B (1998), 'Grain feeding and the dissemination of acid-resistant *Escherichia coli* from cattle', *Science*, **281**, 1666–1668.

DRYSDALE M, MACRAE M, STRACHAN N J C, REID T M S and OGDEN I D (2004), 'The detection of non-O157 *E. coli* in food by immunomagnetic separation', *Journal of Applied Microbiology*, **97**, 220–224.

DUNN J R, KEEN J E and THOMPSON R A (2004), 'Prevalence of Shiga-toxigenic *Escherichia coli* O157:H7 in adult dairy cattle', *Journal of the American Veterinary Medical Association*, **224**, 1151–1158.

EC (2005), 'Commission Regulation (EC) No. 2073/2005 15 November 2005 on microbiological criteria for foodstuffs', *Official Journal of the European Union*, **L338**, 1–2.

ELDER R O, KEEN J E, WITTUM T E, CALLAWAY T R, EDRINGTON T S, ANDERSON R C and NISBET D J (2002), 'Intervention to reduce fecal shedding of enterohemorrhagic *Escherichia coli* O157:H7 in naturally infected cattle using neomycin sulfate', *Journal of Animal Science*, **80**, (Supplement 1), 15.

ERIKSSON E, NERBRINK E, BORCH E, ASPAN A and GUNNARSSON A (2003), 'Verocytotoxin-producing *Escherichia coli* O157:H7 in the Swedish pig population', *Veterinary Record*, **152**, 712–717.

GRAMPIAN HEALTH BOARD (2000), *'Report on the Outbreak of* E. coli *O157 in Children Attending MacDuff Primary School, Aberdeenshire', June 1999.*

GREENQUIST M A, DROUILLARD J S, SARGEANT J M, DEPENBUSCH B E, SHI X, LECHTENBERG K F and NAHARAIA T G (2005), 'Comparison of mucosal swab cultures and fecal cultures for determining prevalence of *Escherichia coli* O157:H7 in feedlot cattle', *Applied and Environmental Microbiology*, **71**, 6431–6433.

HAAS C N, THAYYAR-MADABUSI A, ROSE J B and CHARLES C P (2000), 'Development of a dose-response relationship for *Escherichia coli* O157:H7', *International Journal of Food Microbiology*, **56**, 153–159.

HILBORN E D, MERMIN J H, MSHAR P A, HADLER J L, VOETSCH A, WOJTKUNSKI C, SWARTZ M, MSHAR R, LAMBERT-FAIR M-A, FARRAR J A, GLYNN K and SLUTSKER L (1999), 'A multistate outbreak of *Escherichia coli* O157:H7 infections associated with consumption of mesclun lettuce', *Archives of Internal Medicine*, **159**, 1758–1764.

HOVDE C J, AUSTIN P R, CLOUD K A, WILLIAMS C J and HUNT C W (1999), 'Effect of cattle diet on *Escherichia coli* O157:H7 acid resistance', *Applied and Environmental Microbiology*, **65**, 3233–3235.

HOWIE H, MUKERJEE A, COWDEN J, LEITH J and REID T (2003), 'Investigation of an outbreak of *Escherichia coli* O157 infection caused by environmental exposure at a scout camp', *Epidemiology and Infection*, **131**, 1063–1069.

HRUDEY S E, PAYMENT P, HUCK P M, GILLHAM R W and HRUDEY E J (2003), 'A fatal waterborne disease epidemic in Walkerton, Ontario: comparison with other waterborne outbreaks in the developed world', *Water Science Technology*, **47**, 7–14.

HUPPERTZ H I, BUSCH D, SCHMIDT H, ALEKSIC S and KARCH H (1996), 'Diarrhea in young children associated with *Escherichia coli* non-O157 organisms that produce Shiga-like toxin', *Journal of Pediatrics*, **128**, 341–346.

HUTCHISON M L, WALTERS L D, AVERY S M, SYNGE B A and MOORE A (2004), 'Levels of zoonotic agents in British livestock manures', *Letters in Applied Microbiology*, **39**, 207–214.

ISO (2001), 'ISO16654: 2001, Microbiology of food and animal feeding stuffs – Horizontal method for the detection of *Escherichia coli* O157', Geneva, Switzerland, International Organization for Standardization, http://www.iso.org/iso/en/CatalogueDetailPage.CatalogueDetail?CSNUMBER=29821&scopelist=

JOHNSEN G, WASTESON Y, HEIR E, BERGET O I and HERIKSTAD H (2001), '*Escherichia coli* O157:H7 in faeces from cattle, sheep and pigs in the southwest part of Norway during 1998 and 1999', *International Journal of Food Microbiology*, **65**, 193–200.

KEEN J E, UHLICH G A and ELDER R O (1999), 'Effects of hay and grain-based diets on fecal shedding in naturally-acquired enterohemorrhagic *E. coli* (EHEC) O157 in beef feedlot cattle', in *80th Conference of Research Workers in Animal Diseases*, Chicago, Illinois, USA (Abstract).

KEENE W E and SAZIE E (1997), 'An outbreak of *E. coli* O157:H7 infections traced to jerky made from deer meat', *Journal of the American Medical Association*, **277**, 1229–1232.

KIJIMA-TANAKA M, ISHIHARA K, KOJIMA A, MORIOKA A, NAGATA R, KAWANISHI M, NAKAZAWA M, TAMURA Y and TAKAHASHI T (2005), 'A national surveillance of shiga toxin-producing *Escherichia coli* in food-producing animals in Japan', *Journal of Veterinary Medicine Series B*, **52**, 230–237.

KUDVA I T, HATFIELD P G and HOVDE C J (1997), 'Characterization of *Escherichia coli* O157:H7 and other Shiga toxin-producing *E. coli* serotypes isolated from sheep', *Journal of Clinical Microbiology*, **35**, 892–899.

LAEGREID W W, ELDER R O and KEEN J E (1999), 'Prevalence of *Escherichia coli* O157:H7 in range beef calves at weaning', *Epidemiology and Infection*, **123**, 291–298.

LINDSTEDT B A, VARDUND T and KAPPERUD G (2004), 'Multiple-Locus Variable-Number Tandem-Repeats Analysis of *Escherichia coli* O157 using PCR multiplexing and multi-colored capillary electrophoresis', *Journal of Microbiological Methods*, **58**, 213–222.

LOCKING M E, O'BRIEN S J, REILLY W J, WRIGHT E M, CAMPBELL D M, COIA J E, BROWNING L M and RAMSAY C N (2001), 'Risk factors for sporadic cases of *Escherichia coli* O157 infection: the importance of contact with animal excreta', *Epidemiology and Infection*, **127**, 215–220.

MATHEWS L, McKENDRICK I J, TERNENT H, GUNN G J, SYNGE B and WOOLHOUSE M E (2006), 'Super-shedding cattle and the transmission dynamics of *Escherichia coli* O157', *Epidemiology and Infection*, **134**, 131–142.

McDANIEL T K, JARVIS K G, DONNENBERG M S and KAPER J B (1995), 'A genetic locus of enterocyte effacement conserved among diverse enterobacterial pathogens', *Proceedings of the National Academy of Sciences USA*, **92**, 1664–1668.

McEVOY J M, DOHERTY A M, SHERIDAN J J, THOMSON-CARTER F M, GARVEY P, McGUIRE L, BLAIR I S and McDOWELL D A (2003), 'The prevalence and spread of *Escherichia coli* O157:H7 at a commercial beef abattoir', *Journal of Applied Microbiology*, **95**, 256–266.

MICHINO H, ARAKI K, MINAMI S, TAKAYA S, MIYAZAKI M, ONO A and YANAGAWA H (1999), 'Massive outbreak of *Escherichia coli* O157:H7 infection in school children in Sakai City, Japan, associated with consumption of white radish sprouts', *American Journal of Epidemiology*, **150**, 787–795.

NATARO J P and KAPER J B (1998), 'Diarrheagenic *Escherichia coli*', *Clinical Microbiology Reviews*, **11**, 142–201.

NATIONAL ACADEMY OF SCIENCES (2002), *Escherichia coli O157:H7 in ground beef. Review of a draft risk assessment*, Washington DC, USA, National Academies Press.

NOLLER A C, McELLISTREM M C, PACHECO A G F, BOXRUD D J and HARRISON L H (2003), 'Multilocus Variable-Number Tandem Repeat Analysis distinguishes outbreak and sporadic *Escherichia coli* O157:H7 Isolates', *Journal of Clinical Microbiology*, **41**, 5389–5397.

NURMI E, NUOTIO L and SCHNEITZ C (1992), 'The competitive exclusion concept: development and future', *International Journal of Food Microbiology*, **15**, 237–240.

OGDEN I D, HEPBURN N F and MACRAE M (2001), 'The optimisation of isolation media used in immunomagnetic separation methods for the detection of *E. coli* O157 in foods', *Journal of Applied Microbiology*, **91**, 373–379.

OGDEN I D, MACRAE M and STRACHAN N J C (2004), 'Is the prevalence and shedding concentrations of *E. coli* O157 in beef cattle in Scotland seasonal?' *FEMS Microbiology Letters*, **233**, 297–300.

OGDEN I D, MACRAE M and STRACHAN N J C (2005), 'Concentration and prevalence of *Escherichia coli* O157 in sheep faeces at pasture in Scotland', *Journal of Applied Microbiology*, **98**, 646–651.

OMISAKIN F, MACRAE M, OGDEN I D and STRACHAN N J C (2003), 'Concentration and prevalence of *Escherichia coli* O157 in cattle faeces at slaughter', *Applied and Environmental Microbiology*, **69**, 2444–2447.

PAIBA G A, GIDDENS J C, PASCOE S J S, KIDD S A, BYRNE C, RYAN J B M, SMITH R P and McLAREN I M (2002), 'Faecal carriage of verocytotoxin-producing *Escherichia coli* O157 (VTEC) in cattle and sheep at slaughter in Great Britain', *Veterinary Record*, **150**, 593–598.

PAIBA G A, WILESMITH J W and EVANS S J (2003), 'Prevalence of faecal excretion of verocytotoxigenic *Escherichia coli* O157 in cattle in England and Wales', *Veterinary Record*, **153**, 347–353.

PATON A W, RATCLIFF R M, DOYLE R M, SEYMORE-MURRAY J, LANSER J A and PATON J C (1996), 'Molecular microbiological investigation of an outbreak of hemolytic-uremic syndrome caused by dry fermented sausage contaminated with shiga-like toxin-producing *Escherichia coli*', *Journal of Clinical Microbiology*, **34**, 1622–1627.

PATON A W, SRIMANOTE P, WOODROW M C and PATON J C (2001), 'Characterization of Saa, a novel autoagglutinating adhesin produced by locus of enterocyte effacement-negative Shiga-toxigenic *Escherichia coli* strains that are virulent for humans', *Infection and Immunity*, **69**, 6999–7009.

PEARCE M C, EVANS J, McKENDRICK I J, SMITH A W, KNIGHT H I, MELLOR D J, WOOLHOUSE M E, GUNN G J and LOW C J (2006), 'Prevalence and virulence factors of *Escherichia coli* serogroups O26, O103, O111, and O145 shed by cattle in Scotland', *Applied and Environmental Microbiology*, **72**, 653–659.

PENNINGTON H (1998), *Report on the Circumstances Leading to the 1996 Outbreak of Infection with E. coli O157 in Central Scotland, the Implications for Food Safety and the Lessons to be Learned*, Edinburgh, UK, The Scottish Office (of the UK Government), http://www.scotland.gov.uk/library/documents-w4/pgr-00.htm

POWELL M R, EBEL E, SCHLOSSER W, WALDERHAUG M and KAUSE J (2000), 'Dose response envelope for *E. coli* O157 : H7', *Quantitative Microbiology*, **2**, 141–163.

RANSOM J R, BELK K E, SOFOS J N, SCANGA J A, ROSSMAN M L, SMITH G C and TATUM J D (2003), 'Investigation of on-farm management practices as pre-harvest beef microbiological interventions', *National Cattlemen's Beef Association Research Fact Sheet*, Centennial, CO, USA.

RILEY L W, REMIS R S, HELGERSON S D, McGEE H B, WELLS J G, DAVIS B R, HEBERT R J, OLCOTT E S, JOHNSON L M, HARGRETT N T, BLAKE P A and COHEN M L (1983), 'Hemorrhagic colitis associated with a rare *Escherichia coli* serotype', *New England Journal of Medicine*, **308**, 681–685.

ROE A J, HOEY D E and GALLY D L (2003), 'Regulation, secretion and activity of type III-secreted proteins of enterohaemorrhagic *Escherichia coli* O157', *Biochemical Society Transactions*, **31**, 98–103.

SCHOUTEN J M, VAN DE GIESSEN A W, FRANKENAK D E, JONG M C M and GRAAT E A M (2005), '*Escherichia coli* O157 prevalence in Dutch poultry, pig finishing and veal herds and risk factors in Dutch veal herds', *Preventive Veterinary Medicine*, **70**, 1–15.

SIEGLER R L, OBRIG T G, PYSHER T J, TESH V L, DENKERS N D and TAYLOR F B (2003), 'Response to Shiga toxin 1 and 2 in a baboon model of haemolytic uremic syndrome', *Pediatric Nephrolology*, **18**, 92–96.

SLUTSKER L, RIES A A, MALONEY K, WELLS J G, GREENE K D, GRIFFIN P M (1998), 'A nationwide case-control study of *Escherichia coli* O157 : H7 infection in the United States', *Journal of Infectious Diseases*, **177**, 962–966.

SPROSTON E L, MACRAE M, OGDEN I D, WILSON M J and STRACHAN N J C (2006), 'Slugs: potential novel vectors of *Escherichia coli* O157', *Applied and Environmental Microbiology*, **72**, 144–149.

SRIMANOTE P, PATON A W and PATON J C (2002), 'Characterization of a novel type IV pilus locus encoded on the large plasmid of enterocyte effacement-negative Shiga-toxigenic *Escherichia coli* strains that are virulent for humans', *Infection and Immunity*, **70**, 3094–3100.

STRACHAN N J C, MACRAE M and OGDEN I D (2005a), 'Quantifying the *E. coli* O157 reservoir in Grampian, Scotland', *Veterinary Record*, **156**, 282–283.

STRACHAN N J C, DOYLE M P, KASUGA F, ROTARIU O and OGDEN I D (2005b), 'Dose response modelling of *Escherichia coli* O157 incorporating data from foodborne and environmental outbreaks', *International Journal of Food Microbiology*, **103**, 35–47.

STRACHAN N J C, DUNN G M, LOCKING M E, REID T M S and OGDEN I D (2006), '*E. coli* O157: burger bug or environmental pathogen?' *International Journal of Food Microbiology*, in press.

SYNGE B and PAIBA G (2000), 'Verocytotoxin producing *E. coli* O157' *Veterinary Record*, **147**, 27.

TASK FORCE ON *E. COLI* O157 (2001), http://www.food.gov.uk/multimedia/pdfs/ecolitaskfinreport.pdf

TODD W A, DUNDAS S and COIA J (2001), 'Clinical management of *E. coli* O157 infection', in *Verocytotoxigenic E. coli*, eds Duffy G, Garvey P and McDowell D, Trumbull, CT, USA, Food & Nutrition Press, 393–420.

VAN DONKERSGOED J, GRAHAM T and GANNON V (1999), 'The prevalence of verotoxins, *Escherichia coli* O157:H7, and *Salmonella* in the feces and rumen of cattle at processing', *Canadian Veterinary Journal*, **40**, 332–338.
WHO (1998), 'Zoonotic non-O157 Shiga toxin-producing *Escherichia coli* (STEC)', 23–26 June 1998, Berlin, Germany, Geneva, Switzerland, World Health Organization, W.H.O./CSR/APH/98.8–30.
WILLSHAW G A, SCOTLAND S M, SMITH H R, CHEASTY T, THOMAS A and ROWE B (1994), 'Hybridization of strains of *Escherichia coli* O157 with probes derived from the eaeA gene of enteropathogenic *E. coli* and the eaeA homolog from a Vero cytotoxin-producing strain of *E. coli* O157', *Journal of Clinical Microbiology*, **32**, 897–902.
WILLSHAW G A, SMITH H R, CHEASTY T, WALL P G and ROWE B (1997), 'Vero cytotoxin-producing *Escherichia coli* O157 outbreaks in England and Wales, 1995: phenotypic methods and genotypic subtyping', *Emerging Infectious Diseases*, **3**, 561–565.
ZADIK P M, CHAPMAN P A and SIDDONS C A (1993), 'Use of tellurite for the selection of verocytotoxigenic *Escherichia coli* O157', *Journal of Medical Microbiology*, **39**, 155–158.
ZHAO T, TKALCIC S, DOYLE M P, HARMON B G, BROWN C A and ZHAO P (2003), 'Pathogenicity of enterohemorrhagic *Escherichia coli* in neonatal calves and evaluation of fecal shedding by treatment with probiotic *Escherichia coli*', *Journal of Food Protection*, **66**, 924–930.

4

Faecal indicator organisms for red meat and poultry

G. C. Mead, Consultant, UK

4.1 Introduction

Estimating the number of viable microbes in a food sample, as an index of product safety or quality, is not an exact science. Errors associated with the methods used and the materials examined are due to factors such as clumping of microbial cells, the inability of some cells to proliferate under the prevailing conditions and an uneven distribution of organisms in the food itself. These problems strongly suggest that more importance may be attached to the counts obtained than is justified by their technical or statistical accuracy (Collins *et al.*, 2004). Nevertheless, microbiological criteria of different kinds continue to be used by food businesses and their customers, enforcement agencies, importation authorities, public health laboratories, researchers, etc. In recent years, microbiological analysis has even assumed a new importance in relation to the Hazard Analysis Critical Control Point (HACCP) system, where it has a widely accepted role in validation and verification procedures for process control. Establishing Critical Control Points (CCPs) may involve the enumeration of target organisms and specific levels of reduction to meet food-safety requirements.

Conventional counting techniques are not, of course, the only means available for estimating microbial numbers in foods. Developments have continued to occur in rapid, automated methods, which can offer improvements in reliability and the speed with which results become available. These include a variety of systems, such as those for determining bacterial bioluminescence, changes in impedance or conductivity of microbial cultures and the use of micro-calorimetry, flow cytometry and electronic particle counting. The acceptance of any new method usually depends on statistically analysed comparisons with one or more conventional counting methods, despite the known variability of the latter and that of microbial numbers in the material being analysed. It has been argued that this requirement influenced and even retarded methodological developments (Sharpe, 1980) and may explain partly why the potential of alternative enumeration technology is only slowly being realised.

Although raw foods of animal origin are sometimes contaminated with foodborne human pathogens of one kind or another, their distribution tends to be uneven and numbers are often low. Therefore, the organisms can be difficult to detect and enumerate without recourse to Most Probable Number (MPN) methods, which are laborious and even less accurate than colony counts. As an alternative to seeking a particular pathogen, it is often more appropriate to estimate levels of an appropriate 'indicator' organism instead. Ideally, such an organism (or group of organisms) should be similar in origin and properties to the pathogen in question and must be sufficiently common and abundant to be readily detectable in the majority of samples examined, while being relatively easy to isolate and enumerate. In theory, almost any suitable organism could be used for the purpose (Buttiaux and Mossel, 1961). As a possible candidate, *Bacteroides* spp. are commonly present in large numbers in the faeces of food animals (Williams Smith, 1965 a, b) and thus have the same enteric habitat as pathogens such as *Campylobacter*, *Salmonella*, *Clostridium perfringens* and pathogenic *Escherichia coli.* On the other hand, these organisms have the disadvantage of being obligate anaerobes that would appear to survive poorly outside the gut and they require the use of anaerobic isolation techniques, which are more difficult to apply routinely and on any more than a relatively small scale. Therefore, from a practical viewpoint, *Bacteroides* spp. would be unsuitable as indicators of faecal contamination. Similar considerations would apply to the proposed use of bifidobacteria as faecal indicators for meats and meat products (Beerens, 1998). More appropriate contenders are generic *E. coli*, coliform bacteria and members of the family Enterobacteriaceae as a whole, all of which are the principal concern of this chapter. Many of these organisms are faecal in origin and have similarities to those genera of the family that are pathogenic. Also, there are well-established methods of enumeration that are suitable for routine use. The relative merits of each category as a food hygiene indicator have been debated in Europe and elsewhere for many years and will be considered here mainly in relation to minimising any contamination of raw meat products with enteric foodborne pathogens.

4.2 Origins and terminology of the indicator concept

For more than a hundred years, *E. coli* and coliform bacteria have been used as bacterial indicators of faecal pollution in water supplies. Their use relates to the occurrence of such organisms in the faeces of man and a wide variety of warm-blooded animals, and the fact that the bacterial pathogens of greatest concern in water, such as *S.* Typhi, are all of enteric origin. When water is polluted initially, the levels of any pathogens present are likely to be low in comparison with the large numbers of coliforms and *E. coli* from faecal matter etc., which will usually outnumber salmonellas by many millions to one (Holden, 1970). However, water is often a harsher environment than food for the survival of organisms whose primary habitat is the gut and so low numbers of *Salmonella* may disappear from the water within a few days. Thus, a more abundant faecal indicator serves to highlight the occurrence of faecal pollution *via*, e.g. sewage, and may provide a safety margin by taking longer than the pathogens to die out completely. Unlike the situation in some foods, neither *Salmonella* nor *E. coli* is usually capable of multiplying in water.

As mentioned above, the use of one type of organism to indicate the possible presence of other, potentially harmful organisms that can contaminate food or water has a long history. It also has a specific terminology. For food control purposes, the organisms in question are often referred to as 'markers' of the microbiological quality of food, and they are seen as a key analytical tool for validating compliance with legislation. Where their occurrence in foods is associated with the possible presence of pathogens that are related to them taxonomically, physiologically and ecologically, they are termed 'index' organisms. Alternatively, when their presence relates solely to the effects of processing, as in tests to determine the efficacy of pasteurisation, the term 'indicator' is preferred (Mossel, 1982). In this chapter, the chosen organisms are considered mainly as 'indicators' of faecal contamination during slaughter and dressing/processing operations, but the term 'marker' is used differently. Here, it refers to any suitable organism that is deliberately added at a pre-determined level to a carcass or item of equipment in a slaughter or processing line in order to determine a possible route of cross-contamination, or to verify that a particular control measure is limiting its spread (Mead *et al.*, 1994; Hudson *et al.*, 1998). In this case, the marker organism is one that is readily distinguished from all others present or can be isolated specifically and enumerated on a selective agar medium. The topic is discussed further in Section 4.6.

4.3 Enterobacteriaceae, coliforms and *E. coli*

The Enterobacteriaceae includes some 29 genera, examples of which are shown in Table 4.1. All species are facultative anaerobes that ferment glucose and are oxidase-negative. More recent taxonomic changes in the organisms have been summarised by García-López *et al.* (1998). Some species are enteric in origin and others are primarily environmental. Strains of species belonging to several genera can multiply below 5 °C and have been found on meats and meat products (Table 4.1), where they may

Table 4.1 Some genera included in the Enterobacteriaceae: faecal origin and occurrence of psychrotrophic strains associated with meat

Genus	Faecal origin	Psychrotrophic meat strains
Citrobacter	–[a]	+
Edwardsiella	+	–
Enterobacter	–[a]	+
Erwinia	–[a]	–
Escherichia	+	–[c]
Hafnia	–[a]	+
Klebsiella	–[a]	+
Kluyvera	–[b]	+
Proteus	–[a]	+
Providencia	–	+
Salmonella	+	–
Serratia	–	+
Shigella	+	–
Yersinia	+	–[c]

[a] Sometimes found, but more common in other habitats.
[b] Likely to be mainly environmental.
[c] Reported occasionally (García-López *et al.*, 1998).

be associated with product spoilage during chill storage. The family also includes some important pathogens, such as certain types of *E. coli*, *Salmonella* and *Shigella* spp., and *Yersinia enterocolitica*. As indicator organisms, the Enterobacteriaceae have been used since the 1950s on the basis that they are a better defined taxon than the coliforms (see below) and include more organisms of public health significance (Mossel and Struijk, 1995).

Among the Enterobacteriaceae, those strains that ferment lactose, usually with acid and gas production, are together termed 'coliforms' and include species of *Citrobacter*, *Enterobacter*, *Escherichia* and *Klebsiella*. However, some medical microbiologists extend the definition to include species of *Edwardsiella*, *Hafnia* and *Serratia*, despite their usual inability to ferment lactose. All of these organisms are oxidase-negative, show a fermentative type of carbohydrate dissimilation and grow on relatively simple culture media in the presence of bile. Some strains are psychrotrophic, growing well at chill temperatures, but showing little inhibition at 37 °C (Mossel *et al.*, 1986). Certain strains of *Enterobacter*, for example, can be involved in the spoilage of red meat and poultry, especially under conditions of extended shelf-life (García-López *et al.*, 1998). Another sub-set of the coliform group are the so-called 'faecal coliforms' that ferment lactose at 44.5 ± 0.2 °C within 48 hours and are sometimes referred to as 'thermotolerant'. These comprise *Escherichia* and *Klebsiella* spp., but not all strains isolated are enteric in origin, which reduces the value of the test as an indication of faecal contamination (Mossel and Struijk, 2004).

Of the coliforms that occur in the faeces of warm-blooded animal hosts, >90 % are usually *E. coli*. Other strains of this group can be associated with environmental sources, such as soil, water and vegetation, and tend to persist longer than *E. coli* in contaminated water. They may even be capable of multiplication in this milieu. For practical purposes, it is clear that the coliforms are (i) a rather ill-defined group of organisms and (ii) less specific than *E. coli* alone as an indicator of faecal contamination in water and foods. In the food industry, therefore, coliforms are used more generally as a hygiene indicator and in characterising certain processes. Also, they may be used to demonstrate post-processing contamination of food and microbial growth during storage (Brown and Baird-Parker, 1982).

Although *E. coli* is mainly a gut commensal in man and other warm-blooded hosts, it also includes the less common pathogenic strains and those with resistance to widely-used antimicrobials that may be transferable to other organisms, including pathogens. The most typical strains are motile, produce acid and gas from lactose at 44 °C and below, form indole at both 37 °C and 44 °C, and fail to utilise citrate. They are positive in the methyl red test and negative in the Voges–Proskauer reaction, while H_2S is not produced. Lysine is usually decarboxylated, but the organisms fail to hydrolyse urea, grow in KCN broth or liquefy gelatin. Typical strains that produce indole are sometimes referred to as biotype 1 or 'faecal coli' and are more specific than either coliform bacteria in general or total Enterobacteriaceae as direct or indirect indicators of faecal pollution in water or faecal contamination of foods, such as raw meat.

4.3.1 Isolation and enumeration

Many different culture media and methods have been developed for detecting and enumerating *E. coli* and other Enterobacteriaceae in foods (Manafi, 2003), and no single method is likely to be superior in all circumstances to other, relevant methods. Basically, cultural methods involve elective or selective isolation media and include one or more differential criteria, such as lactose fermentation or indole formation. They divide broadly into methods that involve plating of a sample dilution on or in an agar medium and those requiring the inoculation of multiple tubes of a liquid medium for MPN determination. Plating methods allow colonies to be counted and are relatively rapid and inexpensive. MPN methods are more sensitive, allowing lower numbers of the target organism(s) to be recovered, and they can be used to detect gas as well as acid from carbohydrate fermentation. In contrast, such methods are more costly and laborious and less accurate than colony counts.

Among the available liquid media for MPN determinations, selectivity is often provided by surface-active agents, such as bile salts or lauryl sulphate, either alone or in combination with another selective ingredient, e.g. brilliant green. Lactose broth, however, contains no selective agents and is only considered suitable for testing material that is normally sterile (AOAC, 1990). Minerals-modified glutamate (MMG) broth is unusual in being a chemically-defined elective medium; this and some of the other liquid media utilise the production of acid and gas from lactose as a differential criterion. In solid media, too, bile salts are used to inhibit unwanted organisms, and selectivity may involve certain dyes, sometimes in conjunction with the bile salts or sodium sulphite. Most agar media contain a fermentable carbohydrate and pH indicator to differentiate between target and non-target organisms but, of course, lack any means of detecting gas production. Other differential systems have also been exploited, such as blackening of colonies due to formation of iron sulphide in a medium containing iron and sulphite.

The usual solid medium for isolating and enumerating Enterobacteriaceae is violet–red–bile–glucose (VRBG) agar. On this medium, the organisms typically produce reddish colonies surrounded by zones of a similar colour. Use of an agar overlay for surface-inoculated or pour plates inhibits the growth of other Gram-negative bacteria. Depending on the incubation temperature, the medium will isolate psychrotrophs (4 °C), mesophiles (35–37 °C) or thermotrophs (42–44 °C), although some psychrotrophic strains can also grow at 42 °C (Zeitoun *et al.*, 1994). Typical colonies are confirmed with the oxidase test, which should be negative, and a test for glucose fermentation.

Similarly, coliform bacteria can be isolated on violet–red–bile agar, containing lactose rather than glucose as the fermentable substrate. Again, the strains isolated depend on the temperature of incubation. A temperature of 35/37 °C is sometimes used but, if the full range of coliforms is being sought, plates should be incubated at 30 °C for 24 hours and all colonies counted, regardless of colour. Psychrotrophic strains can be enumerated specifically, following incubation at 4 °C for 10 days. On this medium, lactose-positive bacteria grow as reddish colonies, each of which may be surrounded by a reddish zone. Lactose-negative strains produce pale colonies often surrounded by greenish zones.

For *E. coli*, an effective method of colony counting was developed by Anderson and Baird-Parker (1975), and modified by Holbrook *et al.* (1980) to include a resuscitation step for sublethally injured cells. The method involves the use of cellulose membranes, each of which is laid on the surface of a plate containing MMG agar and inoculated with an appropriate dilution of the test sample. The plate is then incubated at 37 °C for four hours before the membrane is transferred to a tryptone-bile agar (TBA) plate, with incubation at 44.5 °C for a minimum of 18 hours. Membranes showing growth after this period are treated with indole reagent and the reaction allowed to develop under an ultra-violet (UV) lamp or in bright sunlight. Indole-positive colonies appear pink and are counted as *E. coli.*

Another membrane system is the Hydrophobic Grid Membrane Filter (HGMF) technique (Entis, 1984). In this case, the membrane has hydrophobic lines printed in a grid pattern to act as a barrier and prevent colonies from spreading. A suitable dilution of the food sample is filtered through the membrane, which is then placed on the surface of an agar medium, as before. For enumeration purposes, the number of squares that become occupied by colonies is counted and a formula used to convert the results to an MPN. With red-meat carcasses, the HGMF technique permits detection of one cell of *E. coli* per $100\,cm^2$, when all the homogenate from a swab sample is filtered (Gill *et al.*, 1996).

With both solid and liquid isolation media, an increasingly popular approach is the use of enzymatic tests for detecting and enumerating target bacteria. Most strains of *E. coli* produce the enzyme β-D-glucuronidase, which can be detected by incorporating different chromogenic and fluorogenic substrates, including 5-bromo-4-chloro-3 indolyl-β-D-glucuronide (XGLUC). When XGLUC is present, positive colonies show a discrete blue colour. Tryptone-bile-glucuronic (TBX) agar contains XGLUC and is now used with or without membrane filters in standard methods for enumerating *E. coli* in foods and animal feeding stuffs (ISO, 2001a, b). Use of TBX agar allows the detection and enumeration of β-glucuronidase-positive *E. coli* within 24 hours and without the need for confirmatory tests. Detection is limited to strains growing at 44 °C and β-glucuronidase-negative strains, such as *E coli* O157, are not detected on this medium.

The possible use of alternative, rapid methods for detecting and enumerating the indicator (and index) groups within the Enterobacteriaceae should not be overlooked. Notable among them are methods relating to the Petrifilm™ Coliform Count Plate and the *E. coli* Count Plate (3M Corporation, St Paul, MN, USA) described in Chapter 11. There are also electrical methods, involving impedance and conductance, and discussed by Gibson (2003). These are widely used and accepted now for both raw and processed foods, and can obviate the need for MPN determinations. However, the equipment needs to be re-calibrated for each type of food being examined (Wawerla *et al.*, 1999).

Taking account of the various isolation media and methods that are available for *E. coli*, the US Food Safety Inspection Service (FSIS) does not specify exactly the method to be used in conjunction with the HACCP system for meat plants. Instead, FSIS requires a method that is approved by the Association of Official Analytical Chemists or any method that has been validated scientifically in collaborative trials against the three-tube MPN method (USDA-FSIS, 1996). Other major bodies, such as the International Commission on Microbiological Specifications for Foods and the International Standards Organisation, also allow a choice of methods.

4.3.2 The need for resuscitation of injured cells

Any microbial population that is subjected to environmental stress is likely to include a proportion of cells that are viable, but sublethally injured. Injury can arise from processes such as heat treatment, exposure to antimicrobial agents, refrigeration and freezing, all of which can occur during abattoir operations. Structural and physiological effects of stress on microbial cells were reviewed by Stephens and Mackey (2003), who highlight three important manifestations of cellular injury: (i) inhibition by the selective agents used in culture media; (ii) sensitivity to the low concentrations of reactive oxygen species that occur; and (iii) an extended lag period in the growth cycle, during which cell damage is repaired. Injured cells cannot be ignored, since they retain any potential to cause spoilage of foods or illness in man, and resuscitation may occur either in the food or following ingestion. To ensure that injured cells are included in microbial counts, it is necessary to use a resuscitation procedure before selective enumeration is attempted.

Following exposure to stressful conditions, the proportion of injured cells can be relatively high, and Ray and Speck (1973) found that 90 % of cells surviving a freeze treatment were injured and unable to form colonies on VRB agar or deoxycholate–lactose agar. Moreover, exposure to hot agar during plate pouring reduced counts of freeze-injured *E. coli* on a selective medium by 80 %, while those enumerated on a non-selective medium were less affected. The time needed for complete repair of injury appears to vary widely and can be in excess of 20 hours (Stephens *et al.*, 1997). Such a long period is provided by the type of detection method that is commonly used for *Salmonella*, in which a non-selective, liquid medium (buffered peptone water) is employed for pre-enrichment and is followed by selective enrichment and plating. For enumeration purposes, the MPN method can embody a similar approach, with sample dilutions being added to multiple tubes of a non-selective medium prior to subculture into a selective medium for final incubation. However, the precision of the MPN method is low, unless the number of replicate tubes per dilution is extended. Therefore, a plate test, with resuscitation on a solid medium, is generally preferred, as in membrane-filter methods.

Use of suitable resuscitation methods for all faecal indicators probably would increase the counts of these organisms from abattoir samples. In practice, however, no resuscitation step is normally included, unless it is an integral part of the enumeration method. Resuscitation of injured cells is most important in relation to foods processed for safety in which no surviving cells of Enterobacteriaceae would be anticipated. In this case, a resuscitation step is essential, when checking that survivors are truly absent.

4.3.3 Relationship to the total viable count (TVC)

The TVC is used as an indicator of the general microbiological condition of the product and the equipment with which it is produced. It is sometimes referred to as the 'aerobic plate count' or 'aerobic colony count', because the test involves incubation in air. The growth medium is one such as plate count agar, used either for surface inoculation by manual spreading of the inoculum or spiral plating, or as pour-plates, and incubated at 30 °C for up to three days or 35/37 °C for two days. (The organisms being sought have no special growth requirements and therefore supplementation of the medium with blood, for example, is unnecessary.) Incubation at 30 °C allows growth of both aerobic spoilage bacteria that tend to be inhibited above 34 °C and facultative anaerobes from the alimentary tract, such as *E. coli*. Any high counts obtained will need to be interpreted in the light of knowledge of the process and the product. For example, they could indicate excessive faecal contamination or mishandling of the product and consequent multiplication of spoilage organisms. On the other hand, TVCs up to 10^6 per gram do not necessarily indicate an unacceptable product, if the meat is vacuum-packed to extend shelf-life and stored chilled (Brown and Baird-Parker, 1982). Because TVCs at 30 °C allow a higher recovery of microbes than the counts that are usually obtained with selective isolation media, the test has been regarded as highly suitable for HACCP verification in red-meat production (Mackey and Roberts, 1993). It appears that meat produced under well-controlled conditions can readily achieve TVCs of 10^4 per cm^2 or less (Roberts *et al.*, 1980; Hudson *et al.*, 1996). However, this test gives no indication of possible contamination of meat with pathogenic organisms.

Since any microbial testing based on cultural methods is relatively time-consuming, various rapid methods for TVC have been developed and are discussed by Gill (2000). One such method, which avoids the need to culture the organisms, is the Direct Epifluorescent Filter Technique (DEFT). This involves an enzyme treatment of the sample, capture of the organisms on a membrane filter and staining with acridine orange. Cells that show an orange-red fluoresence are viable and are counted with an epifluorescence microscope. While the method could be used for both red meat and poultry, its suitability for the latter was found to depend on the method of sampling (Shaw *et al.*, 1987). DEFT could not be used for neck skin sampled by shaking, because particulate material interfered with counting.

A comparative study by Williams Smith (1965a) showed that $\log_{10}$ median counts of *E. coli* from the faeces of different food animals varied from 4.3 to 6.8 per gram, with higher numbers occurring in pigs and poultry than sheep or cattle. However, carriage-rate is only one factor that is likely to affect subsequent levels of carcass contamination, and others with a major influence include the nature of the slaughter process and its control. For red meat, in particular, the most critical factors appear to be the skill of the operatives and the hygiene control measures that are applied (Gill, 1998). Table 4.2 shows microbial counts from beef carcasses leaving the processing line at each of ten abattoirs. These counts differed between establishments by up to 2.9 log units, suggesting a varying standard of hygiene control. Sometimes, only relatively small numbers of *E. coli* were recovered, but

Table 4.2 Arithmetic means ($\log_{10}$) of microbial counts obtained from sides of beef sampled prior to chilling at ten different abattoirs (adapted from Gill *et al.*, 1998)

Abattoir	TVC (cfu/cm^2)	Coliforms (cfu/100 cm^2)	*E. coli* (cfu/100 cm^2)
1	3.4	2.0	2.1
2	3.1	2.0	2.0
3	4.3	3.0	2.0
4	3.6	2.5	1.7
5	4.9	2.9	1.3
6	3.7	1.9	0.8
7	2.8	1.4	0.7
8	2.2	0.8	0.7
9	3.0	1.6	0.6
10	2.0	–*	–

* Recovery too low for inclusion.

this organism was considered to be a better measure of hygiene performance than TVC in relation to product safety (Gill *et al.*, 1998).

4.4 Monitoring faecal contamination of meat

Despite being less specific than *E. coli* as faecal indicators and a broader grouping than the coliforms, the presence of Enterobacteriaceae on fresh meat and poultry is generally indicative of faecal contamination, and high numbers obtained from abattoir samples would usually suggest unhygienic processing. In this situation, the inclusion of pathogens, such as *Salmonella* and *Shigella*, is likely to have little effect on the count because, if present at all, the pathogens would be greatly outnumbered by *E. coli*. Also, the presence of other non-lactose fermenters could be due mainly to environmental contamination. However, high counts of psychrotrophic Enterobacteriaceae on meat would indicate poor storage conditions. Psychrotrophic strains isolated from red meat have included species of *Citrobacter*, *Enterobacter*, *Erwinia*, *Klebsiella*, *Kluyvera* and *Serratia* (Kleeberger *et al.*, 1980). For poultry, Table 4.3 shows the incidence of Enterobacteriaceae as a whole, after incubation of VRBG agar at 30 °C, in samples taken at different stages of processing (Mead *et al.*, 1982). Caeca were also sampled at slaughter and gave a mean count of log 8.2, while counts from carcass neck skin varied between log 5.2 and 6.4. When isolates were identified, 99 % of those from caecal samples were *E. coli*, as were 72–98 % of strains isolated from carcasses. Psychrotrophic strains were not detected in the caeca or on freshly-slaughtered broilers but, subsequently, they accounted for 15–18 % of all Enterobacteriaceae occurring on car-

Table 4.3 Incidence of psychrotrophic coliforms at a poultry processing plant (adapted from Mead *et al.*, 1982)

Sample	Visit	Enterobacteriaceae*	Psychrotrophic strains (%)
Neck skin: after slaughter	1	4.5	0
	2	5.0	0
	3	4.0	0
after evisceration	1	5.4	41
	2	5.9	38
	3	5.7	51
after water chilling	1	4.6	33
	2	5.0	35
	3	4.4	40
Caecal content	1	8.3	0
	2	7.9	0
	3	6.8	0
Chiller water	1	3.5	0
	2	3.7	23
	3	3.2	41

* Geometric mean of three samples in each case ($\log_{10}$ cfu/g or ml).

casses and appear to have been acquired during processing. The strains studied belonged to the genera *Citrobacter, Enterobacter, Hafnia* and *Serratia*, while Cox *et al.* (1975) also found psychrotrophic *Klebsiella* and *Providencia* spp. in poultry samples.

It is clear that all of the faecal indicators considered here can be found on carcasses of red-meat animals and poultry without any of the associated pathogens necessarily being present. Thus, whatever test is used, there is merely an indication of the possible presence of enteric pathogens. With poultry, there is the added problem that scalding of carcasses below 60 °C allows many of the original faecal contaminants to survive, so that any further contamination during processing may be impossible to distinguish (Notermans *et al.*, 1977). This is a serious obstacle to using any of the usual faecal organisms for monitoring process hygiene effectively in poultry meat production. After water-immersion chilling of faecally-contaminated chicken halves, Cason *et al.* (2004) found no significant differences between counts of *E. coli*, coliforms and Enterobacteriaceae, and demonstrated that counts obtained post-chill showed no relationship at all to pre-chill faecal contamination.

At any stage of an abattoir operation, the comparability of counts obtained for *E. coli*, coliforms and Enterobacteriaceae will depend upon the nature of the component organisms. When *E. coli* is the predominant species on meat, counts of all three are likely to be similar, while a predominance of other related organisms would obviously change this relationship. A study of European poultry processing plants (CEC, 1979) found a very high correlation between counts of coliforms and Enterobacteriaceae at different stages of processing, but levels of Enterobacteriaceae were generally higher. In this particular study, an incubation temperature of 37 °C was used in both cases. Knowledge of the incubation temperature is important in interpreting the results of tests for Enterobacteriaceae. It has been emphasised that counts at 30–37 °C are unsuitable for assessing the bacteriological safety of raw carcass meat, because of the likelihood of recovering psychrotrophic strains over this temperature range (McEvoy *et al.*, 2004; Struijk and Mossel, 2005). Therefore, it is advisable to restrict the test to thermotrophic strains, i.e. those growing at 42.5 °C. Many psychrotrophic Enterobacteriaceae grow relatively uninhibited at 37 °C and they are environmental rather than faecal in origin. The proportion of mesophilic to thermotrophic Enterobacteriaceae on beef increased from log 0.1 ± 0.0 immediately after dressing to 1.8 ± 0.2 after chilling and subsequent mincing in a cold room (van de Moosdijk *et al.,* 1989).

Despite the pitfalls, the organisms described here continue to be used for regulatory purposes. In the USA, where use of the HACCP system in abattoirs is mandatory, *E. coli* performance criteria are specified for chilled carcasses of each of the main meat species, including poultry (USDA-FSIS, 1996). The organism was chosen partly for its specificity as a faecal indicator and partly because of its similarity to enteric pathogens, such as *E. coli* O157:H7 and *Salmonella.* Also, there are reliable, well-established methods for detection and enumeration. The criteria used are not enforceable regulatory standards, but are intended to assist slaughter establishments and the authorities in ensuring that establishments meet their obligation to prevent and reduce contamination of carcasses with faecal material, ingesta and associated bacteria (USDA-FSIS, 1996). The initial criteria were based on the results of a national survey of abattoir performance and not set at levels that would be attainable commercially with rigorous implementation of the HACCP system (Gill, 2000). However, it is intended that numerical targets will be subject to change as new data become available (USDA-FSIS, 1996).

In the European Union (EU), on the other hand, the hygiene of slaughter and dressing operations is being monitored by testing red-meat carcasses for aerobic colony counts and counts of Enterobacteriaceae (EU, 2005). Thus, carcasses are tested after dressing, but prior to chilling, and the limit values are discussed in Chapters 2 and 6. The method used for Enterobacteriaceae is based on ISO 7402:1985, involving a colony count technique (ISO, 1985). In the case of minced meat, mechanically separated meat and meat preparations, however, the required faecal indicator is *E. coli*, with sampling at the end of the manufacturing process. For this purpose, the method is based on BS EN ISO 16649-2: 2001, involving pour-plates of TBX agar or filtration of the diluted sample through cellulose membranes that are placed on the agar surface (ISO, 2001a, b). Following incubation of the plates, colonies of β-glucuronidase-positive *E. coli* are counted.

The testing of poultry carcasses during processing for faecal indicators is not required in the new EU legislation. This is because studies undertaken to assess the relevance of tests for indicator bacteria suggest that there is a poor relationship between bacterial levels and process hygiene. Therefore, the use of microbiological analysis for process verification purposes is likely to be of limited value in this case. From a study of three UK processing plants by Hutchison *et al.* (2006), coefficients of variation (CV) were determined for a range of bacterial indicators, including the three main faecal indicators discussed here. TVCs and counts of Enterobacteriaceae were among those with the lowest CV, but uncertainty measurements for most organisms tended to be high, especially when the counts were low (<1.7 log). It was apparent that the measurement error could easily exceed the measurement itself. Also, only weak relationships were found between bacterial indicator levels on carcasses and the duration of processing, although the cleanliness of the processing environment clearly diminished with time.

4.5 Other indicator organisms

The use of *Aeromonas* and *Listeria* spp. as indicators of potentially pathogenic strains occurring in abattoirs is described in Chapter 6. For frozen meat, there is also the question of whether the usual faecal indicators survive well enough during freezing and frozen storage. When frozen meat is imported, it would be desirable to have a type of indicator that survives as long or longer than the relevant pathogens, and a possible candidate is the enterococci, sometimes referred to loosely as 'faecal streptococci'. Not all of these Gram-positive organisms are associated with the faeces of warm-blooded animals, but some species, including *Enterococcus faecalis* and *Enterococcus. faecium*, are commonly present in the gut and therefore can be found on finished carcasses (Wilkerson *et al.*, 1961). Unfortunately, there is little data on levels of carcass contamination, but more could be obtained readily with available isolation media. An example of a medium used for isolating and enumerating enterococci is the M-Enterococcus agar of Slanetz and Bartley (1957). The medium, which contains sodium azide as the selective agent, was originally used with membrane filters, but is also appropriate for direct plating. Selectivity is increased by incubating plates at 37 °C for four hours, followed by 44 ± 1 °C for 44 hours (Corry *et al.*, 2003).

4.6 Use of *E. coli* for determining cross-contamination

Inoculation of carcasses or selected items of equipment with an easily identifiable marker organism, in order to determine its subsequent spread in the abattoir, is a useful tool in assessing hygiene control, improving existing practices and training staff (Hudson *et al.*, 1998). The organism most commonly used for this purpose is a non-pathogenic strain of *E. coli* K12, with high resistance to nalidixic acid. The resistance is chromosomal rather than plasmid-borne and therefore unlikely to be transferred to other organisms in the abattoir environment. The strain may be enumerated specifically on MacConkey agar no. 3 containing 200 µg/ml of nalidixic acid (Mead *et al.*, 1994). Carcasses are inoculated from a culture grown at 37 °C for 24 hours and diluted as required. The organism is conveniently applied to the carcass surface with a 37 mm-wide paint brush, covering an area of 250 cm^2 for sheep and 500 cm^2 for cattle (Hudson *et al.*, 1998). Poultry carcasses can be dealt with more readily, because they can be removed from the processing line for inoculation. In this case, a method of spray-inoculation has been used (Mead *et al.*, 1994), but carcasses must be inoculated away from the immediate processing area to avoid inadvertent contamination. To recover the marker at the end of the trial, carcasses are sampled in the usual way and sample dilutions are surface-plated on the specific selective medium. All operations must involve aseptic precautions, with the use of sterile sampling instruments and disposable gloves.

Application of the marker in a beef abattoir showed that bagging the excised anus reduced, but did not prevent, the spread of the organism from an inoculum applied in the anal region before the hide was removed. Inoculation of sheep carcasses at a single site in another abattoir led to the marker being recovered from other sites. However, the contamination was significantly reduced ($P < 0.001$) when the operative responsible for flaying had cleaned his hands, arms and apron before and during the handling of each carcass, and used a clean knife on several occasions (Hudson *et al.*, 1998). In a poultry processing plant, use of the marker showed that the spread of organisms during manual cleaning of neck flaps could be reduced when the operative concerned rinsed her hands briefly in chlorinated water between carcasses (Mead *et al.*, 1994).

These are merely examples of marker usage, from which improved practices could develop. Because the techniques involved are relatively simple and do not require sophisticated laboratory facilities, they may be used 'in-house' for training staff and making them aware of microbiological hazards and their control in the abattoir.

4.7 Sources of further information

The classification and properties of organisms that are included in the family Enterobacteriaceae are given in the ninth edition of *Topley and Wilson's Microbiology and Microbial Infections* (Collier *et al.*, 1998). Conventional media and methods for isolating and enumerating indicator organisms in general are described in Corry *et al.* (2003), and the chapter by Manafi in that book covers the Enterobacteriaceae. The use and misuse of the Enterobacteriaceae test in food microbiology are considered by Cox *et al.* (1988). Details of rapid methods that apply to indicators can be found in McMeekin (2003). The role of microbiological testing in food safety management is discussed by the International Commission on Microbiological Specifications for Foods (ICMSF, 2002).

4.8 References

ANDERSON J M and BAIRD-PARKER A C (1975), 'A rapid direct plate method for enumerating *Escherichia coli* biotype I in foods', *Journal of Applied Bacteriology*, **39**, 111–117.

AOAC (1990), *Official Methods of Analysis*, 14th edition, Arlington, VA, USA, Association of Official Analytical Chemists.

BEERENS H (1998), 'Bifidobacteria as indicators of faecal contamination in meat and meat products: detection, determination of origin and comparison with *Escherichia coli*', *International Journal of Food Microbiology*, **40**, 203–207.

BROWN M H and BAIRD-PARKER A C (1982), 'The microbiological examination of meat', in *Meat Microbiology*, ed. Brown M H, London, UK, Applied Science Publishers, 423–520.

BUTTIAUX R and MOSSEL D A A (1961), 'The significance of various organisms of faecal origin in foods and drinking water', *Journal of Applied Bacteriology*, **24**, 353–364.

CASON J A, BERRANG M E, BUHR R J and COX N A (2004), 'Effect of prechill fecal contamination on numbers of bacteria recovered from broiler chicken carcasses before and after immersion chilling', *Journal of Food Protection*, **67**, 1829–1833.

CEC (1979), *Microbiological Methods for Control of Poultry Meat*, Study P203, VI/5021/79-EN, Brussels, Belgium, Commission of the European Communities.

COLLIER L, BALOWS A and SUSSMAN M (eds) (1998), *Topley and Wilson's Microbiology and Microbial Infections*, 9th edition, vol 2, *Systematic Bacteriology*, London, UK, Arnold.

COLLINS C H, LYNE P M, GRANGE J M and FALKINHAM J O III (eds) (2004), *Collins & Lyne's Microbiological Methods*, 8th edition, London, UK, Arnold, 144–155.

CORRY J E L, CURTIS G D W and BAIRD R M (eds) (2003), *Handbook of Culture Media for Food Microbiology*, Amsterdam, Netherlands, Elsevier Science.

COX L J, KELLER N and VAN SCHOTHORST M (1988), 'The use and misuse of quantitative determinations of Enterobacteriaceae in food microbiology', *Journal of Applied Bacteriology*, **65**, Supplement, 237S–249S.

COX N A, MERCURI A J, JUVEN B J and THOMSON J E (1975), 'Enterobacteriaceae at various stages of poultry chilling', *Journal of Food Science*, **40**, 44–46.

ENTIS P (1984), 'Enumeration of total coliforms, faecal coliforms and *Escherichia coli* in foods by hydrophobic grid membrane filter: collaborative study', *Journal of The Association of Analytical Chemists*, **67**, 812–823.

EU (2005), Commission Regulation (EC) No. 2073/2005 of 15 November 2005 on microbiological criteria for foodstuffs', *Official Journal of the European Union*, **L338**, 1–26.

GARCÍA-LÓPEZ M L, PRIETO M and OTERO A (1998), 'The physiological attributes of Gram-negative bacteria associated with spoilage of meat and meat products', in *The Microbiology of Meat and Poultry*, eds Davies A and Board R, London, UK, Blackie Academic and Professional, 1–34.

GIBSON D (2003), 'Electrical methods', in *Detecting Pathogens in Food*, ed. McMeekin T A, Cambridge, UK, Woodhead Publishing, 147–164.

GILL C O (1998), 'Microbiological contamination of meat during slaughter and butchering of cattle, sheep and pigs', in *The Microbiology of Meat and Poultry*, eds Davies A and Board R, London, UK, Blackie Academic and Professional, 118–157.

GILL C O (2000), 'HACCP in primary processing: red meat', in *HACCP in the Meat Industry*, ed. Brown M, Cambridge, UK, Woodhead Publishing, 81–122.

GILL C O, McGINNIS J C and BADONI M (1996), 'Use of total or *Escherichia coli* counts to assess the hygienic characteristics of a beef carcass dressing process', *International Journal of Food Microbiology*, **31**, 181–196.

GILL C O, DELANDES B, RAHN K, HOUDE A and BRYANT J (1998), 'Evaluation of the hygienic performances of the processes for beef carcass dressing at 10 packing plants', *Journal of Applied and Environmental Microbiology*, **84**, 1050–1058.

HOLBROOK R, ANDERSON J M and BAIRD-PARKER A C (1980), 'Modified direct plate method for counting *Escherichia coli* in foods', *Food Technology, Australia*, **32**, 78–83.

HOLDEN W S (ed.) (1970), *Water Treatment and Examination,* London, UK, J & A Churchill.

HUDSON W R, MEAD G C and HINTON M H (1996), 'Relevance of abattoir hygiene assessment to microbial contamination of British beef carcasses', *Veterinary Record*, **139**, 587–589.

HUDSON W R, MEAD G C and HINTON M H (1998), 'Assessing abattoir hygiene with a marker organism', *Veterinary Record*, **142**, 545–547.

HUTCHISON M L, WALTERS L D, MEAD G C, HOWELL M and ALLEN V M (2006), 'An assessment of sampling methods and microbiological hygiene indicators for process verification in poultry slaughterhouses', *Journal of Food Protection*, **69**, 145–153.

ICMSF (2002), *Microorganisms in Foods 7. Microbiological Testing in Food Safety Management*, eds Tompkin R B, Roberts T A, van Schothorst M, Cole M B, Gram L, Buchanan R L and Dahms S, International Commission on Microbiological Specifications for Foods, New York, USA, Kluwer Academic/Plenum Publishers.

ISO (1985), ISO 7402:1985, 'Microbiology – General guidance for the enumeration of Enterobacteriaceae without resuscitation – MPN technique and colony count technique', Geneva, Switzerland, International Organization for Standardization.

ISO (2001a), ISO 16649-1:2001, 'Microbiology of food and animal feeding stuffs – Horizontal method for the enumeration of beta-glucuronidase-positive Escherichia coli – Part 1: Colony-count technique at 44 degrees C using membranes and 5-bromo-4-chloro-3-indolyl beta-D glucuronide', Geneva, Switzerland, International Organization for Standardization.

ISO (2001b), ISO 16649-2:2001, 'Microbiology of food and animal feeding stuffs – Horizontal method for the enumeration of beta-glucuronidase-positive Escherichia coli – Part 2: Colony-count technique at 44 degrees C using 5-bromo-4-chloro-3-indolyl beta D-glucuronide', Geneva, Switzerland, International Organization for Standardization.

KLEEBERGER A, SCHAEFER K and BUSSE M (1980), 'The ecology of enterobacteria on slaughterhouse meat', *Fleischwirtschaft*, **60**, 1529–1531.

MACKEY B M and ROBERTS T A (1993), 'Improving slaughtering hygiene using HACCP and monitoring', *Fleischwirtschaft International*, **2**, 40–45.

MANAFI M (2003), 'Media for detection and enumeration of "total" Enterobacteriaceae, coliforms and *Escherichia coli* from water and foods', in *Handbook of Culture Media for Food Microbiology*, eds Corry J E L, Curtis G D W and Baird R M, Amsterdam, Netherlands, Elsevier Science, 167–193.

McEVOY J M, SHERIDAN J J, BLAIR I S and McDOWELL D A (2004), 'Microbial contamination on beef in relation to hygiene assessment based on criteria used in EU decision 2001/471/EC', *International Journal of Food Microbiology*, **92**, 217–225.

McMEEKIN T A (ed.) (2003), *Detecting Pathogens in Food,* Cambridge, UK, Woodhead Publishing.

MEAD G C, ADAMS B W and HAQUE Z (1982), 'Studies on the incidence, origin and spoilage potential of psychrotrophic Enterobacteraceae occurring on processed poultry', *Fleischwirtschaft*, **62**, 1140–1144.

MEAD G C, HUDSON W R and HINTON M H (1994), 'Use of a marker organism in poultry processing to identify sites of cross-contamination and evaluate possible control measures', *British Poultry Science*, **35**, 345–354.

MOSSEL D A (1982), 'Marker (index and indicator) organisms in food and drinking water. Semantics, ecology, taxonomy and enumeration', *Antonie van Leeuwenhoek*, **48**, 609–611.

MOSSEL D A and STRUIJK C B (1995), 'Escherichia coli, other Enterobacteriaceae and additional indicators as markers of microbiological quality of food: advantages and limitations', *Microbiologia*, **11**, 75–90.

MOSSEL D A A and STRUIJK C B (2004), 'Assessment of the microbial integrity, sensu G. S. Wilson, of piped and bottled drinking water in the condition as ingested', *International Journal of Food Microbiology*, **92**, 375–390.

MOSSEL D A, VAN DER ZEE H, HARDON A P and VAN NETTEN P (1986), 'The enumeration of thermotrophic types amongst Enterobacteriaceae colonizing perishable foods', *Journal of Applied Bacteriology*, **60**, 289–295.

NOTERMANS S, VAN LEUSDEN F M and VAN SCHOTHORST M (1977), 'Suitability of different bacterial groups for determining faecal contamination during post scalding stages in the processing of broiler chickens', *Journal of Applied Bacteriology*, **43**, 383–389.

RAY B and SPECK M L (1973), 'Enumeration of *Escherichia coli* in frozen samples after recovery from injury', *Applied Microbiology*, **25**, 499–503.

ROBERTS T A, MACFIE H J H and HUDSON W R (1980), 'The effect of incubation temperature and site of sampling on assessment of the numbers of bacteria on red meat carcases at commercial abattoirs', *Journal of Hygiene, Cambridge*, **85**, 371–380.

SHARPE A N (1980), *Food Microbiology. A Framework for the Future*, Springfield, Illinois, USA, Thomas.

SHAW B G, HARDING C D, HUDSON W H and FARR L (1987), 'Rapid estimation of microbial numbers on meat and poultry by the Direct Epifluorescent Filter Technique', *Journal of Food Protection*, **50**, 652–657.

SLANETZ L W and BARTLEY C H (1957), 'Numbers of enterococci in water, sewage and feces, determined by the membrane filter technique with an improved medium', *Journal of Bacteriology*, **74**, 591–595.

STEPHENS P J and MACKEY B M (2003), 'Recovery of stressed microorganisms', in *Handbook of Culture Media for Food Microbiology*, eds Corry J E L, Curtis G D W and Baird R M, Amsterdam, Netherlands, Elsevier Science, 25–48.

STEPHENS P J, JOYNSON J A, DAVIES K W, HOLBROOK R, LAPPIN-SCOTT H M and HUMPHREY T J (1997), 'The use of an automated growth analyser to measure recovery times of single, heat-injured *Salmonella* cells, *Journal of Applied Microbiology*, **83**, 445–455.

STRUIJK C B and MOSSEL D A A (2005), 'Letter to the Editor of the International Journal of Food Microbiology', *International Journal of Food Microbiology*, **99**, 113–114.

USDA-FSIS (1996), Part II, Department of Agriculture, Food Safety and Inspection Service, 9 CFR Part 304, *et al.* Pathogen Reduction; Hazard Analysis and Critical Control Point (HACCP) Systems; Final Rule, *Federal Register*, **61**(144), 38806–38989.

VAN DE MOOSDIJK A, MOSSEL D A A and VAN NETTEN P (1989), 'Use of the ecology determinant ("epsilon") factor in establishing relationships between thermotrophic Enterobacteriaceae and *Salmonella* species', *Journal of Applied Bacteriology*, **67**, xviii.

WAWERLA M, STOLLE A, SCHALCH B and EISGRUBER H (1999), 'Impedance microbiology: applications in food hygiene', *Journal of Food Protection*, **62**, 1488–1496.

WILKERSON W B, AYRES J C and KRAFT A A (1961), 'Occurrence of enterococci and coliform organisms on fresh and stored poultry', *Food Technology, Champaign*, **15**, 286–292.

WILLIAMS SMITH H (1965a), 'Observations on the flora of the alimentary tract of animals and factors effecting its composition', *Journal of Pathology and Bacteriology*, **89**, 95–122.

WILLIAMS SMITH H (1965b), 'The development of the flora of the alimentary tract in young animals', *Journal of Pathology and Bacteriology*, **90**, 495–513.

ZEITOUN A A M, DEBEVERE J M and MOSSEL D A A (1994), 'Significance of Enterobacteriaceae as index organisms for hygiene on fresh untreated poultry, poultry treated with lactic acid and poultry stored in a modified atmosphere', *Food Microbiology*, **11**, 169–176.

5

Spoilage organisms of red meat and poultry

J. E. L. Corry, University of Bristol, UK

5.1 Introduction

'Red meat' is usually defined as meat derived from cattle, sheep, pigs, goats, deer and horses, but could include meat from many other mammals. Most microbiological studies have been on cattle, sheep and pigs. 'Poultry', besides chickens, is often taken to include other birds, particularly turkeys, ducks and geese. The microbiology of chicken carcasses has been studied most intensively and, where studied for other birds, seems to be similar for all poultry meat processed in a comparable manner.

The microbiology of red meat and poultry meat is generally similar, and the differences that do occur are mostly related to the different slaughter and processing conditions used. These include: (i) red-meat carcasses are split before evisceration; (ii) red-meat carcasses are skinned (except, usually, pigs), while poultry are processed almost entirely by mechanical means, without removing the skin, and eviscerated without splitting the carcass; (iii) poultry processing usually involves the use of copious quantities of water, while water is used sparingly during red-meat processing, with greater emphasis on avoiding contamination of the meat from hides and gut contents. An example of the different systems used is illustrated by the contrasting methods used to defeather poultry or dehide cattle or sheep. Removal of chicken feathers is accomplished by immersing the carcasses in warm water (52–59 °C), to loosen the feathers, and then flailing or scouring the carcasses with banks of revolving rubber fingers, whilst rinsing with plenty of water. Skinning of cattle and sheep is carried out in the absence of water, preferably with dry, clean hides, and using a careful technique, with or without a mechanical hide-puller, while minimising contact between the dirty outer surface of the skin and the clean underlying tissues. Thus, the microbes present on both red-meat and poultry carcasses are almost entirely on the surface or skin, not in the deep muscle.

For both red meat and poultry, this chapter will consider the nature of the microbes present on processed carcasses and development of the spoilage microflora during storage of meat under different conditions. Also described are media and methods for detecting, isolating and identifying spoilage organisms, including bacteria, yeasts and moulds.

5.2 Microflora of raw meat immediately after slaughter (red meat and poultry)

The microbes present on carcasses and raw meat originate from four main sources: (i) the skin, feathers or hair of the animal or bird (and any faecal matter present); (ii) the gut contents; (iii) the abattoir environment (equipment, air, water) and (iv) the knives, other hand tools and hands of the operatives. Thus, a very wide variety of microbes can be found, including Gram-negative and Gram-positive bacteria, yeasts and moulds, which can encompass pathogens as well as spoilage organisms. Groups reported on red meat prior to storage were reviewed by Dainty *et al.* (1983) and included, in order of frequency: *Micrococcus*, *Pseudomonas*, *Moraxella*/*Acinetobacter*, *Lactobacillus*, *Flavobacterium*, coryneforms, yeasts, Enterobacteriaceae, *Staphylococcus*, *Kurthia*, *Streptococcus*, *Bacillus* and *Brochothrix thermosphacta*. Organisms reported on chicken carcasses before storage have included *Micrococcus*, *Corynebacterium*, *Lactobacillus*, Enterobacteriaceae, *Flavobacterium*, *Acinetobacter*/*Moraxella*, *Pseudomonas*, *Shewanella putrefaciens* and yeasts (Daud *et al.*, 1979; Gallo *et al.*, 1988). Numbers of microbes found on red-meat carcasses vary according to the part of the carcass examined, with higher numbers being found on e.g. the brisket, rump and neck, but are generally $<10^4/\text{cm}^2$ (Mackey and Roberts, 1993). Poultry carcasses tend to carry higher numbers of microbes ($>10^4/\text{cm}^2$ skin), because the skin is not removed, and to show little variation between different parts of the carcass due to the method of processing (Daud *et al.*, 1979). Similarly, the outer surface of unskinned pig carcasses often has higher numbers of microbes than carcasses of sheep or cattle.

Contrary to commonly-held belief, there is strong evidence that migration of microbes from the gut to the rest of the carcass does not occur significantly, either during or just after slaughter (Corry, 1978; Gill, 1979; Nottingham, 1982) and, even if highly contaminated knives or pithing rods are used, surprisingly low numbers of microbes can be detected *post mortem* in the deep muscle (Mackey and Derrick, 1979). Uneviscerated and unbled carcasses keep well at refrigeration temperatures (Corry, 1978). Studies on uneviscerated poultry have shown that they have a longer shelf-life than eviscerated carcasses and spoilage is usually biochemical rather than microbiological (Barnes and Shrimpton, 1957; Mead *et al.*, 1974).

5.3 Factors affecting flora development and changes leading to spoilage

Unless raw meat is stored frozen (below about −10 °C), microbes will inevitably multiply. The composition of the microflora will also change during storage, since only a proportion of the initial contaminants will be able to multiply or even survive. The conditions of storage will influence the types of microbe that can grow and determine how rapidly they can do so. At the end of microbiological shelf-life, the microflora of the meat will be dominated by only one or a few types of microbe. The shelf-life of raw meat is affected by a few main factors (considered in more detail later): (i) the initial number of microbes present (particularly those with spoilage potential); (ii) the species/strains present; (iii) the temperature of storage (the lower the better – meat can be chilled to about −2 °C before it freezes); (iv) the gaseous atmosphere. The pH of the meat *post rigor mortis*, which is related to the glucose content, can also affect shelf-life, particularly for meat stored in a vacuum pack or modified atmosphere. The relative humidity of the storage atmosphere is important only for meat held in the unwrapped state.

The pH of living muscle is normally about 7.4 but, after death, respiration becomes anaerobic, so that glucose in the muscle is converted to lactic acid rather than carbon dioxide and, as a result, the pH falls. Normally, this process ceases before all the glucose in the muscle has been consumed, and the pH stabilises at around 5.5–5.8. If the animal was fatigued and/or stressed immediately prior to slaughter, the glucose concentration in the muscles is likely to be low. In that case, the amount of lactic acid produced is reduced, there is very little residual glucose present, and the final pH will be 6.0 or above. Red meat of high pH is known as dark, firm, dry (DFD), because it tends to look darker than normal and is better at retaining moisture than meat of normal pH. Conversely, some meat, usually pork, may achieve an unusually low pH, and is called pale, soft, exudative (PSE) because it loses more moisture ('drip', 'purge' or 'exudate') and is paler and softer than normal. DFD meat tends to have a shorter shelf-life, especially in a modified atmosphere or vacuum pack. The pale breast meat of chickens has a pH similar to that of normal red meat (5.7–5.9), but the dark leg meat normally has a higher pH (6.2–6.4), making it more perishable than breast meat (Barnes *et al.*, 1979). Also, poultry skin has a pH similar to that of leg meat.

5.3.1 Normal spoilage flora in aerobic conditions

Red meat

At chill temperatures, the most numerous spoilage organisms are *Pseudomonas* spp., and meat usually spoils when numbers of bacteria (almost all *Pseudomonas* spp.) reach 10^7–10^8/cm^2. *Pseudomonas* spp. are obligately aerobic, Gram-negative, oxidase-positive, rod-shaped bacteria, most of which are motile by means of a polar flagellum, although some are non-motile. Species most commonly found are two distinct groups of *Ps. fragi* (some of which are non-motile), *Ps. lundensis*, *Ps. fluorescens* and several groups of fluorescent pseudomonads that are closely related to *Ps. fluorescens* (Dainty *et al.*, 1983; Dainty and Mackey, 1992). Other microbes occurring in lower numbers include *Acinetobacter* and *Psychrobacter* spp., Enterobacteriaceae and sometimes (especially in lamb and pork) *Brochothrix thermosphacta*. The Enterobacteriaceae are psychrotrophic species, particularly *Serratia liquefaciens, Enterobacter agglomerans* and *Hafnia alvei*, rather than the better-known mesophilic species (Dainty and Mackey, 1992).

If red meat is stored as carcasses, hanging in a cold store, yeasts and moulds can form a significant part of the microflora, owing to their tolerance of the relatively dry conditions at the meat surface. Similarly, meat stored frozen, whether as carcasses or deboned and packed, but not maintained at a sufficiently low temperature (<–12 °C), is liable to spoil from growth of yeasts and moulds (Gill *et al.*, 1981; Ismail *et al.*, 1995; Corry, unpublished observations). This occurs because such organisms are much more tolerant of low temperature and a_w than bacteria.

Poultry meat

Pseudomonas spp. similar to those found on red meat also tend to be the most important spoilage bacteria in aerobically stored poultry meat (Arnaut-Rollier *et al.*, 1999), and other groups mentioned for red meat (e.g. Enterobacteriaceae and *Br. thermosphacta*) can also be found. However, among the more common Gram-negative bacteria on poultry are *Acinetobacter* and *Psychrobacter* spp. and sometimes *Shewanella putrefaciens* (previously called *Pseudomonas putrefaciens*, later *Alteromonas putrefaciens*). Nevertheless, there is a direct relationship between initial numbers of pseudomonads on poultry meat and its aerobic shelf-life at chill temperatures (Barnes *et al.*, 1979). Yeasts also multiply on raw poultry and can be found in relatively high numbers (up to about 10^4/ml of carcass rinse or per gram, by the end of shelf-life), with *Candida* spp., especially *C. zeylanoides*, and *Yarrowia lipolytica* being most common (Gallo *et al.*, 1988; Ismail *et al.*, 2000; Hinton *et al.*, 2002). Considering that a yeast cell is about 100 times larger than a bacterium (Ingram, 1971), at 10^4/cm^2, they may well contribute significantly to spoilage. Dark (leg) meat tends to have a shorter shelf-life than breast meat, and to support larger numbers of *Sh. putrefaciens*, and other Gram-negative bacteria, including *Acinetobacter* spp. (Barnes *et al.*, 1979).

Nomenclature of Gram-negative spoilage organisms

Older literature concerning the bacteria found on chill-stored raw meat can be quite confusing, since the nomenclature has changed considerably. For instance, Gram-negative, aerobic, short rods/cocci, forming colourless colonies, were previously called '*Achromobacter*' (from their lack of pigmentation), which is no longer a valid name (Dainty *et al.*, 1983). Later they were divided into motile and non-motile groups. Those that were non-motile and oxidase-positive were classified as *Moraxella* spp. and '*Moraxella*-like', while the remaining non-motile, oxidase-negative strains were *Acinetobacter* spp. The '*Moraxella*-like' strains are now *Psychrobacter* spp., the most common being *P. immobilis* and *P. phenylpyruvica*. Most, if not all, *Moraxella* spp. have now been re-classified as non-motile *Pseudomonas fragi* (Dainty and Mackey, 1992; García-Lopéz and Maradona, 2000; Kämpfer, 2000). The most common *Acinetobacter* spp. found on meat are *A. johnsonii* and *A. lwoffi*; unlike *Psychrobacter* spp., they do not produce acid from glucose. However, none of these three groups of obligate aerobes (*Psychrobacter*, *Acinetobacter* and *Moraxella*) seems to be as important as *Pseudomonas* spp. in the spoilage of poultry or red meat. There is also some doubt as to whether *Moraxella* spp. have any relevance to food (Santos *et al.*, 2000). These authors state: 'Since several researchers have claimed that, with few exceptions, strains formerly identified as *Moraxella* or *Moraxella*-like are almost certainly *Psychrobacter* spp., the real importance of moraxellae in food is not known with certainty.' Using phenotypic tests, *Psychrobacter* spp. are difficult to distinguish from non-motile strains of *Pseudomonas fragi*, since both are oxidase-positive, but their DNA G + C ratio is distinctly different (García-Lopéz and Maradona, 2000). A rapid method of differentiation could be to check for the ability to grow on CFC agar (see Table 5.2 on p. 113). On this medium, pseudomonads can grow, but not *Psychrobacter* spp. Both *Psychrobacter* and *Acinetobacter* are sometimes difficult to de-stain during Gram-staining (García-Lopéz and Maradona, 2000; Kämpfer, 2000), and might be mistaken for Gram-positive organisms.

5.3.2 Spoilage flora under modified atmospheres or in vacuum packaging

Red meat is sometimes stored for several months at –1 °C to +1 °C in vacuum packs; then, the meat may be cut up and sold immediately or packed in a modified atmosphere, which gives a shelf-life of about seven days.

The gas mixture used for modified atmosphere packaging (MAP) of red meat is usually 20–30 % CO_2 and 70–80 % O_2. The high concentration of oxygen is used to maintain an attractive red colour in the muscle myoglobin for as long as possible, while the carbon dioxide inhibits multiplication of *Pseudomonas* spp., the most important spoilage bacteria. As a result, the group of microbes that becomes predominant is the lactic acid bacteria (LAB), with a small proportion of *Pseudomonas* spp., Enterobacteriaceae and sometimes *Br. thermosphacta.* The product deteriorates slowly, due to multiplication of the LAB, including *Carnobacterium* spp., which produce sour and/or cheesy-type odours, rather than the unpleasant, putrid odours produced by pseudomonads. However, shelf-life is usually limited by loss of meat colour, rather than by microbial activity.

Red meat stored in vacuum packs at low temperature acquires a similar flora to meat in MAP, provided that a film with suitably low gas-permeability is used. Analysis of the gaseous atmosphere usually reveals a high level of CO_2, as a result of residual respiration in the meat and from metabolic activity of the microflora. Vacuum-packed meat has a longer shelf-life than meat in the MAP atmosphere described above, because the muscle myoglobin is kept in a reduced state and shelf-life is not limited by such pronounced changes in meat colour. When a vacuum pack is opened, the normal, oxygen-dependent meat colour is regained. By the end of shelf-life, there are usually about $10^8/cm^2$ LAB, with numbers of pseudomonads and Enterobacteriaceae (and occasionally *Br. thermosphacta*) being between 10^2 and $10^4/cm^2$. Packs containing 100 % CO_2 are sometimes used for meat stored prior to retail sale, and these appear to confer the longest shelf-life, because they retard the growth of all the organisms present, including the LAB (Blickstad and Molin, 1984).

Raw pork is not commonly vacuum packed. The reason for this may be that it does not benefit from storage to improve texture ('maturation'). Shelf-life in vacuum packs has also been reported to be shorter for pork than for beef or lamb, due to earlier depletion of glucose (Boers cit. Borch *et al.,* 1996).

Apart from the usual spoilage organisms, *Aeromonas* spp. can be found sometimes at high levels on meat stored in vacuum packs (Isonhood and Drake, 2002; Holley *et al.*, 2004), and high numbers of non-pathogenic *Yersinia enterocolitica* have also been reported (Hanna *et al.*, 1976; Gill and Newton, 1979).

High pH meat in modified atmosphere or vacuum packaging

Red meat with pH > 6.0 should not be vacuum packed, because it is likely to have a short shelf-life due to multiplication of *Br. thermosphacta, Sh. putrefaciens* and/or Enterobacteriaceae, and the low level of glucose present, see below (Gill and Newton, 1979; Newton and Gill, 1981; Dainty and Mackey, 1992). Apparently *Br. thermosphacta* does not grow below pH 5.8 and *Sh. putrefaciens* fails to multiply below pH 6.0 (Gill, 1986). Vacuum-packaged lamb does not keep as long as beef, and this may be related to the greater proportion of surface fat with raised pH, allowing faster multiplication of the organisms in question (Gill and Newton, 1979; Shaw *et al.*, 1980; Dainty *et al.*, 1983). The shelf-life of DFD meat in vacuum packs can be extended by adding glucose and acidifying with lactic and/or citric acid (Gill and Penney, 1985).

Poultry meat in modified atmosphere or vacuum packs

Vacuum packaging does extend the shelf-life of poultry meat, but to a lesser extent than red meat and, therefore, is not widely used. As might be expected, breast meat keeps better in vacuum packs than leg meat, owing to its lower pH (Jones *et al.*, 1982: Patterson *et al.*, 1984; ICMSF, 1998). MAP is being used increasingly for chicken, particularly for breast meat, often with 20–30 % CO_2 and 70–80 % O_2. However, oxygen-containing MAP has been found to impart an unpleasant flavour to turkey breast meat (Mead, 2004). As with red meat, poultry in MAP with high CO_2 or vacuum packed develops a microflora with large numbers of LAB, including *Carnobacterium* spp., although often accompanied by significant numbers of pseudomonads, Enterobacteriaceae and *Br. thermosphacta* (Jones *et al.*, 1982; Mead *et al.*, 1986; Studer *et al.*, 1988).

In earlier studies, 'atypical lactobacilli' were frequently reported on red meat and poultry stored in MAP and vacuum packs (Barnes *et al.*, 1979; Dainty *et al.*, 1983). These have since been classified as a new genus, *Carnobacterium* (Collins *et al.*, 1987). *Cb. divergens* and *Cb. piscicola* are the species most often encountered (Dainty and Mackey, 1992), but *Cb. piscicola* has recently been renamed as *Cb. maltaromaticum* (Mora *et al.*, 2003). These species are also common in MAP and vacuum-packed fish and dairy products, and have attracted considerable attention in recent years as potential sources of bacteriocins and as 'protective' cultures, particularly against *Listeria monocytogenes* (Laursen *et al.*, 2005).

Other LAB found on vacuum-packed meat include *Lactobacillus* spp. (*sake*, *curvatus*, *bavaricus* and *delbreuckii*), *Leuconostoc* spp. (*carnosum*, *gelidum* and *mesenteroides*), *Pediococcus damnosus* and *Lactoccoccus raffinolyticus* (Dainty and Mackey, 1992; Jones, 2004; Laursen *et al.*, 2005). Strains of *Carnobacterium* and *Leuconostoc* spp. are heterofermentative, producing mainly CO_2, lactic and acetic acids and ethanol, and are less acid-tolerant than the homofermentative *Lactobacillus*, *Pediococcus* and *Lactococcus* spp., which produce lactic acid and no CO_2.

Very few detailed studies have been made of the LAB growing in vacuum-packed meat. Jones (2004) investigated the LAB present during storage of vacuum-packed beef at −1.5 °C for up to 16 weeks. Replicate striploins from each of five steers and five bulls, all of normal pH, were studied. Differences in microflora were observed between steer and bull meat, and also between replicate samples of the same type. In general, there was a succession of strains from less to more acid-tolerant, with a gradual decrease in pH value for the drip from pH 5.5. In the steer meat, *Cb. divergens* was replaced from week four onwards by successive strains of *Leuc. mesenteroides*, up to week 16. In the bull meat, *Cb. divergens* predominated up to week eight but, by week 12, it had been replaced by an unidentified LAB (4/5 replicates) or *Leuc. mesenteroides* (1/5 replicates), and by week 16, a strain of *Lactobacillus delbreuckii* predominated, with a lower number of *Pediococcus damnosus.* The pH of the steer meat fell increasingly rapidly (mean 5.17 by week 16) and more so than the bull meat (mean 5.38 by week 16). Numbers of LAB also increased more rapidly in the bull meat. Numbers of LAB stabilised at about $\log_{10}$ 8.5 in both types of meat (after four weeks in bull meat and eight weeks in steer meat). These results indicate that the predominant microflora during storage is influenced not only by the composition of the meat and its storage conditions, but also by the strains of bacteria present initially.

5.4 Spoilage of vacuum-packed red meat of normal pH due to psychrophilic or psychrotrophic *Clostridium* spp.

Vacuum packaging of red meat had been in widespread use for many years. In the late 1980s, however, a new type of spoilage emerged as a serious economic problem and involving vacuum-packed beef from southern Africa (Dainty *et al.*, 1989) and the USA (Kalchayanand *et al.*, 1989). The bacteria responsible for these two outbreaks were later identified as a new species of psychrophilic *Clostridium*: *Cl. estertheticum*, so-named because it produced a complex array of esters (Collins *et al.*, 1992; Spring *et al.,* 2003). The meat had a normal pH, was processed in abattoirs with excellent standards of hygiene and had been stored at well-controlled chill temperatures. The vacuum packs were often grossly distended with gas (a mixture of $H_2 + CO_2$, with a smaller proportion of N_2) and produced a highly unpleasant, vomit-like odour.

Since then, at least two other species of *Clostridium* have been found that are capable of spoiling chilled, vacuum-packed red meat, and *Cl. gasigenes* and *Cl. frigidicarnis* have been reported to spoil meat from New Zealand (Broda *et al.*, 1996, 1999, 2000b; Kalinowski and Tompkin, 1999), although *Cl. estertheticum* is the commonest cause of spoilage. Following the first reports from Southern Africa and the USA, spoilage of meat due to *Cl. estertheticum* has been observed in many countries worldwide, including Ireland, the UK, New Zealand and Brazil, and in lamb and venison as well as beef (Broda *et al.*, 2002; Corry, unpublished observations). *Cl. estertheticum* is a strictly psychrophilic, anaerobic sporeformer, with a temperature range for growth from about −1 °C to 14 °C. Its normal habitat is uncertain, but clearly the organism cannot originate from the intestinal flora, since it does not grow at body temperature, and is most likely to be of soil origin. Fortunately, neither *Cl. estertheticum* nor any of the other cold-tolerant spoilage clostridia produce any toxin, so are unlikely to cause food poisoning (Broda *et al.*, 1998a; Corry, unpublished observations). The most likely mode of entry of *Cl. estertheticum* into the abattoir is via soil-contaminated hides, but the organism could also be ingested through feed contaminated with soil, and hence be present in low numbers in faeces (Broda *et al.*, 2002; Boerema *et al.*, 2003).

Once the clostridium begins to spoil meat from a particular abattoir/cutting plant, special cleaning and sporicidal disinfection regimes are needed to eliminate it. Control depends on careful attention to hygiene during de-hiding, ensuring that air movement is not from the slaughter area and lairage towards the cleaner parts of the line, and effective cleaning and disinfection. The last should involve a sporicidal disinfectant, concentrating on parts of the cutting room where clostridia could colonise, e.g. points of contact between meat and equipment, the interior of the vacuum-packaging machine and the refrigeration units and drip trays in the chillers. Effective disinfection requires, on a regular basis, careful removal of all dirt, dismantling and cleaning of machinery and application of a peroxyacetic acid-containing disinfectant by fogging in order to access all contaminated sites. It is also a useful precaution to use only the minimum temperature and time necessary to heat-shrink the wrapping film, since there is evidence that this process may stimulate any *Clostridium* spores present to germinate and subsequently grow and spoil the meat (Bell *et al.*, 2001; Dr Dorota Broda personal communication).

5.5 Bone taint

Bone taint is a type of spoilage that occurs occasionally in the deep parts of red meat carcasses, near to the bone, in the synovial fluid or bone marrow (Callow and Ingram, 1952; Gardner, 1982; Nottingham, 1982; Roberts and Mead, 1986). The deepest hindquarter joints (particularly stifle – tibia/femur) are most often involved, and also sometimes forequarter joints. Nothing appears amiss until the carcass is cut up and an unpleasant odour is detected. Microbiological examination has sometimes failed to detect any causative organisms. In other cases, mesophilic *Clostridium* spp., *Cl. putrefaciens* (a psychrotrophic species) or enterococci have been found in and around the joint (Callow and Ingram, 1952; Roberts and Mead, 1986).

Slow and/or inadequate chilling is thought to be a factor in the appearance of this problem (Roberts and Mead, 1986). During life, the bacteria may gain entry to lymph nodes through abrasions in the skin. Bone taint still occurs (Corry personal observation), but less often, possibly because chilling has improved. Recent investigations by workers in New Zealand have indicated that cold-tolerant clostridia may also be implicated in bone taint, which could explain why earlier investigations, involving only culture at higher temperatures, often failed to find a causative organism (DeLacey *et al.*, 1998; Boerema *et al.*, 2002). Investigating this problem can be difficult, because samples need to be taken in and near intact joints, using an aseptic technique that avoids contaminating the sample from the surface microflora. This is done by sterilising the outer surface of the meat with a searing (red-hot iron) technique and taking a sample with a modified cork borer.

5.6 Spoilage of uneviscerated poultry

As mentioned previously, uneviscerated poultry, with or without feathers, e.g. game birds, such as pheasants, versus 'New York dressed' chickens, keep longer than uneviscerated carcasses (Mead *et al.*, 1974). Spoilage sometimes occurs due to diffusion of metabolic products produced by the microflora multiplying in the intestines. These can form a greenish discolouration in the (still sterile) muscles and skin (Barnes and Shrimpton, 1957; Mead *et al.*, 1974). In pheasants hung at 15 °C, the causative bacteria were mostly mesophilic clostridia and Enterobacteriaceae (Mead *et al.*, 1973) but, at 1° or 10 °C, the clostridia were strains resembling *Cl. putrefaciens*, which was also found occasionally in the muscle (Mead *et al.*, 1974). This species has been little studied (Ross, 1965; Roberts and Derrick, 1975), but it does not appear to be important in the spoilage of vacuum-packed red meat, possibly because it grows more slowly than *Cl. esterthetícum* at chill temperatures.

5.7 Effect of storage temperature on spoilage

The lower the temperature, the longer the lag phase and generation time of the spoilage flora, and the longer the shelf-life that can be achieved (Dainty and Mackey, 1992). This is particularly so for MAP or vacuum-packed meat, because the inhibitory effect of CO_2 is related to its solubility, which increases as the temperature is reduced (Gill, 1988). The storage temperature also determines which microbes will predominate, for instance, the minimum growth temperature of most psychrotrophic Enterobacteriaceae is >1 °C (Borch *et al.*, 1996; Riddel and Korkeala, 1997), so these are unlikely to grow in meat stored <0 °C.

Regez *et al.* (1988) observed that the proportion of *Pseudomonas* spp. in the spoilage flora of poultry kept in sealed plastic bags diminished as the storage temperature was raised from 0 to 20 °C, with *Aeromonas* spp., Enterobacteriaceae and acinetobacters predominating at 10° and 20 °C. Storage of vacuum-packed red meat at temperatures above the chill range is likely to select a microflora containing mesophilic clostridia and Enterobacteriaceae (Ingram and Dainty, 1971), which would be highly hazardous in the unlikely event of the meat being eaten!

5.8 Metabolic products responsible for spoilage

Table 5.1 summarises information on metabolites produced by the major spoilage organisms. In general, the predominant microflora in vacuum packs or MAP produces lower levels of products with unpleasant odours. However, it is often impossible to identify exactly the organism(s) responsible for producing a specific chemical, and there may well be interactions between organisms in the production of some spoilage compounds. For a detailed account of metabolic products formed during aerobic and anaerobic spoilage of meat and poultry, see Viehweg *et al.* (1989), Dainty and Mackey (1992) and Nychas *et al.* (1998).

Table 5.1 Substrates used for growth and production of metabolites by major meat spoilage bacteria

Bacterium	Substrates used for growth		Major end products of metabolism	
	Aerobic	Anaerobic	Aerobic	Anaerobic
Pseudomonas	Glucose^{1} Amino acids^{2} Lactic acid^{3}		Slime, suphides, esters, acids, amines	
Acinetobacter/ Psychrobacter	Amino acids^{1} Lactic acid^{2}		Esters, nitriles, oximes, sulphides	
Shewanella putrefaciens	Glucose^{1} Amino acids^{2} Lactic acid^{3}	Glucose^{1} Amino acids^{1}	Volatile sulphides	H_2S
Brochothrix thermosphacta	Glucose^{1} Ribose^{2}	Glucose^{1}	Acetic acid, acetoin, isovaleric acid, isobutyric acid	Lactic acid, volatile fatty acids, ethanol
Enterobacter	Glucose^{1} Glucose-6-phosphate^{2} Amino acids^{3} Lactic acid	Glucose^{1} Glucose-6-phosphate^{2} Amino acids^{3}	Sulphides, amines	Lactic acid, CO_2, H_2, H_2S, amines
Lactobacillus		Glucose^{1} Amino acids^{1}		Lactic acid, volatile fatty acids
Leuconostoc		Glucose^{1}		Lactic acid, volatile fatty acids, butanoic acid
Carnobacterium		Glucose^{1}		Lactic acid, diacetyl, acetate
Clostridium estertheticum		Glucose Amino acids		CO_2, H_2, butanol, ethanol, butanoic acid, esters, volatile sulphur-containing compounds

The superscript number indicates the order of substrate utilisation, where known.
Source: adapted from ICMSF (1996), with additions from Collins *et al.* (1992) and Jones (2004).

5.9 Isolation and identification of spoilage organisms

Table 5.2 summarises the plating media, method of inoculation, incubation conditions and confirmatory tests for the major spoilage organisms, except *Sh. putrefaciens* and cold-tolerant *Clostridium* species.

When examining stored meat, decimal dilutions may be prepared in MRD (maximum recovery diluent: 0.1 % peptone, 0.9 % NaCl) and surface-plated on the media listed below. If spoilage due to *Cl. estertheticum* is suspected, the author uses a polymerase chain reaction (PCR)-based method of detection.

Since Enterobacteriaceae are sought as spoilage organisms, not as indicators, violet–red–bile–glucose (VRBG) plates should be incubated at 25 ° or 30 °C, because some of the species important in spoilage are unable to grow at 37 °C. Selection depends on inclusion of bile salts and crystal violet to inhibit Gram-positive bacteria, and the medium contains an indicator system of phenol red and glucose in which Enterobacteriaceae produce acid from glucose, yielding red–pink colonies >1 mm in diameter. The medium can be surface-inoculated, without overlaying, for ease of use, although this does not follow recognised practice.

CFC agar contains the quaternary ammonium compound cetrimide to inhibit Gram-positive bacteria and yeasts, as well as the antibiotics fucidin and cephaloridine to inhibit unwanted Gram-negative bacteria (Mead and Adams, 1977; Mead, 1985). Since pseudomonads are strictly aerobic bacteria, inoculation must be on the surface, not by pour-plate. Many laboratories count all colonies on CFC as 'presumptive' *Pseudomonas* spp., but some oxidase-negative bacteria are able to grow on the medium, so it is recommended that all well-spaced colonies be tested for oxidase. This can be done by flooding the plate with the test reagent, but a neater method is to 'blot' the plate with a piece of filter paper to give a mirror image of the colonies, and then flood the paper with the oxidase reagent. Stanbridge and Board (1994) modified CFC agar so that they could differentiate between *Pseudomonas* spp. and any Enterobacteriaceae that could grow on the medium; however, the author did not find this to be effective (Corry, unpublished observations).

Mead (1985) reported that most strains of *Sh. putrefaciens* failed to grow on CFC agar, and the present author has not isolated the organism from chicken or red meat on this medium. However, Tryfinopoulou *et al.* (2001) found, when examining fish stored in modified atmospheres, that *Sh. putrefaciens* from this source does grow, and cannot be differentiated easily from *Pseudomonas* spp., because it is oxidase-positive. Nevertheless, there is a possible distinction in that *Sh. putrefaciens* produces salmon-pink colonies (Mead, 1985). The organism grows readily on non-selective media, such as plate count agar, and on Lyngby iron agar, which is not selective, but detects H_2S producers, including *Sh. putrefaciens*, by yielding black colonies, due to the reaction of H_2S with ferric citrate (Gram *et al.*, 1987). Pour

Table 5.2 Plating media commonly used to enumerate spoilage organisms of red meat and poultry

Group selected	Medium	References	Inoculation	Incubation: temperature/time/ atmosphere	Confirmatory test(s)
Pseudomonas spp.	CFC (cephaloridine-fucidin-cetrimide) agar	Mead and Adams (1977), ISO (1996b)	Surface	25 °C, 48 h, aerobic	Count oxidase-positive colonies
Lactic acid bacteria	MRS (de Man, Rogosa and Sharpe) agar	Anon. (2003d), ISO (1996a)	Surface or pour-plate	25 °C, 5 days, anaerobic	Count catalase-negative colonies
Brochothrix thermosphacta	STAA (streptomycin-thallium acetate-actidione) agar	Gardner (1966, 2003),ISO (1997)	Surface	25 °C, 48 h, aerobic	Count catalase-positive, oxidase-negative colonies
Yeasts and moulds	RBC (rose bengal-chloramphenicol) agar	Jarvis (1973), Anon. (2003c)	Surface	25 °C, 5 days in dark, lids up, aerobic	Differentiate bacteria from yeasts by microscopy
Yeasts and moulds	DRBC (dichloran-rose bengal-chloramphenicol) agar	King *et al.* (1979) Anon. (2003e)	Surface	25 °C, 5 days in dark, lids up, aerobic	Differentiate bacteria from yeasts by microscopy
Yeasts and moulds	OGY (oxytetracycline-glucose-yeast extract) agar	Mossel *et al.* (1970) Anon. (2003f)	Surface	25 °C, 5 days in dark, lids up, aerobic	Differentiate bacteria from yeasts by microscopy
Enterobacteriaceae	VRBG (violet–red–bile–glucose) agar	Mossel *et al.* (1978), ISO (2004)	Pour-plate or surface, with or without overlay	30 °C, 48 h, aerobic	Count all pink or red colonies >1 mm diameter

plating and aerobic incubation at 20–25 °C for three days is recommended for Lyngby agar.

LAB are usually enumerated on deMan, Rogosa and Sharpe (MRS) agar pH 6.5 (deMan *et al.*, 1960), which will also grow *Carnobacterium* spp. MRS agar is not selective, but is nutritionally rich, and often incubated anaerobically to inhibit aerobic competitors. MRS agar at pH 5.7 with sorbic acid as a selective agent has been recommended for the examination of refrigerated meats for LAB (Reuter, 1985; Anonymous, 2003a). This might exclude carnobacteria, which are considered less acid-tolerant than other LAB (Schillinger and Holzapfel, 2003), although Jones (2004) successfully isolated *Carnobacterium divergens* using this medium. Cresol red-thallium acetate (CTAS) agar at pH 9.0 (Anonymous, 2003b) contains thallium acetate, nalidixic acid and sodium citrate as selective agents in addition to the high pH, with sucrose and cresol red as indicators, and also triphenyl tetrazolium chloride. On this medium, *Carnobacterium maltaromaticum* produces small bronze-metallic, shiny, yellowish to pinkish colonies and changes the medium from red to yellow due to acid production from sucrose, as well as clearing the precipitate originally present in the medium. By contrast, *Cb. divergens* produces only pinprick-sized colonies, often without any colour change in the medium. *Leuconostoc* spp. and some enterococci will also grow on CTAS (Anonymous, 2003b; Schillinger and Holzapfel, 2003). In a further development, Wasney *et al.* (2001) formulated cresol red–thallium acetate–sucrose–inulin (CTSI) medium from CTAS by reducing the concentrations of thallium acetate and sucrose, and adding inulin, while increasing the level of manganese sulphate, adding vancomycin and nisin to inhibit unwanted organisms, and including thiamine as a growth factor.

Various media have been suggested for enumerating yeasts and moulds in high a_w foods (Table 5.2, Beuchat, 2003). The most commonly used is rose bengal–chloramphenicol (RBC) agar, which originally contained chlortetracycline to inhibit bacteria, but now includes chloramphenicol, which is more stable. The rose bengal has the advantage of retarding the spread of mould colonies so that both yeasts and moulds can be counted. As with all media designed to select yeasts, it is wise to check yeast-like colonies by microscopy, because some bacteria are able to grow on these media. The oxytetracycline–glucose–yeast extract (OGY) medium of Mossel *et al.* (1970) originally contained gentamicin, in addition to oxytetracycline. It was developed specifically for the examination of meat, especially raw minced meat, which can contain high numbers of yeasts, but may be less suitable for meat with moulds present, as these might spread and obscure the yeasts colonies. DRBC agar is a development from RBC agar, and contains dichloran in addition to rose bengal. This combination is more efficient at retarding the spread of mould colonies.

There is no selective medium that is suitable for cold-tolerant clostridia in general, or for *Cl. estertheticum* in particular. The organisms can be isolated sometimes by treating samples with ethanol or heat, in order to inactivate vegetative cells of other organisms, leaving bacterial spores (Broda *et al.*, 1998b and c). Occasionally it has been possible to isolate *Cl. estertheticum* by using ethanol treatment and plating on blood agar incubated anaerobically at 10 °C for 2–3 weeks (Corry, unpublished observations). Some strains of *Cl. estertheticum* are β-haemolytic, and therefore easy to detect on blood agar. A much more rapid and reliable method of detecting this and other clostridial species is to use PCR-based methods (Helps *et al.*, 1999; Broda *et al.*, 2000a; Boerema *et al.*, 2002).

5.10 Future trends

Transportation of vacuum-packed, chilled red meat over large distances, especially from countries in the Southern Hemisphere, such as Brazil, Argentina, Southern Africa, Australia and New Zealand, to Europe and the Middle and Far East is likely to continue for the foreseeable future. Losses due to spoilage from psychrophilic clostridia (and possibly due to high-pH meat) are kept confidential for commercial reasons, but appear to have been increasing in the last 10 years and must run into millions of US dollars. The way forward is likely to include improved methods of hygiene control in the abattoir, better methods of detection for cold-tolerant clostridia and possibly novel methods of in-pack control, such as the use of strains of LAB that are antagonistic to clostridia. Better attention to welfare of animals pre-slaughter will reduce the incidence of high-pH meat, and stricter control of temperature combined with good hygiene will maximise meat shelf-life in vacuum packs.

The trend for poultry to be raised cheaply in some countries, such as Brazil and Thailand, and exported to the richer areas, such as Europe and the USA, is likely to continue. Currently, most is transported raw and deboned in frozen form, either in aerobic packs or vacuum packed, and sometimes the meat is cooked before vacuum packaging and freezing. In future, there could be developments to transport whole carcasses in 100 % CO_2 or vacuum packed, but attention would need to be given to the spoilage implications of these practices.

5.11 Sources of further information and advice

Useful books on practical food microbiology include Bell *et al.* (2005), Downes and Ito (2001), Mossel *et al.* (1995) and Corry *et al.* (2003). AgResearch of Hamilton, New Zealand at http://www.agresearch.co.nz, formerly Meat Research Institute of New Zealand (MIRINZ), has produced an on-line manual of methods for the microbiological examination of meat, which has replaced the hard copy previously available (MIRINZ, 1991).

5.12 References

ANONYMOUS (2003a), 'de Man, Rogosa and Sharpe agar with sorbic acid (MRS-S agar)', in *Handbook of Culture Media for Food Microbiology*, eds Corry J E L, Curtis G D W and Baird R M, Progress in Industrial Microbiology vol. 37, Oxford, UK, Elsevier, 514–516.

ANONYMOUS (2003b), 'Cresol red thallium acetate (CTAS) agar', in *Handbook of Culture Media for Food Microbiology*, eds Corry J E L, Curtis G D W and Baird R M, Progress in Industrial Microbiology vol. 37, Oxford, UK, Elsevier, 446–448.

ANONYMOUS (2003c), 'Rose bengal chloramphenicol agar', in *Handbook of Culture Media for Food Microbiology*, eds Corry J E L, Curtis G D W and Baird R M, Progress in Industrial Microbiology vol. 37, Oxford, UK, Elsevier, 586–588.

ANONYMOUS (2003d), 'de Man, Rogosa and Sharpe agar', in *Handbook of Culture Media for Food Microbiology*, eds Corry J E L, Curtis G D W and Baird R M, Progress in Industrial Microbiology vol. 37, Oxford, UK, Elsevier, 511–513.

ANONYMOUS (2003e), 'Dichloran rose bengal chloramphenicol (DRBC) agar', in *Handbook of Culture Media for Food Microbiology*, eds Corry J E L, Curtis G D W and Baird R M, Progress in Industrial Microbiology vol. 37, Oxford, UK, Elsevier, 456–458.

ANONYMOUS (2003f), 'Oxytetracycline glucose yeast extract (OGY) agar', in *Handbook of Culture Media for Food Microbiology*, eds Corry J E L, Curtis G D W and Baird R M, Progress in Industrial Microbiology vol. 37, Oxford, UK, Elsevier, 541–543.

ARNAUT-ROLLIER I, DE ZUTTER L and VAN HOOF J (1999), 'Identities of the *Pseudomonas* species in flora from chilled chickens', *International Journal of Food Microbiology*, **48**, 87–89.

BARNES E M and SHRIMPTON D A (1957), 'Causes of greening of uneviscerasted poultry carcasses during storage', *Journal of Applied Bacteriology*, **20**, 273–285.

BARNES E M, MEAD G C, IMPEY C S and ADAMS B W (1979), 'Spoilage organisms of refrigerated poultry meat', in *Cold Tolerant Microbes in Spoilage and the Environment*, eds Russell A D and Fuller R, Society for Applied Bacteriology Technical Series no. 13, London, UK, Academic Press, 101–116.

BELL C, NEAVES P and WILLIAMS A P (2005), *Food Microbiology and Laboratory Practice*, Oxford, UK, Blackwell.

BELL R G, MOORHEAD S M and BRODA D M (2001), 'Influence of heat shrink treatments on the onset of clostridial "blown pack" spoilage of vacuum packed chilled meat', *Food Research International*, **34**, 271–275.

BEUCHAT L R (2003), 'Media for detecting and enumerating yeasts and moulds', in *Handbook of Culture Media for Food Microbiology*, eds Corry J E L, Curtis G D W and Baird R M, Progress in Industrial Microbiology vol. 37, Oxford, UK, Elsevier, 369–385.

BLICKSTAD E and MOLIN G (1984), 'Growth and end-product formation in fermenter cultures of *Brochothrix thermosphacta* ATCC 11509 and two psychrotrophic *Lactobacillus* spp. in different gaseous atmospheres', *Journal of Applied Bacteriology*, **57**, 213–220.

BOEREMA J A, BRODA D M and BELL R G (2002), 'PCR detection of psychrotolerant clostridia associated with deep tissue spoilage of vacuum-packed chilled meats', *Letters in Applied Microbiology*, **35**, 446–450.

BOEREMA J A, BRODA D M and BELL R G (2003), 'Abattoir sources of psychrophilic clostridia causing blown pack spoilage of vacuum-packed chilled meats determined by culture-based and molecular detection procedures', *Letters in Applied Microbiology*, **36**, 406–411.

BORCH E, KANT-MUERMANS M L and BLIXT Y (1996), 'Bacterial spoilage of meat and cured meat', *International Journal of Food Microbiology*, **33**, 103–120.

BRODA D M, BOEREMA J A and BELL R G (1998a), 'A PCR survey of psychrotrophic *Clostridium botulinum*-like isolates for the presence of BoNT genes', *Letters in Applied Microbiology*, **27**, 119–223.

BRODA D M, DE LACEY K M and BELL R G (1998b), 'Efficacy of heat and ethanol spore treatments for the isolation of psychrotrophic *Clostridium* spp. associated with the spoilage of vacuum-packed chilled meats', *International Journal of Food Microbiology*, **39**, 61–68.

BRODA D M, DE LACEY K M and BELL R G (1998c), 'Influence of culture media on the recovery of psychrotrophic *Clostridium* spp. associated with the spoilage of vacuum-packed chilled meats', *International Journal of Food Microbiology*, **39**, 69–78.

BRODA D M, MUSGRAVE D R and BELL R G (2000a), 'Use of restriction fragment length polymorphism analysis to differentiate strains of psychrophilic and psychrotrophic clostridia associated with "blown pack" spoilage of vacuum-packed meats', *Journal of Applied Microbiology*, **88**, 107–116.

BRODA D M, LAWSON P A, BELL R G and MUSGRAVE D R (1999), '*Clostridium frigidicarnis* sp nov., a psychrotolerant bacterium associated with "blown pack" spoilage of vacuum-packed meats', *International Journal of Systematic Bacteriology*, **49**, 1539–1550.

BRODA D M, BELL R G, BOEREMA J A and MUSGRAVE D R (2002), 'The abattoir source of culturable psychrophilic *Clostridium* spp. causing "blown pack" spoilage of vacuum-packed chilled venison', *Journal of Applied Microbiology*, **93**, 817–824.

BRODA D M, DELACEY K M, BELL R G, BRAGGINS T J and COOK R L (1996), 'Psychrotrophic *Clostridium* spp. associated with "blown pack" spoilage of chilled vacuum-packed red meats and dog rolls in gas-impermeable plastic casings', *International Journal of Food Microbiology*, **29**, 335–352.

BRODA D M, SAUL D J, LAWSON P A, BELL R G and MUSGRAVE D R (2000b), '*Clostridium gasigenes* sp. nov., a psychrophile causing spoilage of vacuum-packed meat', *International Journal of Systematic and Evolutionary Microbiology*, **50**, 107–118.

CALLOW E H and INGRAM M (1952), 'Bone taint', *Food*, **24**, 52–55.

COLLINS M D, FARROW J A E, PHILLIPS B A, FERUSU S and JONES D (1987), 'Classification of *Lacytobacillus divergens*, *Lactobacillus piscicola* and some catalase negative asporogenous, rod-shaped bacteria from poultry in a new genus *Carnobacterium*', *International Journal of Systematic Bacteriology*, **37**, 310–316.

COLLINS M D, RODRIGUES U M, DAINTY R H, EDWARDS R A and ROBERTS T A (1992), 'Taxonomic studies on a psychrophilic *Clostridium* from vacuum-packed beef: description of *Clostridium estertheticum* sp.', *FEMS Microbiology Letters*, **96**, 235–240.

CORRY J E L (1978), 'Possible sources of ethanol ante- and post-mortem: its relationship to the biochemistry and microbiology of decomposition', *Journal of Applied Bacteriology*, **44**, 1–56.

CORRY J E L, CURTIS G D W and BAIRD R M (eds) (2003), *Handbook of Culture Media for Food Microbiology*, Progress in Industrial Microbiology vol. 37, Oxford, UK, Elsevier.

DAINTY R H and MACKEY B M (1992), 'The relationship between the phenotypic properties of bacteria from chill-stored meat and spoilage processes', *Journal of Applied Microbiology*, **73**, 103S–114S.

DAINTY R H, SHAW B G and ROBERTS T A (1983), 'Microbial and chemical changes in chill-stored red meats', in *Food Microbiology: Advances and Prospects*, eds Roberts T A and Skinner F A, Society of Applied Bacteriology Symposium Series no. 11, London, UK, Academic Press, 151–178.

DAINTY R H, EDWARDS R A and HIBBARD C R (1989), 'Spoilage of vacuum-packed beef by a *Clostridium* sp.', *Journal of the Science of Food and Agriculture*, **49**, 473–486.

DAUD H B, McMEEKIN T A and THOMAS C J (1979), 'Spoilage association of chicken skin', *Applied and Environmental Microbiology*, **37**, 399–401.

DELACEY K M, BRODA D M and BELL R G (1998), 'In vitro assessment of psychrotrophic *Clostridium* spp. as possible causative agents of bone-taint in beef', *Food Microbiology*, **15**, 583–589.

DE MAN J C, ROGOSA M and SHARPE M E (1960), 'A medium for the cultivation of lactobacilli', *Journal of Applied Bacteriology*, **23**, 130–135.

DOWNES F P and ITO K (2001), *Compendium of Methods for the Microbiological Examination of Foods*, 4th edition, Washington, DC, USA, American Public Health Association.

GALLO L, SCHMITT R E and SCHMIDT-LORENZ W (1988), 'Microbial spoilage of refrigerated fresh broilers. Part I. Bacterial flora and growth during storage', *Lebenmittel-Wissenschaft und Technologie*, **21**, 216–223.

GARCÍA-LOPÉZ M-L and MARADONA M P (2000), '*Psychrobacter*', in *Encyclopaedia of Food Microbiology*, eds Robinson R K, Batt C A and Patel P D, vol. 3, London, UK, Academic Press, 1875–1882.

GARDNER G A (1966), 'A selective medium for the enumeration of *Microbacterium thermosphactum*', *Journal of Applied Bacteriology*, **29**, 455–460.

GARDNER G A (1982), 'Microbiology of processing: bacon and ham', in *Meat Microbiology*, ed. Brown M H, London, UK, Applied Science Publishers, 129–178.

GARDNER G A (2003), 'Culture media for non-sporulating Gram positive, catalase positive food spoilage bacteria', in *Handbook of Culture Media for Food Microbiology*, eds Corry J E L, Curtis G D W and Baird R M, Oxford, UK, Elsevier, 141–145.

GILL C O (1979), 'A review: intrinsic bacteria in meat', *Journal of Applied Bacteriology*, **47**, 367–378.

GILL C O (1986), 'The control of microbial spoilage in fresh meats', in *Advances in Meat Research Vol. 2 Meat and Poultry Microbiology*, eds Pearson A M and Dutson T R, Westport, CT, USA, 49–88.

GILL C O (1988), 'The solubility of carbon dioxide in meat', *Meat Science*, **22**, 65–71.

GILL C O and NEWTON K G (1979), 'Spoilage of vacuum-packed dark firm dry meat at chill temperatures', *Applied and Environmental Microbiology*, **37**, 362–364.

GILL C O and PENNEY N (1985), 'Modification of in-pack conditions to extend the storage life of vacuum-packaged lamb', *Meat Science*, **14**, 43–60.

GILL C O, LOWRY P D and DIMENNA M E (1981), 'A note on the identities of organisms causing black spot spoilage of meat', *Journal of Applied Bacteriology*, **51**, 183–187.

GRAM L, TROLLE G and HUSS H H (1987), 'Detection of specific spoilage bacteria from fish stored at low (0 °C) and high (20 °C) temperatures', *International Journal of Food Microbiology*, **4**, 65–72.

HANNA M O, ZINK D L, CARPENTER Z L and VANDERZANT C (1976), '*Yersinia enterocolitica*-like organisms from vacuum-packaged beef and lamb', *Journal of Food Science*, **41**, 1254–1256.

HELPS C R, HARBOUR D A and CORRY J E L (1999), 'PCR-based 16S ribosomal DNA detection technique for *Clostridium estertheticum* causing spoilage in vacuum-packed chill-stored beef', *International Journal of Food Microbiology*, **52**, 57–65.

HINTON A, CASON J A and INGRAM K D (2002), 'Enumeration and identification of yeasts associated with commercial poultry processing and spoilage of refrigerated broiler carcasses', *Journal of Food Protection*, **65**, 993–998.

HOLLEY R A, PEIRSON M D, LAM J and TAN K B (2004), 'Microbial profiles of commercial, vacuum-packaged, fresh pork of normal or short storage life', *International Journal of Food Microbiology*, **97**, 53–62.

ICMSF (2005), 'Poultry products', in *Microorganisms in Foods 6. Microbial Ecology of Food Commodities*, 2nd edition, International Commission on Microbiological Specifications for Foods, New York USA, Kluwer Academic/Plenum Publishers, 107–173.

INGRAM M (1971), 'Microbial changes in foods – general considerations', *Journal of Applied Bacteriology*, **34**, 1–8.

INGRAM M and DAINTY R H (1971), 'Changes caused by microbes in spoilage of meats', *Journal of Applied Bacteriology*, **34**, 21–39.

ISMAIL M A, ABOUELALA A H, NASSAR A and MICHAIL D G (1995), 'Fungal contamination of beef carcasses and the environment in a slaughterhouse', *Food Microbiology*, **12**, 441–445.

ISMAIL S A S, DEAK T, ABD EL-RAHMAN H A, YASSIEN M A M and BEUCHAT L R (2000), 'Presence and changes in populations of yeasts on raw and processed poultry products stored at refrigeration temperature', *International Journal of Food Microbiology*, **62**, 113–121.

ISO (1996a), ISO 13720: 1996, 'Meat and meat products – enumeration of lactic acid bacteria – colony count technique at 30 °C', Geneva, Switzerland, International Organization for Standardization.

ISO (1996b), ISO 13721: 1996, 'Meat and meat products – enumeration of *Pseudomonas* spp.', Geneva, Switzerland, International Organization for Standardization.

ISO (1997), ISO 13722: 1997, 'Meat and meat products – enumeration of *Brochothrix thermosphacta* – colony count technique', Geneva, Switzerland, International Organization for Standardization.

ISO (2004), ISO 21528-2: 2004, 'Microbiology of food and animal feeding stuffs – horizontal methods for the detection and enumeration of Enterobacteriaceae. Part 2: Colony count method', Geneva, Switzerland, International Organization for Standardization.

ISONHOOD J H and DRAKE M (2002), '*Aeromonas* species in foods', *Journal of Food Protection*, **65**, 575–582.

JARVIS B (1973), 'Comparison of an improved rose-bengal-chlortetracycline agar with other media for the selective isolation of moulds and yeasts in food', *Journal of Applied Bacteriology*, **36**, 723–727.

JONES J M, MEAD G C, GRIFFITHS N M and ADAMS B W (1982), 'Influence of packaging on microbiological, chemical and sensory changes in chill-stored turkey portions', *British Poultry Science*, **23**, 25–40.

JONES R J (2004), 'Observations on the succession dynamics of lactic acid bacteria populations in chill-stored vacuum-packaged beef', *International Journal of Food Microbiology*, **90**, 273–282.

KALCHAYANAND N, RAY B, FIELD R A and JOHNSON M C (1989), 'Spoilage of vacuum-packaged refrigerated beef by clostridium', *Journal of Food Protection*, **52**, 424–426.

KALINOWSKI R M and TOMPKIN R B (1999), 'Psychrotrophic clostridia causing spoilage in cooked meat and poultry products', *Journal of Food Protection*, **62**, 766–772.

KÄMPFER P (2000), '*Acinetobacter*', in *Encyclopedia of Food Microbiology*, eds Robinson R K, Batt C A and Patel P D, vol. 1, London, UK, Academic Press, 7–16.

KING D A, HOCKING A D and PITT J I (1979), 'Dicloran rose bengal medium for the enumeration and isolation of moulds from foods', *Applied and Environmental Microbiology*, **37**, 959–964.

LAURSEN B G, BAY L, CLEENWERCK I, VANCANNEYT M, SWINGS J, DALGAARD P and LEISNER J J (2005), '*Carnobacterium divergens* and *Carnobacterium maltaromaticum* as spoilers or protective cultures in meat and seafood: phenotypic and genotypic characterisation', *Systematic and Applied Microbiology*, **28**, 151–164.

MACKEY B M and DERRICK C (1979), 'Contamination of the deep tissues of carcasses by bacteria present on the slaughter instruments or in the gut', *Journal of Applied Bacteriology*, **46**, 355–366.

MACKEY B M and ROBERTS T A (1993), 'Improving slaughter hygiene using HACCP and monitoring', *Fleischwirtschaft*, **73**, 58–61.

MEAD G C (1985), 'Enumeration of pseudomonads, using cephaloridine-fucidin-cetrimide agar (CFC)', *International Journal of Food Microbiology*, **2**, 21–26.

MEAD G C (2004), 'Shelf-life and spoilage of poultry meat', in *Poultry Meat Processing and Quality*, ed. Mead G C, Cambridge, UK Woodhead Publishing, 283–303.

MEAD G C and ADAMS B W (1977), 'A selective medium for the rapid isolation of pseudomonads', *Journal of General Microbiology*, **128**, 1249–1264.

MEAD G C, CHAMBERLAIN A M and BORLAND E D (1973), 'Microbial changes leading to the spoilage of hung pheasants, with special reference to the clostridia', *Journal of Applied Bacteriology*, **36**, 279–287.

MEAD G C, BARNES E M and IMPEY C S (1974), 'Microbiological changes in the uneviscerated bird hung at 10 °C with particular reference to the pheasant', *British Poultry Science*, **15**, 381–390.

MEAD G C, GRIFFITHS N M, GREY T C and ADAMS B W (1986), 'The keeping quality of chilled duck portions in modified atmosphere packs', *Lebensmittel-Wissenschaft und Technologie*, **16**, 142–146.

MIRINZ (1991), '*Microbiological Methods for the Meat Industry*', 2nd edition, ed. Cook R L, publication no. 873, ISSN 0465-4390, Hamilton, New Zealand, Meat Research Institute of New Zealand.

MORA D, SCARPELLINI M, FRANZETTI L, COLOMBO S and GALLI A (2003), 'Reclassification of *Lactobacillus maltaromicus* (Miller *et al.* 1974) DSM 20342^T and DSM 20344 and *Carnobacterium piscicola* (Collins *et al.* 1987) DSM 20730^T and DSM 20722 as *Carnobacterium maltaromaticum* comb. nov.' *International Journal of Systematic and Evolutionary Microbiology*, **53**, 675–678.

MOSSEL D A A, EELDERINK I, KOOPMANS M and VAN ROSSEM F (1978), 'Optimisation of a MacConkey-type medium for the enumeration of Enterobacteriaceae', *Laboratory Practice*, **27**, 1049–1050.

MOSSEL D A A, KLEYNEN-SEMMELING A M C, VINCENTIE H M, BEERENS H and CATSARAS M (1970), 'Oxytetracycline-glucose-yeast-extract agar for selective enumeration of moulds and yeasts in foods and clinical material', *Journal of Applied Bacteriology*, **33**, 454–457.

MOSSEL D A A, CORRY J E L, STRUIJK C B and BAIRD R M (1995), *Essentials of the Microbiology of Foods. A Textbook for Advanced Studies*, Chichester, UK, John Wiley.

NEWTON K G and GILL C O (1981), 'The microbiology of DFD fresh meats: a review', *Meat Science*, **5**, 223–232.

NOTTINGHAM P M (1982), 'Microbiology of carcass meats', in *Meat Microbiology*, ed. Brown M H, London, UK, Applied Science Publishers, 13–65.

NYCHAS G-J, DROSINOS E H and BOARD R G (1998), 'Chemical changes in stored meat', in *The Microbiology of Meat and Poultry*, eds Davies A and Board R G, London, UK, Blackie, 288–326.

PATTERSON J T, GILLESPIE C W and HOUGH B (1984), 'Aspects of the microbiology of vacuum and gas-packaged chicken, including pre-treatments with lactic acid and potassium sorbate', *British Poultry Science*, **25**, 457–465.

REGEZ P, GALLO L, SCHMITT R E and SCHMIDT-LORENZ W (1988), 'Microbial spoilage of refrigerated fresh broilers. 3 Effect of storage-temperature on microbial association of poultry carcasses', *Lebensmittel-Wissenschaft und Technologie*, **21**, 229–233.

REUTER G (1985), 'Elective and selective media for lactic acid bacteria', *International Journal of Food Microbiology*, **2**, 55–68.

RIDDEL J and KORKEALA H (1997), 'Minimum growth temperatures of *Hafnia alvei* and other Enterobacteriaceae isolated from refrigerated meat determined with a temperature gradient incubator', *International Journal of Food Microbiology*, **35**, 287–292.

ROBERTS T A and DERRICK C M (1975), 'Sporulation of *Clostridium putrefaciens* and the resistance of the spores to heat, γ-radiation and curing salts', *Journal of Applied Bacteriology*, **38**, 33–37.

ROBERTS T A and MEAD G C (1986), 'Involvement of intestinal anaerobes in the spoilage of red meats, poultry and fish', in *Anaerobic Bacteria in Habitats other than Man*, eds Barnes E M and Mead G C, Society for Applied Bacteriology Symposium Series no. 13, Oxford, UK, Blackwell Scientific, 333–349.

ROSS H E (1965), '*Clostridium putrefaciens*: a neglected anaerobe', *Journal of Applied Bacteriology*, **28**, 49–51.

SANTOS J-A, GARCIA-LOPEZ M-L and OTERO A (2000), '*Moraxella*', in *Encyclopedia of Food Microbiology*, eds Robinson R K, Batt C A and Patel P D, vol. 2, London, UK, Academic Press, 1487–1492.

SCHILLINGER U and HOLZAPFEL W H (2003), 'Culture media for lactic acid bacteria', in *Handbook of Culture Media for Food Microbiology*, eds Corry J E L, Curtis G D W and Baird R M, Progress in Industrial Microbiology vol. 37, Oxford, UK, Elsevier, 127–145.

SHAW B G, HARDING C D and TAYLOR A A (1980), 'The microbiology and storage stability of vacuum-packed lamb', *Journal of Food Technolology*, **15**, 397–405.

SPRING S, MERKHOFFER B, WEISS N, KROPPENSTEDT R M, HIPPE H and STACKEBRANDT E (2003), 'Characterization of novel psychrophilic clostridia from an Antarctic microbial mat: description of *Clostridium frigoris* sp nov., *Clostridium lacusfryxellense* sp nov., *Clostridium bowmanii* sp nov and *Clostridium psychrophilum* sp nov and reclassification of *Clostridium laramiense* as *Clostridium esterthеticum* subsp *laramiense* subsp nov.', *International Journal of Systematic and Evolutionary Microbiology*, **53**, 1019–1029.

STANBRIDGE L H and BOARD R G (1994), 'A modification of the *Pseudomonas* selective medium, CFC, that allows differentiation between meat pseudomonads and Enterobacteriaceae', *Letters in Applied Microbiology*, **18**, 327–328.

STUDER P, SCHMITT R E, GALLO L and SCHMIDT-LORENZ W (1988), 'Microbial spoilage of refrigerated fresh broilers. 2 Effect of packing on microbial association of poultry carcasses', *Lebenmittel-Wissenschaft und Technologie*, **21**, 224–228.

TRYFINOPOULOU P, DROSINOS E H and NYCHAS G-J E (2001), 'Performance of *Pseudomonas* CFC-selective medium in fish storage ecosystems', *Journal of Microbiological Methods*, **47**, 243–247.

VIEHWEG S H, SCHMITT R E and SCHMIDT-LORENZ W (1989), 'Microbial spoilage of refrigerated fresh broilers. Part VII Production of off odours from poultry skin by bacterial isolates,' *Lebensmittel-Wissenschaft und Technologie*, **22**, 356–367.

WASNEY M, HOLLEY R A and JAYAS D S (2001), 'Cresol Red Thallium Acetate Inulin (CTSI) agar for the selective recovery of *Carnobacterium* spp.', *International Journal of Food Microbiology*, **64**, 167–174.

6

Sampling of red meat

C. O. Gill, Lacombe Research Centre, Canada

6.1 Introduction

Raw red meats from healthy animals inevitably will be contaminated with bacteria (Nottingham, 1982). Those bacteria will include organisms that can grow on the product to cause spoilage, if the meat is not frozen, and may include enteric pathogens that can cause disease in humans. Nevertheless, routine microbiological testing of red meats was relatively uncommon until the last decade of the twentieth century. Instead, the microbiological safety and storage stability of meat was decided by reference to visible contamination on the product, the general appearance of the meat and possibly its odour (Hathaway and McKenzie, 1991). However, traditional meat inspection evidently failed to control the contamination of meat with enteric pathogens (Mackey and Roberts, 1993), and the storage stability of raw meats in preservative packaging that could greatly extend shelf-life at chill temperatures could not be determined reliably from the appearance and odour of the meat alone (Gill, 1996). Therefore, routine microbiological testing of red meats at various stages of processing and distribution has become a common practice. In this chapter, the objectives of microbiological sampling of raw red meats and the appropriate procedures used in sampling meat for different purposes will be discussed.

6.2 Purposes of routine microbiological sampling

Routine microbiological testing of meat may be undertaken to decide on the acceptance or rejection of batches of product, when produced at a plant or received from a supplier, for the development, validation and verification of Hazard Analysis Critical Control Point (HACCP) or Quality Management (QM) systems, or for regulatory purposes.

The purpose of acceptance testing is to obtain data from which to decide, with reasonable certainty, whether or not a batch of product is microbiologically safe or has an adequate shelf-life before the onset of microbial spoilage, given the intended use and expected treatment of the product. A batch of product may, of course, be subject to microbiological testing for both safety and storage stability. That might involve the collection of different samples for the two purposes if, for example, testing for a pathogen involved enrichment of the whole of each sample.

What is regarded as a reasonable chance that an acceptance/rejection decision is correct will depend not only on the certainty that would be theoretically desirable, but also on the practical limitations of sampling. Sample processing capability, cost and convenience, rather than theoretical considerations, are likely to determine the number and size of the samples and the type of testing that can be used (Jarvis and Malcolm, 1986). As well as other factors, the choice of sampling procedure will be constrained by the time required for the completion of testing, since many forms of raw meat cannot be held for lengthy periods without incurring costs and risking unacceptable loss of quality and/or storage life.

Testing to decide on the acceptance or rejection of product batches will not in itself prevent unacceptable batches being produced or delivered (Marquardt, 1984). While some action might be taken to avoid further misprocessing or mishandling following the identification of unacceptable batches, with current high-speed processing and large-scale distribution, much product could be rendered unacceptable before test results were available. Therefore, systems for the control of production, storage and distribution processes that avoid the generation of unacceptable product are to be preferred to procedures for its subsequent detection and rejection (Mayes, 1993).

QM and HACCP systems are implemented to ensure that acceptable products emerge consistently from any process. QM systems are operated to control the commercially important quality attributes of products, while HACCP systems are used to assure product safety. With respect to the microbiological condition of raw meat, QM and HACCP systems should function respectively to control spoilage and pathogenic organisms on the product. Even so, in some circumstances, the same microbiological data might be used for both purposes. However, formal separation of the two types of system should be maintained, because quality specifications can be changed as and when necessary to meet different commercial circumstances and customer expectations, while safety should not be compromised, irrespective of the commercial environment. Nevertheless, in practice, QM and HACCP systems often seem to be confused.

For process control purposes, microbiological sampling of the product may be undertaken when control systems are being developed, in order to determine where in a process the numbers of target bacteria are increasing or decreasing substantially (Gill and Jones, 1997). The operations that result in substantial increases or decreases in bacterial numbers are then Critical Control Points (CCPs) in the process. Such objective determination of CCPs is desirable, because subjective assessment of processes, as recommended for HACCP-system development (USDA, 1992), may not identify properly the microbiological effects of the various operations on the raw product. If the microbiological effects of individual operations are uncertain, CCPs may be misidentified and the system developed for control of microbiological contamination will then fail.

Microbiological sampling can also be undertaken to validate the control exerted at CCPs and to verify objectively the performance of a control system. Validation of the control achieved at the CCPs in individual processes is necessary, because it cannot be assumed safely that actions or treatments that are effective in controlling contamination of meat under test conditions, or at other facilities, are equally effective in commercial circumstances or in other processes (Gill *et al.*, 1998a; Gill and Landers, 2003a). Verification of control by determining the microbiological condition of the product obtained from a particular process is desirable, since control failures may not be detected by inspection of either the relevant documentation or the process itself, or both (Gill *et al.*, 2003). In addition, verification by appropriate testing can permit the microbiological condition of the product to be confirmed readily and objectively by third parties, such as customers and regulatory authorities.

Failure to meet verification criteria when evidently Standard Operating Procedures (SOPs) are being maintained should trigger sampling of the product at various stages of processing to identify the operation(s) at which there has been loss of control and to validate any corrective action. Sampling the product before and after any new or modified operation in a process to determine the microbiological effects of the operation is also appropriate. Otherwise, microbiological testing for process control purposes should diminish as the SOPs for controlling contamination become firmly established and effective control, as a result of maintenance of the SOPs, is confirmed.

Since the mid-1990s, regulatory authorities in most developed countries have promulgated regulations that require the implementation of HACCP systems at red meat packing and processing plants. It might then be expected that microbiological testing associated with meat inspection would be such as to ensure the implementation and maintenance of effective HACCP systems. That would seem to have been the intention behind requirements for routine sampling of red-meat carcasses, ground beef and beef products intended for grinding because, if performance criteria are not met, the HACCP system associated with production of the failed product must be reviewed and, where necessary, improved (USDA, 1996; EC, 2001). However, the performance criteria that have been advanced have the form of acceptance tests and, in some cases, failure to meet a test criterion results in rejection of the lot (CFIA, 2005). Thus, in the microbiological testing required by current regulations, there appears to be a degree of confusion between testing for HACCP purposes and for the acceptance of batches of product.

6.3 Target organisms

The organisms targeted in microbiological testing of meat may be groups of organisms that are usually present among the aerobic or anaerobic spoilage flora, specific organisms of high spoilage potential or which cause relatively unusual forms of spoilage, specific pathogens or groups of organisms indicative of the possible presence of pathogenic or spoilage organisms.

Testing of meat for spoilage organisms is usually undertaken only when the reasons for unexpected spoilage of the product or means of extending product storage life are being sought. Spoiled product may be examined for groups of organisms, such as pseudomonads or enterobacteria, or for specific organisms, such as *Shewanella putrefaciens*, *Brochothrix thermosphacta* or *Clostridium estertheticum* (Gill, 2003). Products undergoing processing may then be sampled to enumerate the organisms found to be involved in the unexpected or inconveniently early spoilage, in order to determine whether unusual or heavy contamination of the product with those organisms might be the reason for such spoilage (Boerema *et al.*, 2003; Holley *et al.*, 2004).

Routine testing of raw red meats for specific bacterial pathogens is mostly to detect *Salmonella* or *Escherichia coli* O157:H7 (Evrendilek *et al.*, 2001), an enterohaemorrhagic strain of *E. coli* that is carried by, but does not cause, disease in cattle and which has been involved in outbreaks of human disease associated with the consumption of ground beef in North America and Europe (Low *et al.*, 2005). Testing for other meatborne pathogens, such as *Yersinia enterocolitica* or *Listeria monocytogenes*, is likely to be undertaken only when outbreaks of disease caused by these agents are being investigated, or in response to commercial concerns about the possibility of ready-to-eat foods prepared from the raw meat being contaminated with pathogens from the raw product (Johnson *et al.*, 1990; Frederiksson-Ahomaa *et al.*, 2001).

In addition to testing for pathogenic bacteria, some meats may also be examined for the parasites that cause trichinellosis. Larvae of *Trichinella* spp. infect muscle tissue and are transmitted to new hosts when the infected meat is consumed. The classical domestic cycle of trichinellosis involves the acquisition of *T. spiralis* by humans and pigs as a result of the consumption of infected pork (Campbell, 1988). In regions where *T. spiralis* infection of domestic pigs persists, suspect carcasses may be examined for the organism.

In addition, muscle tissue from every horse carcass intended for human consumption in countries of the European Union (EU) must be examined microscopically for *Trichinella* larvae (Gamble *et al.*, 1996), since human outbreaks of trichinellosis have been caused by infected horse meat (Ancelle, 1998).

Most microbiological sampling of red meat involves the enumeration of indicator organisms. These are groups of organisms that include species capable of compromising the safety or storage stability of a food, are related to them, behave similarly or are derived from the same source. Most or all of the organisms within an indicator group may be non-pathogenic or play little or no part in spoilage processes.

The indicator organisms for which meat is probably most often examined are those recovered on non-selective agars, incubated in air at a specified temperature for a specified time, with all visible colonies being counted to obtain a total viable count (TVC). Counts of viable aerobic organisms are commonly referred to as aerobic plate counts (APCs). Different temperatures and times of incubation may be used. Typical conditions would be 35 °C for 48 hours (Messer *et al.*, 2000), although lower temperatures and longer times are often preferred in relation to the spoilage flora, since some spoilage organisms grow poorly if at all at temperatures above 30 °C (Jay, 2002).

The APC can be used for enumerating not only the aerobic spoilage flora but also the spoilage flora that develops under anaerobic conditions, such as those usually found in vacuum packs, because most of the organisms that multiply in that case are aerotolerant or facultatively anaerobic (Stanbridge and Davies, 1998). The incubation conditions may be set to restrict the types of organisms recovered and so obtain counts of organisms other than those of total aerobes. For example, the incubation temperature can be reduced to 7 °C for a period of 10 days to enumerate psychrotrophs or incubation can be under anaerobic conditions to exclude strictly aerobic organisms.

APCs are widely used to assess the extent to which general contamination is being controlled during the dressing of carcasses and the subsequent processing of meat, or to assess the stage of development of the spoilage flora on the product. However, there is no necessary relationship between the total numbers of bacteria and the numbers of pathogens on meat at any stage of processing or storage (Jericho *et al.*, 1996; Arthur *et al.*, 2004). Also, there is no necessary relationship between total numbers on the meat and the number of spoilage organisms (Gill, 1995), until the spoilage organisms have grown to predominate (Fig. 6.1). Thus, the relevance of APCs to the safety or storage stability of the meat is often uncertain.

It is generally thought that the most hazardous organisms found on meat are associated with faecal contamination of the product (USDA, 1996). It is then considered appropriate to assess the microbiological safety of meat by reference to the numbers of bacteria indicative of contamination with

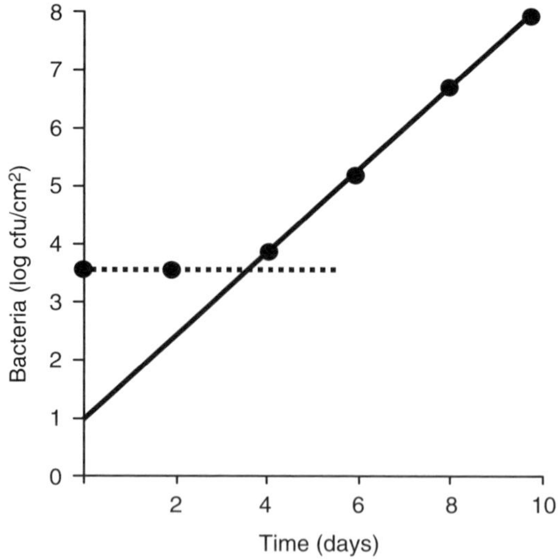

Fig. 6.1 Delayed increase in aerobic plate count (•) during development of a meat spoilage flora at chill temperature; psychrotrophs (—), mesophiles (---).

faecal organisms. The groups of organisms that may be enumerated for that purpose are Enterobacteriaceae, coliforms, faecal coliforms and generic *E. coli* (Brenner, 1992; Hitchins *et al.*, 1998). The relationship between the numbers of organisms in the first two groups and faecal organisms can be uncertain, since both groups include environmental and spoilage organisms, which can be more numerous than faecal organisms on the meat. Faecal coliforms, i.e. coliforms that grow and produce acid from lactose at 44.5 °C, include some species other than *E. coli*. Since methods for the direct enumeration of *E. coli* are readily available, this organism, rather than the less specific faecal coliforms, is now usually enumerated in meat samples (Fung, 2002). The topic is discussed further in Chapter 4.

Other indicator organisms that have been suggested as possibly useful in assessing the microbiological condition of meat are generic *Aeromonas* and *Listeria* spp. Aeromonads can grow to high numbers in water remaining on or in meat-plant equipment, while listerias appear to be common in the flora associated with detritus and filth that persists in equipment, waste systems or other fixtures and fittings in meat plants (Gobat and Jemmi, 1991; Isonhood and Drake, 2002). Thus, contamination of meat with those organisms can be indicative of sources of contamination that may harbour the pathogens *Aeromonas hydrophila* or *L. monocytogenes* (Gill *et al.*, 1999b, 2005).

Although the microbiological effects of meat chilling, storage, transport and display can be assessed by recovering bacteria from the product, they can also be determined, often more conveniently, by calculating the possible growth of pathogenic and/or spoilage organisms from product temperature histories (Gill *et al.*, 2002; Gill and Landers, 2003b). The addition of bacteria to meat during these four types of process can be largely avoided, or wholly prevented, if the product is contained in a protective pack. However, growth of psychrotrophic spoilage organisms will occur, growth of psychrotolerant pathogens is likely to occur and growth of mesophilic pathogens may occur, particularly during product chilling. The degree of growth of the three groups of organisms depends on the temperature conditions experienced by the product and can be calculated from the product temperature history and an equation relating the growth rate of each indicator organism to temperature. Since temperature histories usually can be collected with less effort, and processed more rapidly than carrying out microbiological tests, such temperature function integration procedures offer a relatively rapid means of assessing the microbiological effects of controlling meat temperature (Gill and Landers, 2005). If the calculated bacterial growth is found to be excessive, then inspection of temperature history data can usually indicate the stage of the process where temperature control was deficient, and so direct attention to appropriate action to resolve the problem.

6.4 Sampling plans

A sampling plan can be as simple as the sampling of all units of product produced by a process or received from a supplier. Such sampling is required by some authorities at small packing plants, where testing for *E. coli* O157:H7 is needed to validate control of contamination of beef carcasses (CFIA, 2005). Also, some manufacturers of hamburger patties undertake testing of all bulk containers (combo bins) of manufacturing beef received at their plants for *E. coli* O157:H7. However, microbiological testing usually involves the sampling of relatively few items selected at random from all those in a lot.

Acceptance testing can involve the use of either attributes or variables sampling plans (Hildebrandt and Weiss, 1994). When bacteria are detected, but not enumerated, samples can be scored only for the attributes of being positive or negative for the target organism. Therefore, such data must be collected and interpreted by reference to attributes sampling plans. When the numbers of bacteria in samples are determined by plating or Most Probable Number (MPN) procedures, the data may be collected and interpreted by reference to attributes or variables sampling plans.

For bacterial numbers to be considered as attributes, criteria for acceptable and unacceptable levels must be set. In a two-class attributes acceptance plan, numbers below a stated value (m) are regarded as acceptable, while numbers equal to or more than m are regarded as unacceptable (Hildebrandt *et al.*, 1995). The plan specifies the number of samples (n) that must be analysed when a decision is to be made on the acceptability of a lot and the number of samples (*c*) yielding unacceptable numbers that can be tolerated in an acceptable lot. Acceptance plans for sampling that involve only the detection of certain organisms are necessarily two-class attributes acceptance plans (Hildebrandt and Böhmer, 1998).

In a three-class attributes acceptance plan, the three classes are defined by two stated values for bacterial numbers (Legan *et al.*, 2001). Numbers equal to or less than the lesser value (m) are regarded as wholly acceptable. Numbers equal to or greater than the larger value (M) are regarded as wholly unacceptable. Values $> m$ but $< M$ are regarded as conditionally acceptable in that a lot will be acceptable if no sample yields bacteria at levels $\geq M$ and, in the number of samples (n) required for a decision, the samples that yield bacteria at levels $> m$ but $< M$ do not exceed a stipulated number (c).

For variables sampling plans, the numerical data are considered as such, with the assumption that the distribution of bacteria in or on a product conforms to a recognised model (Corradini *et al.*, 2001). In the case of meat, the distribution of bacteria is assumed to be log normal. That is, for a set of microbial counts collected from the product at random, the log values of the counts will be normally distributed (Kilsby and Pugh, 1981). Therefore, statistical analysis of the data can be based on this normal distribution and used in determining product acceptability.

For acceptance plans, the number of samples (n) required for decision, the tolerable proportion of samples (p) yielding bacteria at log numbers that are not wholly acceptable and the lowest probability (P) of rejecting unacceptable product are established (Kilsby *et al.*, 1979). A factor (k_1) is calculated from those specified values and the mean ($\bar{x}$) and standard deviation (s) are calculated for each set of log values. Then, when log numbers greater than a value, V, are not wholly acceptable, the product is unacceptable when $\bar{x} + k_1 s$ is more than V.

In sampling for purposes of process control, it is necessary to assume that the product passing through a process forms a single population and that the distribution of bacteria in or on the product conforms to a recognised model. Again, for meat, the log normal distribution can be assumed. Then, the population of any group of bacteria on the product at any stage of processing is described fully by the population parameters for the mean of the log numbers (mean log) and the standard deviation of those numbers. Values for the mean log and standard deviation can be estimated with reasonable confidence from the numbers of bacteria recovered from a relatively small group of samples collected at random (Gill, 2000). Then, the estimated population parameters can be used to compare the numbers of bacteria on the product at any stage of the process as, for example, before and after each CCP. However, the numbers of bacteria on or in the product cannot be compared properly by reference to estimated mean log values ($\bar{x}$) (Brown and Baird-Parker, 1982). That is because, as a product moves

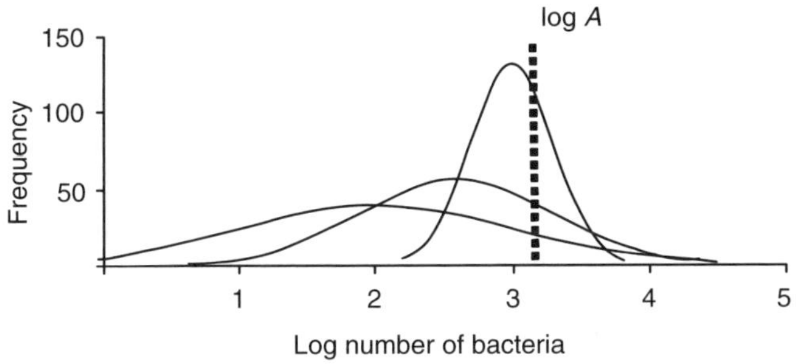

Fig. 6.2 Different distributions of bacteria at the same log mean number ($\log A$) on meat.

through the process, bacteria on or in it are redistributed, with a reduction in the variance between the numbers of bacteria associated with individual samples. Thus, values estimated for the standard deviation of the log values (s) are likely to decrease at successive stages of any process used for the preparation of a raw meat product. The decreases in variance will result in increased values for $\bar{x}$, even though no bacteria are added to the product (Fig. 6.2). Therefore, comparison of values for $\bar{x}$ can lead to the erroneous conclusion that numbers of bacteria on the product have increased during processing when, in fact, they have remained unchanged.

If indeed the numbers have remained unchanged, then the mean numbers of bacteria on the product will not change during processing. The log of the mean numbers ($\log A$) can be calculated from the formula $\log A = \bar{x} + \log_n 10 \, (s^2/2)$ (Kilsby and Pugh, 1981). Then, bacterial numbers on the product at different stages of processing can be compared properly by reference to $\log A$ values.

6.5 Sampling sites

Although it may be possible sometimes to take the whole of a product item as a sample, for example an entire hamburger patty, the sampling method and/or the size of the item often precludes sampling of the whole product. Then, appropriate sampling sites must be decided. An item can be sampled at one or several sites, which may be chosen as being representative of all sites on or in the item, as likely to be more heavily contaminated than most other sites or taken at random. With smaller items, such as primal or consumer cuts, sampling usually would be at one site regarded as representative although, if fat cover is extensive, such products might be sampled at both a fat and a lean surface in each case.

Large items, for which there are no sites where heavy contamination is considered particularly likely, may be sampled at representative sites or at random. For example, a bulk container of manufacturing beef may be sampled by cutting cores through the meat at the centre and two or more corners of the bulked meat; or it may be sampled by cutting portions from pieces of meat selected at random from product in the container (Siragusa, 1998). In contrast, it has become the usual practice to sample carcasses at sites that are known to be often heavily contaminated, such as the hock, rump, brisket and neck (Roberts *et al.*, 1980); and cartons of hot-boned meat are often sampled at their geometric centres, since the most extensive bacterial growth, and thus the highest numbers, could be expected at that slowest cooling site (AQIS, 2001).

The rationale for sampling only selected sites that are expected to be relatively heavily contaminated is that, if such sites are microbiologically acceptable, then the rest of the product will be too. However, for items like beef carcasses, the part of the carcass that is examined is only a small fraction of the total surface area (Jericho *et al.*, 1997). The microbiological condition of the rest of the carcass may then be far better, or only marginally so, than the condition of the selected sites. Thus, assessments of the microbiological condition of the product based on the same data may differ considerably, depending on the relationship that is assumed to exist between the microbiological condition of the selected sites and the condition of the rest of the carcass.

For estimating the mean numbers of bacteria in or on a product, random sampling of sites and product units is necessary by, for example, reference to a grid (Fig. 6.3). Otherwise, the redistribution of bacteria from the more heavily contaminated to less contaminated sites may give the illusion of numbers being reduced, if only initially heavily-contaminated sites are sampled (Bell, 1997). However, a site or sites may be selected for sampling when an operation that affects only part of an item, such as a carcass, is being investigated. Then, it is assumed that the selected site is representative of all the sites in the affected area (Gill *et al.*, 1998b). Differences in the log mean numbers at the site before and after the operation therefore indicate the microbiological effects of the operation on the site; but they do not indicate the microbiological effects of the operation on the item as a whole, unless the total surface area or mass of the item is taken into account.

6.6 Methods of sampling

Since muscle and fat tissues from healthy animals are usually sterile, meat items composed of intact tissues will carry bacteria only on their surfaces (Gill, 1979). Therefore, microbiological sampling of such tissues should be directed at the recovery of bacteria from surfaces. On the other hand, deep tissues, as well as the surfaces of some organs, such as the liver, can be contaminated with bacteria (Gill, 1988), and bacteria will be carried into

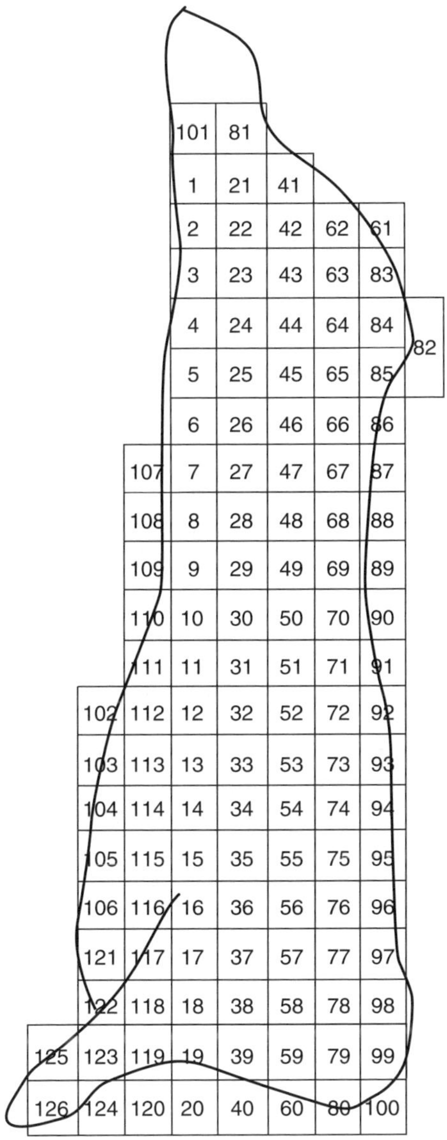

Fig. 6.3 Reference grid for the random selection of sites to be sampled on beef carcass sides.

muscle tissues that are subjected to mechanical tenderising or pumping with brine (Gill and McGinnis, 2004). However, the numbers of bacteria in the deep tissues of such products will always be less than those on the surface, and the composition of the flora at the surface and in the deep tissue will be the same, at least initially. Therefore, unless the numbers of contaminants in the deep tissues are to be determined, microbiological testing of these products need involve only the recovery of bacteria from surfaces.

Surfaces may be sampled by excision of a portion of surface tissue or by swabbing part or all of a meat surface. It is generally believed that larger numbers of bacteria per unit area will be recovered by excision than by swabbing (Sharpe *et al.*, 1996). However, it appears that, if swabs of an appropriate material are used, then the numbers recovered by swabbing or excision will be similar (Table 6.1). Generally, mildly abrasive and absorbent materials, such as synthetic sponges, can be used to recover bacteria from all types of meat surface (Pearce and Bolton, 2005). Mildly abrasive, but non-absorbent materials, such as medical gauze or cheese cloth, may be equally effective for recovering bacteria from carcass surfaces, but not from cuts of meat (Gill and Jones, 2000). Absorbent, but non-abrasive materials, such as cotton or alginate wools, may be reasonably effective for recovering bacteria from the surfaces of meat cuts, but not from carcasses (Gill *et al.*, 2001a). The differences between materials for recovering bacteria from carcasses and cut surfaces probably arise because carcass surfaces include few areas of cut muscle and are often dry, so an abrasive action is required to remove some surface material with the accompanying bacteria. Meat cuts, however, usually have extensive areas of cut-muscle tissue, with disrupted muscle fibres between which bacteria may penetrate to depths of 1–2 mm. Moreover, both the muscle and fat surfaces are usually wet. Apparently, recovery of bacteria from such surfaces is better with absorbent than non-absorbent materials.

It is usual to delimit the area to be sampled by swabbing with a sterile template or to use a coring device to cut a disc of fixed size for excision. However, bacterial numbers are properly assessed as log values, since the numbers of viable organisms change exponentially during their growth or inactivation, and some variation in the area sampled will have little effect

Table 6.1 Log mean numbers of total aerobic bacteria recovered from beef or pork carcasses or cuts by excision or swabbing with cellulose acetate sponge, medical gauze or cotton wool buds (Gill and Jones, 2000; Gill *et al.*, 2001a)

Sampling method	Log mean numbers (log cfu/cm^2)			
	Beef		Pork	
	Carcasses	Cuts	Carcasses	Cuts
Excision	2.90	4.27	2.37	3.37
Sponge	3.02	4.08	2.35	2.78
Gauze	3.21	3.41	2.38	2.31
Cotton wool	2.61	4.26	2.12	3.01

on the log numbers of bacteria recovered. Then, there is usually little or nothing to be gained by defining precisely the area to be sampled and, in some circumstances, such as sampling carcasses on a high-speed line, defining the area to be sampled by means of a template may be wholly impractical (Gill *et al.*, 1996). Therefore, in general, the exact definition of the area to be sampled is an unnecessary and largely pointless exercise.

With excision of tissue, the area sampled is usually limited to no more than 100 cm^2. With swabbing, however, the sampled area can be of any size that is convenient and practical (Gill *et al.*, 2001b). Although the sample size may be of little consequence for bacteria that are relatively numerous, such as total aerobes, large samples are desirable when relatively rare indicator organisms or specific pathogens are being enumerated or detected. Moreover, swabbing can be performed more rapidly than excision sampling, using less equipment and without damage to the items sampled. Thus, in most circumstances, sampling of meat surfaces by swabbing must be preferred to sampling by excision.

If the flora of the deep tissue of non-comminuted meat is to be investigated, it is vital to sterilise the surface before sampling to avoid contaminating the excised deep tissue with organisms from the surface. Searing of the surface to a depth of 3 mm, using a heated template, may be sufficient to avoid contamination by surface flora (Gill and Penney, 1977). However, if the meat surfaces are wet and contamination of the sample must be wholly avoided, it may be necessary to excise a portion of meat, place it in boiling water and then dry it before searing the surface and excising the deep-tissue sample (Gill and McGinnis, 2004). The adequacy of an aseptic sampling technique can be judged by sampling intact muscle, from which sterile tissue should be obtained.

Since bacteria are dispersed throughout comminuted meat, samples of such a product are necessarily portions of the meat. With product that is to be comminuted, the portions of meat collected by coring or selecting pieces at random are usually ground, and some of the ground meat obtained is taken for microbiological testing.

6.7 Detection methods and enumeration of bacteria

The samples obtained by swabbing, excision of tissue or collection of comminuted product are macerated with appropriate volumes of diluent. Some mechanical disruption of the sample is usual as, for example, with the use of a stomacher, in which a strong plastic bag containing both sample and diluent is pounded by paddles (Jay and Margitic, 1979). The diluent is usually a buffering or nutrient solution, such as Butterfield's phosphate diluent or peptone water. However, a sample may be macerated with a pre-enrichment or enrichment broth, when detection of a pathogen rather than enumeration of indicator organisms is intended.

With tissue samples, the amount of diluent added in ml is usually equivalent to nine times the weight of the sample in grams. Thus, the initial preparation is a ten-fold dilution of the sample. With swab samples, the same, convenient volume of diluent can be used with a swab of a particular type, whatever the size of the swabbed area. Thus, the numbers of bacteria present in 1 ml of fluid may be those recovered from 100 cm^2 or more. With a swab composed of an absorbent material, such as a sponge, a substantial proportion of the total fluid usually will be retained by the swab. When a swab composed of a non-absorbent material, such as medical gauze, is used, the amount of fluid retained by the swab will be trivial, if it is compressed to expel fluid, when being removed from its container following maceration.

Bacteria may be enumerated by spread, pour or spiral plating on suitable agar media, using undiluted or appropriately diluted sample suspensions; by inoculating fluids into commercially prepared plates containing dry nutrients and gelling agents and, in some instances, selective agents; by MPN procedures, using tubes containing non-selective or selective broths; or by hydrophobic grid membrane filtration (HGMF) procedures, which involve filtering fluids through membrane filters with upper surfaces divided into squares by lines of hydrophobic material, and incubating the filters on appropriate agar media (Peeler *et al.*, 1992; Swanson *et al.*, 1992).

All the plating methods are suitable for enumerating bacteria when their numbers in sample suspensions are relatively high. However, usually no more than 1 ml of fluid can be added to an agar plate. Consequently, when the bacteria being sought are few, MPN procedures are to be preferred, since all the sample fluid can be distributed, if necessary, into MPN tubes. The HGMF procedure is an MPN method, because the number of squares containing colonies following incubation of the filters are counted and any square may contain more than one colony. With fluids from macerated tissues, the sensitivity of an HGMF procedure is not necessarily greater than that of a plating procedure, because blocking of the filter by material suspended in the fluid limits the volume of undiluted fluid that can be filtered to 1 ml or less in most cases. However, with swab samples, it is often possible to filter all the undiluted fluid obtained from the swab (Gill and Badoni, 2004). Then, the sensitivity is determined only by the area of product surface that is available for swabbing and can be sampled in practice.

For the detection of a pathogen, the sample may be incubated first in a non-selective broth that can include materials to mitigate oxidative or other damage of injured cells, in order to allow their resuscitation (McDonald *et al.*, 1983). Subsequently, one or more portions of the pre-enrichment broth can be transferred to at least one enrichment broth in which growth of the target organism will be favoured over that of other bacteria likely to be present in the sample (Harrigan and McCance, 1976). Alternatively, a sample can be incubated directly in an enrichment broth, if the possible presence of injured cells that would be inhibited by selective agents in the broth can be disregarded. After incubation of a selective broth, portions may be spread on one or more selective agars to obtain colonies with characteristics typical of the pathogen. Presumptive pathogens isolated from such colonies must be confirmed, usually by further tests. Equally, a pathogen may be detected in enrichment media by enzyme immunoassay, latex agglutination, immunomagnetic particle assay, detection of specific DNA by polymerase chain reaction (PCR) procedures, or other immunological or genetic methods for which kits or reagents and primers are commercially available (Feng, 1997; Girrafa and Neviani, 2001). The time required by such methods to determine whether or not a pathogen is present in the sample is generally substantially less than that needed to obtain presumptive isolates and confirm their identity by biochemical tests.

6.8 Interpretation of data

Acceptance sampling in relation to the possible presence of enteric pathogens, such as *E. coli* O157:H7 in raw meat, often involves the examination of a single or compound sample from a lot, for detection of the pathogen. What is considered to be a lot can vary from a single carcass or bulk container to the whole of a day's production at a large plant. Since the incidence of enteric pathogens on fresh meat is usually low, there is only a very small probability of recovering a pathogenic organism from a single sample taken from a contaminated lot, when the pathogen is uniformly distributed throughout the product and detection of the organism in a contaminated sample is certain (Acuff, 1999). However, the distribution of bacteria in raw meat is never uniform, and detection of the target organism, whenever it is present in a sample, is never certain. Thus, although detection of a pathogen in a sample shows that a lot is contaminated and might justify rejection of the lot, failure to detect the pathogen may not demonstrate reliably that it is absent from the lot (Table 6.2).

Table 6.2 Numbers of samples required to detect an organism with a probability of 0.90 or better when the organism is present in 10 % or smaller proportions of samples

Proportion of positive samples (%)	Number of samples for detection		
	0.90 probability	0.95 probability	0.99 probability
10.0	23	30	46
1.0	230	300	461
0.1	2303	2996	4605

Unfortunately, in both commercial and regulatory practice, failure to detect a pathogen is often taken as positive proof of its absence from a lot at the time of sampling. Consequently, if the pathogen should be detected subsequently, or cause disease, it is often assumed that contamination occurred after the last sampling, and this may be erroneous. The testing of individual batches of product for pathogens that are infrequently present in raw meat is therefore of little use for discriminating between contaminated and uncontaminated batches. It does, however, give rise to the unwarranted assumption that the product is free from the pathogen for which it was tested, and can misdirect any investigation of the source of a pathogen detected in the product during secondary processing or following consumption.

Although increasing the number of samples per lot must increase the probability of detecting an uncommon pathogen, any increase in sampling within practicable limits will still permit the detection of few contaminated lots and possible acceptance of most of them. Even when a pathogen is present relatively frequently, as is *Salmonella* in some raw meats, and several samples are obtained from lots in accordance with an attributes sampling plan, the probability of accepting a lot that is unacceptably contaminated is still likely to be high (ICMSF, 1974).

Batches of some raw meats, such as chilled, vacuum-packaged products carried by sea to distant markets, may be tested by customers on arrival to decide whether or not the remaining storage life of the product is adequate. Such testing usually involves only the determination of APCs from swab samples, samples of excised tissue or exudate. Apparently, the criterion for acceptance is often that none of the counts obtained per cm^2, g or ml from one or a few samples exceeds a stipulated, maximum value. Obviously, the results of any such test can be related properly to the condition of the batch as a whole only when the relationship between the numbers of bacteria recovered and the remaining storage life of the product is well established, and the distribution of bacteria on or in the product at the time of its receipt is known, or can be reasonably assumed, so that the proportion of product units that may have an unacceptably short storage life can be assessed. However, there is often little or no specific information on those matters. Thus, in general practice, it seems that a batch is deemed unacceptable when the counts from any sample exceed the specified values. With such an approach, the possibility of misjudging the storage life of product in a batch must be very high.

Criteria in the form of attributes acceptance plans for *E. coli* and *Salmonella* have been formulated by the USDA as standards against which the performance of the HACCP system for controlling carcass contamination is assessed (Table 6.3). The criteria for *E. coli* are applied at any time to the n values obtained prior to the data being examined, while the data for *Salmonella* are collected by obtaining one sample per day until a data set is complete. Whether these criteria are relevant to the safety of meat is

Table 6.3 Sampling plans for the microbiological testing of red meat carcasses and ground beef required by the US Department of Agriculture (USDA)

Organisms	Product	Testing frequency	Test parameters			
			n[a]	m[b]	M[c]	c[d]
Generic *Escherichia* coli	Cattle carcasses	1/300[e]	13	5	100	3
	Pig carcasses	1/1000[e]	13	10	10000	3
Salmonella	Beef carcasses	1/day	82	–	–	1
	Cow carcasses	1/day	58	–	–	2
	Pig carcasses	1/day	55	–	–	6
	Ground beef	1/day	53	–	–	5

[a] Number of samples required for decision.
[b] Samples yielding $< m$ cfu/cm^2 or g are wholly acceptable.
[c] Samples yielding $> M$ cfu/cm^2 or g are wholly unacceptable.
[d] Tolerable number of samples yielding bacteria at numbers $> m < M$, or positive for *Salmonella*.
[e] One test for each 300 or 1000 carcasses processed.

questionable, since they are based on data from a survey of chilled carcasses at US plants rather than on Food Safety Objectives (FSOs), which would be more appropriate (Van Schothorst, 1998; Gorris, 2005). Moreover, the criteria reflect the general condition of commercial carcasses before the introduction of decontamination treatments that are now common in North American meat packing plants (Belk, 2001). Also, they were devised on the assumption that enteric organisms are deposited on meat only during the carcass dressing process, whereas at some plants, at least, most of these organisms are acquired during carcass breaking (Gill *et al.*, 1999a). Furthermore, a re-examination of the HACCP system is required in response to failure to meet the criteria, although the data can give no indication of where control may have failed. Finally, compliance with such criteria does not demonstrate that meat is microbiologically safe.

In view of the uncertainty of most conclusions that can be drawn from the application of attributes sampling plans to raw meats, the use of variables sampling plans would seem to be preferable. Variables sampling can be used to identify objectively the CCPs in a meat production process, as described in Section 6.4. Although variables sampling usually involves the enumeration of indicator organisms rather than pathogens *per se*, because of the infrequency of enteric pathogens on meat, it is often possible to enumerate indicator organisms when present in very low numbers, and to deal with some proportion of the values in a set that are censored because they were below the limit of detection in the method of sample analysis (Marks and Coleman, 1998).

For an indicator recovered from a product in a specific process, it might be possible sometimes to determine the proportionate presence of pathogens or spoilage organisms encompassed by the indicator group. However, the only general relationship that can be assumed between the key organisms and those indicative of their possible presence is that the numbers of pathogenic or spoilage organisms do not exceed the levels of the indicator organisms. Even so, the use of numerical data to describe contamination of the product with indicator organisms allows the ranking of process performances and the setting of FSOs or other standards of microbiological quality by reference to the best performances (Brown *et al.*, 2000). Obviously, certainty about the safety or storage stability of the product can be enhanced by proper selection of the indicator organisms to be enumerated. Thus, criteria based on numbers of generic *E. coli* will provide more certain information about possible contamination with enteric pathogens than will criteria based on numbers of Enterobacteriaceae, as is required in EU regulations (Table 6.4), because the latter group of organisms includes many saprophytic species that have no connection with faecal contamination. Similarly, enumeration of pseudomonads or lactic acid bacteria, rather than APCs, will allow better understanding of the extent to which the product is contaminated with organisms that are capable of causing aerobic or anaerobic spoilage.

In practice, the enumeration of several indicator organisms is desirable, for detailed understanding of the extent to which the product is contaminated with various types of bacteria, and the different sources of some of those organisms. Although that may involve extensive examination of the product, this can be spread over a lengthy period, and the amount of sampling required can be greatly reduced once control over microbiological contamination of the product is established.

Table 6.4 Sampling plans for the microbiological testing of red meat carcasses as required by the European Union (EU) and enacted in the UK

Organisms	Carcass type	Test parameters[a]		
		Unacceptable[b]	Acceptable[c]	Satisfactory[d]
Aerobes	Pig	4.3	4.3	3.3
	All other	4.3	4.3	2.8
Enterobacteriaceae	Pig	2.3	2.3	1.3
	All other	1.8	1.8	0.8

Figures relate to sampling of carcasses by sponge swabbing only (for excision sampling, see www.ukmeat.org).

[a] Mean log numbers per five carcasses sampled during one sampling session.
[b] Mean log/number of positives is above value given.
[c] Mean log is below value given.
[d] Mean log/number of positives is equal to or below value given.

6.9 Future trends

For the past few years, it has been widely acknowledged that the microbiological safety of raw meats, and indeed the safety of foods in general, will only be improved if hazardous microorganisms are controlled at all stages of production and processing. It is generally accepted that control can be achieved reliably by the implementation of HACCP systems for all food-related processes (Motarjemi, 2000).

Although the application of HACCP systems at all stages of food production, 'from farm to fork', is desirable in principle, it is now evident that such systems cannot be applied in practice to all the relevant processes (Sperber, 2005). For example, HACCP systems cannot be applied to processes that are inherently and uncertainly variable, such as most farming activities would seem to be, since situations of that type cannot be controlled consistently to maintain all microbiological hazards within known limits. Or again, at small food-handling establishments, such as many commercial kitchens, the introduction and maintenance of formal HACCP systems may be economically impractical. It seems likely that the limitations in applying HACCP systems will be recognised increasingly, with mitigation of microbiological hazards being sought for those processes that are not amenable to HACCP implementation by identification and encouragement of appropriate Good Practices.

For processes in which hazardous microbiological contamination can be controlled by HACCP systems, it is to be hoped that, increasingly, such systems will be implemented or suitably modified and also validated and verified by reference to appropriate microbiological data. However, in the immediate future, it seems likely that subjective assessments will continue to predominate in decisions relating to HACCP systems, while microbiological data will continue to be collected and used mainly for explicit or *de facto* acceptance testing. Despite acceptance testing being wholly at variance with the HACCP approach to assuring meat safety, and often essentially useless for identifying hazardous products, use of this approach is likely to increase, since it gives an illusion of attention to safety concerns, even though little or nothing may be achieved in practice.

The situation with regard to assuring the storage life of raw meats may be similar. Many packers now understand how to achieve consistently a long storage life for a product that is distributed to distant markets. However, the storage life of all raw meat is greatly affected by the temperatures to which it is exposed, and control of product temperature at some stages of distribution can be uncertain. Thus, those receiving consignments of raw meat may continue to apply acceptance tests that do not adequately determine the microbiological condition of the product, but perhaps slightly reduce the risk of accepting meat that is approaching spoilage, without considering the likelihood of rejecting wholly acceptable product.

6.10 Sources of further information

Methods for collecting and processing samples from foods in general, and detecting or enumerating specific bacteria in such samples, are detailed in Downes and Ito (2001) and FDA (1998). Microbiological methods for meat are described in Jensen *et al.* (2004). Sampling plans are discussed in ICMSF (2002) and their use with HACCP systems for meat plant processes is considered in Brown (2000). Both microbiological methods and sampling plans are discussed in Lund *et al.* (2000) and Robinson *et al.* (2000).

6.11 References

ACUFF G R (1999), 'Ground beef', in *The Role of Microbiological Testing in Beef Food Safety Programs*, Kansas City, US, USA, American Meat Science Association.

ANCELLE T (1998), 'History of trichinellosis outbreaks linked to horse-meat consumption', 1975–1998, *Eurosurveillance*, **3**, 86–89.

AQIS (2001), *Hot-boning and warm-boning. Procedures associated with program approval and criteria to be applied to hot and warm boned meat*, Notice No. MEAT 2001/20, Canberra, Australia, Australian Quarantine and Inspection Service.

ARTHUR M T, BOSILEVAC J M, NOU X, SHACKELFORD S D, WHEELER T L, KENT M P, JARONI D, PAULING B, ALLEN D M and KOOHMARAIE M (2004), '*Escherichia coli* O157 prevalence and enumeration of aerobic bacteria, *Enterobacteriaceae* and *Escherichia coli* O157 at various steps in commercial beef processing plants', *Journal of Food Protection*, **67**, 658–665.

BELK K E (2001), 'Beef decontamination technologies', *Beef Facts*, Denver, CO, USA, National Cattlemen's Beef Association.

BELL R G (1997), 'Distribution and sources of microbial contamination on beef carcasses', *Journal of Applied Microbiology*, **82**, 292–300.

BOEREMA J A, BRODA D M and BELL R G (2003), 'Abattoir sources of psychrophilic clostridia causing blown pack spoilage of vacuum-packed chilled meats determined by culture-based and molecular detection procedures', *Letters in Applied Microbiology*, **36**, 406–411.

BRENNER D J (1992), 'Introduction to the family Enterobacteriaceae', in *The Prokaryotes*, 2nd edition, eds Balows A, Trüper H G, Dworkin M, Harder W and Schleifer K H, New York, USA, Springer Verlag, 2673–2695.

BROWN M H (ed.) (2000), *HACCP in the Meat Industry*, Cambridge, UK, Woodhead Publishing.

BROWN M H and BAIRD-PARKER A C (1982), 'The microbiological examination of meat', in *Meat Microbiology*, ed. Brown M H, London, UK, Applied Science Publishers, 423–520.

BROWN M H, GILL C O, HOLLINGSWORTH J, NICKELSON R, SEWARD S, SHERIDAN J J, STEVENSON T, SUMNER J L, THENO D M, USBORNE W R and ZINK D (2000), 'The role of microbiological testing in systems for assuring the safety of beef', *International Journal of Food Microbiology*, **62**, 7–16.

CAMPBELL W C (1988), 'Trichinosis revisited, another look at modes of transmission', *Parasitology Today*, **4**, 83–86.

CFIA (2005), Policy on the control of *E. coli* O157:H7 contamination in raw beef products, *Meat Hygiene Manual of Procedures*, Chapter 4, Annex 0, Nepean, Ontanio, Canada, Canadian Food Inspection Agency.

CORRADINI M G, NORMAND M D, NUSSINOVITCH A, HOROWITZ J and PELEG M (2001), 'Estimating the frequency of high microbial counts in commercial food products using various distribution functions', *Journal of Food Protection*, **64**, 674–681.

DOWNES F P and ITO K (2001), *Compendium of Methods for the Microbiological Examination of Foods*, 4th edition, Washington, DC, USA, American Public Health Association.

EC (2001), European Communities, Commission decision of 8 June 2001, *Official Journal of the European Union*, **L165**, 48–53.

EVRENDILEK G A, RICHTER E R and CHISM G W (2001), 'Concurrent detection of *Escherichia coli* O157:H7 and *Salmonella* from a single enrichment in 24 h', *Journal of Food Safety*, **21**, 75–86.

FDA (1998), Food and Drug Administration, *Bacteriological Analytical Manual*, 8th edition, revision A, Gaithersburg, MD, USA, AOAC International.

FENG P (1997), 'Impact of molecular biology on the detection of foodborne pathogens', *Molecular Biotechnology*, **7**, 267–278.

FREDRIKSSON-AHOMAA N, HALLANVUO S, KORTE T, SIITONEN A and KORKEALA H (2001), 'Correspondence of genotypes of sporadic *Yersinia enterocolitica* bioserotype 4/0:3 strains from human and porcine sources', *Epidemiology Infection*, **12**, 37–47.

FUNG D Y (2002), 'Predictions for rapid methods and automation in food microbiology', *Journal of the AOAC International*, **85**, 1000–100.

GAMBLE H R, GAJADHAR A A and SOLOMON M B (1996), 'Methods for detection of trichinellosis in horses', *Journal of Food Protection*, **59**, 420–425.

GILL C O (1979), 'Intrinsic bacteria in meat', *Journal of Applied Bacteriology*, **47**, 367–376.

GILL C O (1988), 'Microbiology of edible meat byproducts', in *Edible Meat Byproducts*, eds Pearson A M and Dutson T R, London, UK, Elsevier Applied Science, 47–82.

GILL C O (1995), 'MAP and CAP of fresh, red meats, poultry and offals', in *Principles of Modified-Atmosphere and Sous Vide Packaging*, eds Farber J M and Dodds K L, Lancaster, PA, USA, Technomic Publishing Co, 105–136.

GILL C O (1996), 'Extending the storage life of raw chilled meats', *Meat Science*, **43**, S99–S109.

GILL C O (2000), 'HACCP in primary processing: red meat', in *HACCP in the Meat Industry*, ed. Brown M H, Cambridge, UK, Woodhead Publishing, 81–122.

GILL C O (2003), 'Active packaging in practice: meat', in *Novel Food Packaging Techniques*, ed. Ahvenainen R, Cambridge, UK, Woodhead Publishing, 365–383.

GILL C O and BADONI M (2004), 'Effects of peroxyacetic acid, acidified sodium chlorite or lactic acid solutions on the microflora of chilled beef carcasses', *International Journal of Food Microbiology*, **91**, 43–50.

GILL C O and JONES T (1997), 'Assessment of the hygienic characteristics of a process for dressing pasteurized pig carcasses', *Food Microbiology*, **14**, 81–91.

GILL C O and JONES T (2000), 'Microbiological sampling of carcasses by excision or swabbing', *Journal of Food Protection*, **63**, 167–173.

GILL C O and LANDERS C (2003a), 'Microbiological effects of carcass decontamination treatments at four beef packing plants', *Meat Science*, **65**, 1005–1011.

GILL C O and LANDERS C (2003b), 'Effects of spray-cooling processes on the microbiological conditions of decontaminated beef carcasses', *Journal of Food Protection*, **66**, 1247–1252.

GILL C O and LANDERS C (2005), 'Microbiological conditions of horse meat prepared at a North American packing plant, and control of the temperatures of product air freighted to Europe', *Meat Science*, **69**, 501–507.

GILL C O and McGINNIS J C (2004), 'Microbiological conditions of mechanically tenderized beef cuts prepared at four retail stores', *International Journal of Food Microbiology*, **95**, 95–102.

GILL C O and PENNEY N (1977), 'Penetration of bacteria into meat', *Applied and Environmental Microbiology*, **33**, 1284–1286.

GILL C O, BADONI M and JONES T (1996), 'Hygienic effects of trimming and washing operations in a beef-carcass-dressing process', *Journal of Food Protection*, **59**, 666–669.

GILL C O, DESLANDES B, RAHN K, HOUDE A and BRYANT J (1998a), 'Evaluation of the hygienic performances of the processes for beef carcass dressing at 10 packing plants', *Journal of Applied Microbiology*, **84**, 1050–1058.

GILL C O, McGINNIS J C and BRYANT J (1998b), 'Microbial contamination of meat during the skinning of beef carcass hindquarters at three slaughtering plants', *International Journal of Food Microbiology*, **42**, 175–184.

GILL C O, BADONI M and McGINNIS J C (1999a), 'Assessment of the adequacy of cleaning of equipment for breaking beef carcasses', *International Journal of Food Microbiology*, **46**, 1–8.

GILL C O, BAKER L P and JONES T (1999b), 'Identification of inadequately cleaned equipment used in a sheep carcass-breaking process', *Journal of Food Protection*, **62**, 637–643.

GILL C O, BADONI M and McGINNIS J C (2001a), 'Microbiological sampling of meat cuts and manufacturing beef by excision or swabbing', *Journal of Food Protection*, **64**, 325–334.

GILL C O, McGINNIS J C and BRYANT J (2001b), 'Contamination of beef chucks with *Escherichia coli* during carcass breaking', *Journal of Food Protection*, **64**, 1824–1827.

GILL C O, JONES T, RAHN K, HOUDE A, McGINNIS J C, CAMPBELL S, HOLLEY R A and LeBLANC D I (2002), 'Control of product temperatures during the distribution of retail ready beef to stores and vacuum packaged beef to restaurants', *Dairy Food and Environmental Sanitation*, **22**, 422–428.

GILL C O, BRYANT J and LANDERS C (2003), 'Identification of critical control points for control of microbiological contamination in processes leading to the production of ground beef at a packing plant', *Food Microbiology*, **20**, 641–650.

GILL C O, McGINNIS J C, HOUDE A, LAMOUREUX L and LeBLANC D (2005), 'Microbiological conditions of moisture-enhanced pork before and after cooking', *Food Microbiology*, **22**, 321–327.

GIRRAFA G and NEVIANI E (2001), 'DNA-based, culture independent strategies for evaluating microbial communities in food-associated ecosystems', *International Journal of Food Microbiology*, **67**, 19–34.

GOBAT P F and JEMMI T (1991), 'Epidemiological studies on *Listeria* spp. in slaughterhouses', *Fleischwirtsch International*, **1**, 44–49.

GORRIS L G M (2005), 'Food safety objective: an integral part of food chain management', *Food Control*, **16**, 801–809.

HARRIGAN W F and McCANCE M E (1976), 'The detection and enumeration of pathogenic and toxigenic organisms', in *Laboratory Methods in Food and Dairy Microbiology*, ed. Harrison WF, London, UK, Academic Press, 147–160.

HATHAWAY S C and McKENZIE A I (1991), 'Postmortem meat inspection programs; separating science and tradition', *Journal of Food Protection*, **54**, 471–475.

HILDEBRANDT G and BÖHMER L (1998), 'Sampling plans for the assessment of *Salmonella* contamination', *Fleishwirtsch International*, **1/2**, 44–45.

HILDEBRANDT G and WEISS H (1994), 'Sampling plans in microbiological quality control 1. Description of plans in common use', *Fleischwirtsch International*, **2**, 54–57.

HILDEBRANDT G, BÖHMER L and DAHMS S (1995), 'Three class attributes plans in microbiological quality control: a contribution to the discussion', *Journal of Food Protection*, **58**, 784–790.

HITCHINS A D, FENG P, WATKINS W D, RIPPEY S R and CHANDLER L A (1998), '*Escherichia coli* and the coliform bacteria', in Food and Drug Administration, *Bacteriological Analytical Manual*, 8th edition, revision A, Gaithersburg, MD, USA, AOAC International, Chapter 4.

HOLLEY R A, PEIRSON M D, LAM J and TAN K B (2004), 'Microbial profiles of commercial, vacuum-packaged, fresh pork of normal or short storage life', *International Journal of Food Microbiology*, **97**, 53–62.

ICMSF (1986), Sampling plans for situations involving direct hazards from pathogens, *Microorganisms in Foods 2: Sampling for Microbiological Analysis: Principles and Specific Applications*, 2nd edition, eds Roberts T A, Bryan F L, Christian J N B, Kilsby D, Olsun J C Jr and Silliker JH, International Commission on Microbiological Specifications for Foods, Toronto, Canada, University of Toronto Press, 65–75.

ICMSF (2002), *Microorganisms in Foods 7: Microbiological Testing in Food Safety Management*, International Commission on Microbiological Specifications for Foods, New York, USA, Kluwer Academic/Plenum Publishers.

ISONHOOD J H and DRAKE M (2002), 'Aeromonas species in foods', *Journal of Food Protection*, **65**, 575–582.

JARVIS G A and MALCOLM S A (1986), 'Comparison of three-class attributes sampling plans and variables sampling plans for lot acceptance sampling in food microbiology', *Journal of Food Protection*, **49**, 724–728.

JAY J M (2002), 'A review of aerobic and psychrotrophic plate count procedures for fresh meat and poultry products', *Journal of Food Protection*, **65**, 1200–1206.

JAY J M and MARGITIC S (1979), 'Comparison of homogenizing, shaking, and blending on the recovery of microorganisms and endotoxins from fresh and frozen ground beef as assessed by plate counts and the Limulus amoebocyte lysate test', *Applied and Environmental Microbiology*, **38**, 879–884.

JENSEN W K, DEVINE C and DIKEMAN M (2004), *Encyclopedia of Meat Sciences*, Amsterdam, Netherlands, Elsevier.

JERICHO K W F, KOZUB G C, BRADLEY J A, GANNON V P J, GOLSTEYN-THOMAS E J, GIERUS M, NISHIYAMA B J, KING R K, TANAKA E E, D'SOUZA S and DIXON-MACDOUGALL J M (1996), 'Microbiological verification of the control of the processes of dressing, cooling and processing of beef carcasses at a high line-speed abattoir', *Food Microbiology*, **13**, 291–301.

JERICHO K W F, KOZUB G C, GANNON V P J, GOLSTEYN-THOMAS E J, KING R K, BIGHAM R L, TANAKA E E, DIXON-MacDOUGALL J M, NISHIYAMA B J, KIRBYSON M and BRADLEY J A (1997), 'Verification of the level of microbiological control for the slaughter and cooling processes of beef carcass production at a high-line-speed abattoir', *Journal of Food Protection*, **60**, 1509–1514.

JOHNSON J L, DOYLE M P and CASSENS R G (1990), '*Listeria monocytogenes* and other *Listeria* spp. in meat and meat products', *Journal of Food Protection*, **53**, 81–91.

KILSBY D C and PUGH M E (1981), 'The relevance of the distribution of micro-organisms within batches of food to the control of microbiological hazards from foods', *Journal of Applied Bacteriology*, **51**, 345–354.

KILSBY D C, ASPINALL L J and BAIRD-PARKER A C (1979), 'A system for setting numerical microbiological specifications for foods', *Journal of Applied Bacteriology*, **46**, 591–599.

LEGAN J D, VANDEVEN M H, DAHMS S and COLE M B (2001), 'Determining the concentration of microorganisms controlled by attributes sampling plans', *Food Control*, **12**, 137–147.

LOW J C, McKENDRICK I J, McKECHNIE C, FENLON D, NAYLOR S W, CURRIE C, SMITH D G E, ALLISON L and GALLY D L (2005), 'Rectal carriage of enterohemorrhagic *Escherichia coli* O157 in slaughtered cattle', *Applied and Environmental Microbiology*, **71**, 93–97.

LUND B M, BAIRD-PARKER A C and GOULD G W (2000), *The Microbiological Safety and Quality of Food*, Gaithersburg, MD, USA, Aspen Publications.

McDONALD L C, HACKNEY C R and RAY B (1983), 'Enhanced recovery of injured *Escherichia coli* by compounds that degrade hydrogen peroxide or block its formation', *Applied and Environmental Microbiology*, **45**, 360–365.

MACKEY B M and ROBERTS T A (1993), 'Improving slaughtering hygiene using HACCP and monitoring', *Fleischwirtschaft International*, **2**, 40–45.

MARKS H and COLEMAN M (1998), 'Estimating distributions of numbers of organisms in food products', *Journal of Food Protection*, **61**, 1535–1540.

MARQUARDT D W (1984), 'New technical and educational directions for managing product quality', *American Statistician*, **38**, 8–14.

MOTARJEMI Y (2000), 'Regulatory assessment of HACCP: a FAO/WHO consultation on the role of government agencies in assessing HACCP', *Food Control*, **11**, 341–344.

MAYES T (1993), 'The application of management systems to food safety and quality', *Trends in Food Science and Technology*, **4**, 216–219.

MESSER J W, RICE E W and JOHNSON C H (2000), 'Total viable counts; pour plate technique; spread plate technique', in *Encyclopedia of Food Microbiology*, eds Robinson R K, Batt C A and Patel P D, London, UK, Academic Press, 2154–2160.

NOTTINGHAM P M (1982), 'Microbiology of carcass meats', in *Meat Microbiology*, ed. Brown H, London, UK, Applied Science Publishers, 13–65.

PEARCE R A and BOLTON D J (2005), 'Excision vs. sponge sampling – a comparison of methods for the microbiological sampling of beef, pork and lamb carcasses', *Journal of Applied Microbiology*, **98**, 896–900.

PEELER J T, HOUGHTBY G A and RAINOSEK A P (1992), 'The most probable number technique', in *Compendium of Methods for the Microbiological Examination of Foods*, eds Vanderzant C and Splittstoesser D F, Washington, DC, USA, American Public Health Association, 105–120.

ROBERTS T A, MacFIE H J H and HUDSON W R (1980), 'The effect of incubation temperature and site of sampling on assessment of the numbers of bacteria on red meat carcasses at commercial abattoirs', *Journal of Hygiene Cambridge*, **85**, 371–380.

ROBINSON R K, BATT C A and PATEL D P (eds) (2000), *Encyclopedia of Food Microbiology*, San Diego, CA, USA, Academic Press.

SHARPE A N, ISIGIDI BIN KINGOMBE C, WATNEY P, PARRINGTON L J, DUDAS I and DIOTTE M P (1996), 'Efficient non destructive sampler for carcasses and other surfaces', *Journal of Food Protection*, **59**, 757–763.

SIRAGUSA G R (1998), 'Microbiological sampling plans for beef trim combos', in *The Role of Microbiological Testing in Beef Food Safety Programs*, Kansas City, KS, USA, American Meat Science Association.

SPERBER W H (2005), 'HACCP does not work from farm to table', *Food Control*, **16**, 511–514.

STANBRIDGE L H and DAVIES A R (1998), 'The microbiology of chill-stored meat' in *The Microbiology of Meat and Poultry*, eds Davies A and Board R, London, UK, Blackie Academic, 174–219.

SWANSON K M J, BUSTA F F, PETERSON E H and JOHNSON M G (1992), 'Colony count methods', in *Compendium of Methods for the Microbiological Examination of Foods*, eds Vanderzant C and Splittstoesser D F, Washington, DC, USA, American Public Health Association, 75–95.

USDA (1992), US Department of Agriculture, National Advisory Committee on Microbiological Criteria for Foods, 'Hazard analysis and critical control point system', *International Journal of Food Microbiology*, **16**, 1–23.

USDA (1996), US Department of Agriculture, 'Pathogen reduction; hazard analysis and critical control point (HACCP) system; final rule', *Federal Register*, **61**, 38805–38989.

van SCHOTHORST M (1998), 'Principles for the establishment of microbiological food safety objectives and related control measures', *Food Control*, **9**, 379–384.

7

Sampling methods for poultry-meat products

G. C. Mead, Consultant, UK

7.1 Introduction

Microbial contamination of processed poultry has important implications for both product safety and shelf-life, and its determination, whether qualitative or quantitative, requires sampling methods that are sensitive and easy to apply, without causing unacceptable damage to the product. Such methods must also yield results that are readily interpreted by those responsible for monitoring and controlling product quality. Among meat animals, poultry is unusual, because the carcass remains whole throughout the primary processing operation and the abdominal cavity is relatively inaccessible. Although microbial contaminants will be present on all exposed surfaces, their distribution tends to be uneven and many are attached to, or entrapped in, skin and muscle in a manner that resists attempts to remove them by physical means, such as rinsing or swabbing. Numerous sampling methods have been developed and some of them have been compared, but the choice of method for a particular purpose must always take account of the likely numbers and distribution of the organisms being sought.

More recently, microbiological analysis of poultry meat has gained a new impetus from the continued growth in international trade and widespread application of Hazard Analysis Critical Control Point (HACCP) principles in processing plants, whether this has been voluntary, mandatory or a customer requirement. The HACCP system can involve microbiological testing in a number of ways. In particular, it may be used at different stages of processing to validate CCPs or to verify that the control programme as a whole is functioning effectively. Testing may also be needed to demonstrate compliance with certain performance criteria, such as those for *Escherichia coli* and *Salmonella* that are required in the USA (USDA-FSIS, 1996). Any use of a microbiological criterion requires that the method of sampling the product is agreed and clearly described, because it will have an important influence on the results obtained. In addition, there will be a need for a sampling plan to indicate the number of samples to be taken on each occasion and appropriate limit values for the target organism(s).

Firstly, this chapter will consider the nature of microbial contamination on raw poultry and its association with carcass surfaces. Then, it will review the sampling of both chilled and frozen products, and describe those methods that are considered most suitable for routine use. Attention will also be given to sampling further-processed products, in which the nature and distribution of the main contaminants are generally different from those on raw meat.

7.2 Origins and distribution of carcass contaminants

Birds arriving for slaughter at the processing plant carry many millions of different microbes, including some of possible public health significance and others capable of spoiling the processed product during refrigerated storage. The organisms occur on the skin, among the feathers and in the nasopharynx and alimentary tract of the bird. Among those that can be present are the main foodborne pathogens associated with poultry: *Campylobacter jejuni*, non-host-specific salmonellas and *Clostridium perfringens*. All of them may be carried asymptomatically in the alimentary tract and are transmitted to carcass surfaces via faecal contamination. Another potential food-poisoning organism is *Staphylococcus aureus*, which can occur in the nasopharynx and on the skin, although avian strains rarely produce the enterotoxin that causes human illness (Mead and Dodd, 1990). Cold-tolerant spoilage bacteria and yeasts are also found initially, especially among the feathers, and are acquired largely from the environment in which the birds are reared. The principal spoilage bacteria are pigmented and non-pigmented strains of *Pseudomonas* spp. and most are destroyed as carcasses pass through the scalding process to loosen the feathers. However, recontamination occurs during subsequent stages of processing, especially via the hands and gloves of operatives (Holder *et al.*, 1997; Mead, 2004). By contrast, *Listeria monocytogenes* is a cold-tolerant pathogen that is rarely isolated from the live bird, but appears to multiply on processing equipment, from which it can be transferred to carcasses being processed (Genigeorgis *et al.*, 1989; Hudson and Mead, 1989; Ojeniyi *et al.*, 1996).

Generally, microbial contamination occurs over the entire outer surface of the processed carcass and in the abdominal cavity. However, numbers may vary widely, as can the distribution of the organisms. In studying the variability of the microbial load on carcasses sampled at three different processing plants, McNab *et al.* (1993) found that factors operating at the level of the individual bird, the bird-lot and the processing plant all had an effect, being responsible for 56, 26 and 18 %, respectively, of the count variability. When carcass contamination was studied over a five-month period at a single processing plant, Renwick *et al.* (1993) noted that variability between carcasses, lots from a single supplier and lots from different suppliers represented 73, 14 and 13 %, respectively, of the overall variability observed. The microbial load increased with factors such as longer crating and holding periods, and slaughter during winter months. While plant-to-plant differences in processing procedures and hygiene control are to be expected, experience suggests that, for any one plant, microbial counts vary less between sampling sites on the same carcass than they do between individual carcasses (Patterson and Gibbs, 1975). In comparing counts from wing, breast and leg, Lahellec and Meurier (1970) found that no single site was consistently more contaminated than any other. However, it is widely recognised that counts from the breast area are relatively low in most cases, while the neck skin is a more heavily contaminated part of the carcass (Barnes *et al.*, 1973a). This may be due partly to the position of the carcass during processing. When hung by the legs, the carcass drains down into the flap of skin that remains after neck removal.

In relation to carcass sampling, there is an important difference between *Salmonella* and *Campylobacter* spp. Although the proportion of processed carcasses that are contaminated with *Salmonella* has sometimes been above 50 % (Waldroup, 1996), the number of bacterial cells per carcass is usually low, as indicated in Table 7.1. Only occasionally are carcasses found to carry in excess of a thousand salmonellas (Jørgensen *et al.*, 2002). A further study in Finland, where there is an official *Salmonella* control programme that encompasses broilers, confirmed that low numbers occur consistently (Hirn *et al.*, 1992). Regular examination of frozen chicken carcasses over a three-year period revealed that up to 11 % were *Salmonella*-positive, but 70 % of

Table 7.1 Incidence and numbers of *Salmonella* on raw poultry carcasses in studies cited by Waldroup (1996)

Country	Year	Positive carcasses (%)	Count (cfu/carcass)
USA	1969	20.5	<30
The Netherlands	1981	?	63
Japan	1981	25	<100
USA	1983	11.6	<1
USA	1991	17–50	5–34
India	1991	100	300
USA	1992	40.8	1.8
USA	1993	21.8	1.4

the positive carcasses contained less than five viable cells. This low level of contamination would be expected from the correspondingly low carriage-rate in Finnish broiler flocks at the time of slaughter.

The situation is different for *Campylobacter* and, at slaughter, levels of intestinal carriage often exceed 10^6 cfu per gram of caecal content in carrier birds. Hood *et al.* (1988) found up to 1.5×10^6 cfu per carcass in fresh, retail broilers, and surface swabs from different sampling sites, including the abdominal cavity, showed a relatively uniform distribution of the organisms. In another study, Berndtson *et al.* (1992) confirmed the presence of *Campylobacter* on the skin and in the abdominal cavity of processed broilers and showed that the organisms could penetrate the skin, presumably via feather holes. Approximately 3 % of muscle samples were also reported to contain *Campylobacter*, but this could have been due to cross-contamination during sampling. However, deep-muscle contamination was also reported by Luber *et al.* (2005). Both *Campylobacter* and *Salmonella* can sometimes invade certain internal organs of the bird, such as the liver, but their ability to persist in muscle tissue may be limited by the antimicrobial factors that are associated with normal tissues and fluids (Skarnes and Watson, 1957). Traditionally, the deep muscle of processed poultry has been regarded as free from viable organisms and multiplication was not observed in uneviscerated carcasses held at 10 °C or 15 °C for several days (Barnes *et al.*, 1973b), although no account was taken of *Campylobacter*. More work is needed to clarify the situation.

7.3 Microbial attachment to skin and other carcass surfaces

The ability of microbes to become attached to a variety of surfaces is well known. The phenomenon has significant consequences for the persistence of microorganisms, cleaning of carcasses during processing and carcass sampling. However, the biological surfaces involved are complex and the mechanisms by which microbes become attached or entrapped are still poorly understood. Early studies on the attachment of specific bacterial strains to broiler skin showed that only flagellated bacteria attached significantly, and the rate and extent of attachment were influenced by factors such as bacterial strain, time of exposure, temperature, pH and the number of organisms present in the water-film overlying the skin surface (Notermans and Kampelmacher, 1974, 1975a). Organisms present in the water-film were readily removed by washing, but otherwise became attached to the skin, where they were protected from the effects of carcass scalding and washing (Notermans and Kampelmacher, 1975b).

Later work provided a different perspective on the way in which bacteria are retained by chicken skin. Most importantly, McMeekin and Thomas (1978) found that bacterial motility had a negligible effect and many non-motile bacteria were retained equally well. Examination of naturally-contaminated skin by electron microscopy showed that the predominant organisms initially were types resembling *Micrococcus* spp. (Thomas and McMeekin, 1980). These bacteria were located in accumulations of sebum-like substances at the surface of the stratum corneum. Removal of the outer epidermal layer during 'hard' scalding and plucking of carcasses exposed a new surface for microbial contamination, which contained many capillary-size channels and crevices. During subsequent water-immersion chilling, swelling of the skin occurred and was presumed to be due to water absorption. This exposed the channels and crevices to contaminants in the water, a process affected by the time and temperature of immersion (Thomas and McMeekin, 1982). It was suggested that swelling of the skin could trap bacteria already present in channels and crevices and render them even less amenable to physical removal. Alternatively, skin-swelling may provide access to new niches for contaminants to occupy.

Skin, of course, is not the only type of surface on the carcass to which microbes are exposed. Thomas and McMeekin (1981) showed that immersion in water of chicken muscle fascia or muscle perimysium caused collagen associated with the connective tissue to expand and form a dense network of fibres at the surface. Two strains of *Salmonella*, both of which were afimbriate and one non-motile, became attached to the collagen fibres, but only after extended periods of immersion. These findings are relevant to the underside of the neck flap, a key sampling site (see below), which is capable, therefore, of trapping microbial contaminants.

The entrapment of bacteria in skin crevices, following water immersion, was confirmed by Lillard (1985, 1989). Again, 'attachment' was not significantly affected by the presence of flagella or fimbriae; furthermore, electrostatic attraction between the organisms and the surface appeared to play no part. It is noteworthy that bacteria were found to be firmly attached to the skin, even before the bird was processed. In experimental trials on skin portions, attachment began within only 15 seconds and increased with time in a linear manner. The rate of bacterial attachment showed no significant difference for skin from 'soft'- and 'hard'-scalded carcasses.

7.4 Sampling carcasses in the processing plant

Sampling plans are beyond the scope of this chapter but, for commercial purposes, practical considerations tend to dictate how many carcasses will be sampled at any specific stage of the process. In high-rate production, any feasible number will represent only a very small fraction of a single lot. Nevertheless, data generated regularly on this basis are useful to indicate trends in product contamination. By contrast, most scientific studies have involved a minimum of five replicate carcasses although, if *Salmonella* is being sought and the prevalence is likely to be low, a larger number may be necessary to yield any positives. Whatever the level of sampling, carcasses should always be taken for analysis at random and only discarded or avoided if they are incomplete or otherwise visibly damaged.

The timing of sampling may be less critical than was once thought. In a European Commission study (CEC, 1979), sampling was delayed until two hours after the start of processing to allow a period of equilibration in the microbial contamination of processing equipment. This decision appears to have been based on supposition, because subsequent studies have failed to find any consistent relationship between microbial contamination of a lot and the time at which samples were taken after processing commenced (McNab *et al.*, 1991, 1993; Renwick *et al.*, 1993). Clearly, other factors may have a greater influence on the microbial load of a particular lot. On the other hand, sampling of the first birds to enter the process, when the equipment is at its cleanest, may give misleading results.

7.5 Methods of sampling carcasses

Many different methods have been developed for sampling poultry carcasses during and/or after processing to determine their microbiological condition. These involve the following basic techniques: surface swabbing, whole-carcass rinsing, tissue excision and maceration, repeated dipping of a whole carcass in diluent, collection of drip (weepage), high-pressure spraying or scraping of a defined area of skin, spraying the abdominal cavity and lifting skin contaminants with an agar contact-plate or nitrocellulose membrane (Mercuri and Kotula, 1964; Avens and Miller, 1970; Patterson, 1971; Barnes *et al.*, 1973a; Blankenship *et al.*, 1975; Russell *et al.*, 1997). Within the above categories, there are a number of variations, thereby extending the methods available. Differences between them in relation to the recovery of microorganisms can be explained largely by (i) the incidence and distribution of target organisms on the carcass and (ii) the degree of attachment or entrapment and therefore the ease with which the organisms can be removed. Among the most widely used, practical and cost-effective techniques for in-plant sampling are swabbing, whole-carcass rinsing and tissue excision. Their relative merits are as follows.

7.5.1 Swab-sampling

Swabbing was one of the earliest techniques used in experimental studies on carcass contamination. It has the advantage of being non-destructive and can be used for carcasses of any size. It is, however, relatively inefficient at removing organisms from the surface being sampled, as shown in Table 7.2. To explain this disadvantage, Sharpe *et al.* (1996) suggested that microbial release from surfaces is affected by the concentration of microbes already in suspension. With swab-sampling, therefore, the high concentration of microbes at the swab–sample interface may inhibit the release of

Table 7.2 Removal of microorganisms from poultry skin by repeated swab sampling (Patterson, 1971)

Swab no.	Count ($\log_{10}$ cfu/16 cm^2)	Percent recovery*
1	5.6	38
2	5.7	43
3	5.0	9
4	4.9	8
5	4.3	2

* Of total number removed by five swabs.

further organisms. Conventional swabs have cottonwool tips, but cotton gauze has also been used for poultry (Patterson, 1971) and other materials are possible, such as cellulose acetate sponge (Gill and Badoni, 2005). Since organisms collected from the test surface must be released into a diluent for analytical purposes, mechanical agitation of the suspended swab is required. A more convenient alternative is to use calcium alginate swabs, as described, for example, by Notermans *et al.* (1976), Izat *et al.* (1989) and Sveum *et al.* (1992). After sampling, the swab can be dissolved in a sodium hexametaphosphate solution, thereby releasing the organisms into suspension.

For quantitative sampling, a sterile template is used to delineate the required surface area. This may be up to 50 cm^2 and Sveum *et al.* (1992) favoured five such areas per carcass sampled. If a minority organism, such as *Salmonella*, is being sought, it is obviously advantageous to sample as much of the carcass as possible. To obtain microbial counts, on the other hand, a smaller area is suitable, e.g. 10 cm^2, and again more than one site on each carcass may be sampled. The counts are expressed per unit area sampled.

7.5.2 Whole-carcass rinsing

The whole-carcass rinsing technique is widely used, although many variations exist, as indicated in Table 7.3. It is the only technique that samples the entire carcass, including the abdominal cavity. Although essentially non-destructive, carcasses removed from the processing-line for sampling can only be returned if (i) the rinse fluid used is plain tap water or distilled water, (ii) sampling is carried out in the processing area and (iii) no temperature-abuse of the carcasses occurs before re-hanging on the processing line.

The organisms recovered by rinse sampling are those that are readily removed from the carcass and not firmly attached or entrapped. Lillard (1989) showed that, even after 40 successive rinses, relatively high numbers of organisms were still being removed. Nevertheless, because the entire

Table 7.3 Some examples of different methods used for whole-carcass rinse sampling

Reference	Rinse fluid	Amount (ml)	Rinsing time (sec)
Cox and Blankenship, 1975	Lactose broth + 0.6 % Tergitol	500	60
Cox *et al.*, 1981	Water	100	60
Mulder and Bolder, 1981	Peptone-saline	1000	30
Izat *et al.*, 1989	Distilled water	200	60
Stern, 1995	Phosphate-buffered saline	200	120
Li *et al.*, 2002	Buffered peptone water	100	150

surface, inside and out, is subjected to rinsing, the technique is particularly suitable for detecting minority organisms and is mandatory in the US Pathogen Reduction Programme for detecting *Salmonella* and counting *Escherichia coli* (USDA-FSIS, 1996). In the past, it has also found favour in Europe (CEC, 1979).

The basic procedure is as follows.

1. Place each carcass in a sterile, water-tight polythene bag and weigh, if necessary (see below). (A 3500 ml stomacher-type bag is suitable for the purpose.)
2. Add the required amount of diluent (related to carcass size).
3. Seal the bag with, e.g. a cable-tie.
4. Holding the bagged carcass with both hands, shake vigorously, firstly in a vertical direction, then horizontally, for a total of up to one minute. In doing so, one hand supports the carcass through the bottom of the bag, the other holds the legs through the closed bag at the top end.
5. After shaking, drain the carcass thoroughly into the bag and remove the carcass.
6. Transfer some or all of the rinse fluid to an appropriate screw-capped container for subsequent analysis.

Note that any handling of carcasses should involve the use of sterile, disposable gloves.

When carried out properly as a manual operation, whole-carcass rinsing is a tiring activity, especially when there are numerous carcasses to be sampled on any one occasion. Operator fatigue may lead to inadequate shaking and inconsistent results. For this reason, the shaking process was automated and a multi-unit sampler, capable of processing up to six chicken carcasses simultaneously, has been developed by Dickens *et al.* (1985). By automating the process, it becomes more standardised, a source of error is avoided, there is a saving of time and the technique can be applied to larger carcasses, such as those of turkeys, which are not amenable to manual shaking (Dickens *et al.*, 1986). In relation to plate counts and recovery of

Salmonella, no significant differences were observed between manual and automated rinsing (Dickens *et al.*, 1985, 1986).

Microbial counts obtained from rinse sampling may be expressed as cfu per ml of rinse fluid. Alternatively, they may be related to the overall surface area of the carcass and expressed per unit area by using appropriate conversion factors given by Simonsen (1971) and Thomas (1978). For the usual size of broiler (*c.* 1 kg eviscerated weight), the surface area in cm^2 is approximately equal to the carcass weight in grams +500 (Simonsen, 1971). Table 7.3 shows the variation between rinsing methods in the nature and amount of rinse fluid used and the required period of shaking. For broiler carcasses, even as little as 100 ml of fluid is used in one method of *Salmonella* detection (Cox *et al.*, 1981). This is an important practical consideration, because large amounts of rinse fluid are cumbersome to prepare, handle and transport to the processing plant. Also, with only 100 ml of rinse fluid, all of it can be used for sample enrichment. Some studies have shown that *Salmonella* detection is enhanced by incubating the rinsed carcass in the bag as a pre-enrichment step or at least including the skin (Cox and Blankenship, 1975; Jørgensen *et al.*, 2002; Simmons *et al.*, 2003). Inevitably, however, whole carcasses require a considerable amount of incubator space, which is not always available and, in this case, the method is obviously destructive.

7.5.3 Skin excision and maceration

In one respect, maceration of skin samples is a 'gold standard', because it ensures that virtually all the organisms present are released for detection or enumeration, although not all parts of the carcass are sampled. In principle, skin maceration is a destructive technique and may cause unacceptable damage to the carcass if more than one site is sampled. The problem can be avoided by confining attention to the neck flap, and only a small portion (*c.* 5 g) needs to be removed for microbial counts. *Salmonella* detection, however, usually requires a larger sample (10–25 g). A great advantage of this technique is that samples can be taken, using sterile scissors and forceps, without removing carcasses from the processing line, even at line-speeds of 6000 carcasses per hour. This means that the technique is easily the most rapid of those described here. Each skin sample can be collected in a clean plastic bag that is initially inverted over the hand and then used to enclose the cut sample by drawing it upwards. The bag is sealed with a knot. Samples collected thus are usually macerated in a stomacher. Alternatively, a mechanical blender can be used or samples can be shaken vigorously in screw-capped bottles containing glass beads or sharp sand (Patterson, 1971; Cox *et al.*, 1976).

Tests on chicken meat and skin, following maceration for two minutes, have shown that counts obtained with a stomacher are equivalent to those involving a mechanical blender (Emswiler *et al.*, 1977). Not surprisingly, skin maceration usually gives higher recoveries of microorganisms than either swabbing or rinse sampling, but the *Salmonella* detection-rate may be significantly lower (e.g. Cox *et al.*, 1978), because of the small area sampled. The neck-skin method is particularly suitable for detecting the effects of processing on carcass contamination in general (CEC, 1979) and, with this type of sample, coefficients of variation were significantly lower for counts of indicator bacteria than those obtained from carcass rinses (Hutchison *et al.*, 2006). There is no consistent relationship between counts obtained by neck-skin excision and whole-carcass rinsing, but the proportion of organisms recovered by the latter tends to diminish as carcasses pass through the processing operation (CEC, 1979). This may suggest that more organisms become firmly attached to carcass surfaces towards the end of the process and many of these are only recoverable by skin maceration.

Although the stomacher is widely used in food microbiology, Sharpe (2001) has developed the Pulsifier® (Filtaflex Ltd, Ontario, Canada), a device that utilises a combination of shock waves and intense agitation to release microbial contaminants from food samples. Counts thus obtained are said to be as high or higher than those from a sample macerated in a stomacher, and there is less debris in the resultant suspension to interfere with pipetting.

7.6 Sampling frozen meat

Frozen carcasses or other items are normally thawed before sampling by holding at 1–5 °C overnight (Barnes *et al.*, 1973a). As an alternative, Patterson and Gibbs (1975) held carcasses at room temperature for 2–4 hours, which could be barely sufficient to soften the tissues. However, thawing must be controlled carefully to avoid significant increases in microbial numbers.

Frozen blocks of deboned meat present a different challenge and usually need to be tested before thawing and use in product manufacture. In this case, microbial contaminants will be distributed throughout the block. Shallow blocks can be sampled by chipping small pieces of meat from the surface with a meat chopper, chisel and hammer (Paterson and Gibbs, 1975). For larger blocks, use can be made of a plug cutter fitted to a variable-speed electric drill (Barnes *et al.*, 1973a). A sterile, metal disc is fitted between the chuck and the plug cutter to prevent air currents from the drill contaminating the sample. Samples should be taken from several different sites on the block and then macerated.

7.7 Cut portions and other raw products

Cut portions may be prepared with or without skin, but skinless products are not necessarily less contaminated with microbes (Berrang *et al.*, 2001). As with whole carcasses, tissue excision and maceration are likely to give the highest recovery of microbes from portions. For this purpose, *c.* 5 g of cut muscle is needed for analysis, including skin, if present. The size and/or shape of the product may render quantitative swab-sampling difficult, although other methods, such as rinsing or collecting drip, are also possible (Barnes *et al.*, 1973a). In the case of giblets (gizzard, heart, liver and neck), the best method is to excise portions of tissue, for maceration. With mechanically recovered meat (MRM), samples may be collected by the method described previously for neck skin, using a plastic bag inverted over the hand.

7.8 Cooked, ready-to-eat products

There is now a multiplicity of further-processed poultry products (Fletcher, 2004), and sampling any of them must take account of the likely incidence and distribution of microbial contaminants. Commercial cooking processes can be expected to destroy most vegetative bacteria, including those of any pathogens, but not necessarily thermoduric organisms, such as enterococci, or bacterial spores. In practice, cooked products may be susceptible to re-contamination during subsequent handling and could acquire pathogens. However, microbial contamination of cooked items is likely to be at a low level and detection of pathogens, such as *Listeria monocytogenes*, will require relatively large samples. For whole carcasses, Barnes *et al.* (1973a) favoured the removal of meat from one complete half and maceration in twice the weight of diluent. Otherwise, swab sampling can be used and a suitably large area covered, including the abdominal cavity.

Barnes *et al.* (1973a) also describe the sampling of cooked poultry rolls, which are used as a source of sliced meat. A roll comprises pieces of meat that are compressed together and wrapped in a cylindrical casing. Each roll is cooked, cooled and either frozen or stored chilled. Initially, microbial contaminants will be present throughout the roll and, after cooking, sampling will need to detect any possible survivors at the centre. Usually, the roll can be examined in the unfrozen state with the aid of sterile cork borers (*c.* 2 cm diameter). Samples from the centre and other areas are analysed following maceration.

7.9 Pooling of samples

When samples from raw products are taken to determine the presence of *Salmonella*, it is a common practice within the industry for a number of samples to be treated as a composite. Thus, there is a saving in labour and materials, which need not reduce recovery of the target organism at the enrichment stage. In addition, pooling of samples may also be appropriate for microbial counting purposes, since it has the effect of averaging numbers across all samples (Hutchison *et al.*, 2006). The only disadvantage is that between-sample variation is then not apparent and one heavily-contaminated sub-sample could result in an exceptionally high overall count.

7.10 Sample handling and transportation

All samples should be taken aseptically by properly trained personnel, using pre-sterilised instruments and materials. When products are sampled in the production area, the use of glass bottles, pipettes, etc. must be avoided, due to the risk of breakage. The labelling of samples should be clear and unambiguous, and appropriate records kept, detailing the timing and origin of each sample.

In principle, samples should be analysed as soon as possible after collection and, in the meantime, protected from any subsequent contamination or physical damage. Transportation and storage conditions should also aim to minimise any risk of microbial multiplication. This means that unfrozen samples should be transported with ice or freezer packs in insulated containers and, if held until the following day for analysis, storage at 0 °C is advisable. In the author's experience, counts change little overnight at this temperature. At 2–5 °C, counts may be expected to increase by about one log unit overnight (Hutchison *et al.*, 2006). Frozen samples may need to be transported with dry ice to prevent thawing in transit.

Campylobacter is a common contaminant of raw poultry and where the organism is being sought specifically, allowance should be made for its relatively delicate nature. Swab samples must not be allowed to dry out and, if delays are anticipated before analysis, it may be best to use a transport medium, such as Fastidious Anaerobe Broth (Humphrey, 1995).

7.11 Future trends

The demand for microbiological testing of foods continues to grow and is likely to have a considerable impact on the European poultry industry, as the HACCP system becomes more widely used over the next few years in both old and new Member States of the European Union. In particular, testing will be needed to verify that primary processing and product manufacture are suitably hazard-controlled and meet appropriate performance criteria. While methods for the detection of specific pathogens are becoming increasingly rapid and effective, there is little prospect that such methods could be applied to all carcasses on-line in high-rate poultry processing. This means that, for the foreseeable future, sampling will continue to be needed prior to analysis. As the present chapter has indicated, sampling itself has always involved a compromise, both with respect to the number of samples that can be taken from a single lot and the manner in which individual units are treated to remove target organisms. Over the last 30 years, very few new methods of carcass sampling have been developed for poultry and any that are in the future must be relatively simple, practical and cost-effective for large-scale use. In the shorter term, it would seem more important to take the most appropriate of the existing methods and standardise them for international application. Only then can proper comparative studies be carried out to determine the microbiological condition of poultry products in different countries, which may be needed for international trade.

7.12 Sources of further information and advice

Three of the articles cited earlier are useful reviews of sampling methods for raw and cooked poultry products and provide further information on the relative efficacy of such methods. These articles are Barnes *et al.* (1973a), Patterson and Gibbs (1975) and Russell *et al.* (1997). Methods acceptable in the European Union for sampling carcasses and processed meats, including MRM, for statutory control purposes, are given in www.ukmeat.org. Statistical aspects of sampling and the development of sampling plans for foods, including poultry products, are described by the International Commission on Microbiological Specifications for Foods (ICMSF, 1986). A further publication (ICMSF, 2002) gives advice on the risk-based approach to managing food safety and describes appropriate sampling plans; see also Legan and Vandeven (2003). The role of microbiological testing in implementing the HACCP system in the meat industry, especially in relation to validation and verification procedures, is discussed by Sheridan (2000) and Brown (2000).

7.13 References

AVENS J S and MILLER B F (1970), 'Quantifying bacteria on poultry carcass skin', *Poultry Science*, **49**, 1309–1315.

BARNES E M, IMPEY C S and PARRY R T (1973a), 'The sampling of chickens, turkeys, ducks and game birds', in *Sampling – Microbiological Monitoring of Environments*, eds Board R G and Lovelock D W, Society for Applied Bacteriology, Technical Series No 7, London, UK, Academic Press, 63–75.

BARNES E M, MEAD G C and GRIFFITHS N M (1973b), 'The microbiology and sensory evaluation of pheasants hung at 5, 10 and 15 °C', *British Poultry Science*, **14**, 229–240.

BERNDTSON E, TIVEMO M and ENGVALL A (1992), 'Distribution and numbers of *Campylobacter* in newly slaughtered broiler chickens and hens', *International Journal of Food Microbiology*, **15**, 45–50.

BERRANG M E, LADELY S R and BUHR R J (2001), 'Presence and level of *Campylobacter*, coliforms, *Escherichia coli*, and total aerobic bacteria recovered from broiler parts with and without skin', *Journal of Food Protection*, **64**, 184–188.

BLANKENSHIP L C, COX N A, CRAVEN S E and RICHARDSON R L (1975), 'Total rinse method for microbiological sampling of the internal cavity of eviscerated broiler carcasses', *Applied Microbiology*, **30**, 290–292.

BROWN M H (2000), 'Validation and verification of HACCP plans', in *HACCP in the Meat Industry*, ed. Brown M, Cambridge, UK, Woodhead Publishing, 231–272.

CEC (1979), *Microbiological Methods for Control of Poultry Meat*, Study P203, VI/5021/79-EN, Brussels, Belgium, Commission of the European Communities.

COX N A and BLANKENSHIP L C (1975), 'Comparison of rinse sampling methods for detection of salmonellae on eviscerated broiler carcasses', *Journal of Food Science*, **40**, 1333–1334.

COX N A, THOMSON J E and BAILEY J S (1981), 'Sampling of broiler carcasses for *Salmonella* with low volume rinse water', *Poultry Science*, **60**, 768–770.

COX N A, MERCURI A J, THOMSON J E and CHEW V (1976), 'Swab and excised tissue sampling for total and *Enterobacteriaceae* counts of fresh and surface-frozen broiler skin', *Poultry Science*, **55**, 2405–2408.

COX N A, MERCURI A J, TANNER D A, CARSON M O, THOMSON J E and BAILEY J S (1978), 'Effectiveness of sampling methods for *Salmonella* detection on processed broilers', *Journal of Food Protection*, **41**, 341–343.

DICKENS J A, COX N A and BAILEY J S (1986), 'Evaluation of a mechanical shaker for microbiological rinse sampling of turkey carcasses', *Poultry Science*, **65**, 1100–1102.

DICKENS J A, COX N A, BAILEY J S and THOMSON J E (1985), 'Automated microbiological sampling of broiler carcasses', *Poultry Science*, **64**, 1116–1120.

EMSWILER B S, PIERSON C J and KOTULA A W (1977), 'Stomaching vs blending', *Food Technology*, **31**, 40–42.

FLETCHER D L (2004), 'Further processing of poultry', in *Poultry Meat Processing and Quality*, ed. Mead G C, Cambridge, UK, Woodhead Publishing, 108–134.

GENIGEORGIS C A, DUTULESCU D and GARAYZABAL J F (1989), 'Prevalence of *Listeria* spp. in poultry meat at the supermarket and slaughterhouse level', *Journal of Food Protection*, **52**, 618–624.

GILL C O and BADONI M (2005), 'Recovery of bacteria from poultry carcasses by rinsing, swabbing or excision of skin', *Food Microbiology*, **22**, 101–107.

HIRN J, NURMI E, JOHANSSON T and NUOTIO L (1992), 'Long-term experience with competitive exclusion and salmonellas in Finland', *International Journal of Food Microbiology*, **15**, 281–285.

HOLDER J S, CORRY J E L and HINTON M H (1997), 'Microbial status of chicken portions and portioning equipment', *British Poultry Science*, **38**, 505–511.

HOOD A M, PEARSON A D and SHAHAMAT M (1988), 'The extent of surface contamination of retailed chickens with *Campylobacter jejuni* serogroups', *Epidemiology and Infection*, **100**, 17–25.

HUDSON W R and MEAD G C (1989), '*Listeria* contamination at a poultry processing plant', *Letters in Applied Microbiology*, **9**, 211–214.

HUMPHREY T, MASON M and MARTIN K (1995), 'The isolation of *Campylobacter jejuni* from contaminated surfaces and its survival in diluents', *International Journal of Food Microbiology*, **26**, 295–303.

HUTCHISON M L, WALTERS L D, MEAD G C, HOWELL M and ALLEN V M (2006), 'An assessment of sampling methods and microbiological hygiene indicators for process verification in poultry slaughterhouses', *Journal of Food Protection*, **69**, 145–153.

ICMSF (1986), *Microorganisms in Foods 2. Sampling for microbiological analysis; Principles and specific applications*, 2nd edition, International Commission on Microbiological Specifications for Foods, Toronto, Canada, University of Toronto Press.

ICMSF (2002), *Microorganisms in Foods 7. Microbiological testing in food safety management*, International Commission on Microbiological Specifications for Foods, New York, USA, Kluwer Academic/Plenum Publishers.

IZAT A L, COLBERG M, DRIGGERS C D and THOMAS R A (1989), 'Effects of sampling method and feed withdrawal period on recovery of microorganisms from poultry carcasses', *Journal of Food Protection*, **52**, 480–483.

JØRGENSEN F, BAILEY R, WILLIAMS S, HENDERSON P, WAREING D R A, BOLTON F J, FROST J A, WARD L and HUMPHREY T J (2002), 'Prevalence and numbers of *Salmonella* and *Campylobacter* spp. on raw, whole chickens in relation to sampling methods', *International Journal of Food Microbiology*, **76**, 151–164.

LAHELLEC C and MEURIER C (1970), 'Variations dans l'importance quantitative des contaminations bactériennes des carcasses de volailles', *Bulletin d'Information, Station Experimentale d'Aviculture de Ploufragen (Côtes-du-Nord)*, **10**, 118–125.

LEGAN D and VANDEVEN M H (2003), 'Sampling techniques', in *Detecting Pathogens in Food*, ed. McMeekin T A, Cambridge, UK, Woodhead Publishing, 20–51.

LI Y, YANG H and SWEM B L (2002), 'Effect of high-temperature inside-outside spray on survival of *Campylobacter jejuni* attached to prechill chicken carcasses', *Poultry Science*, **81**, 1371–1377.

LILLARD H S (1985), 'Bacterial cell characteristics and conditions influencing their adhesion to poultry skin', *Journal of Food Protection*, **48**, 803–807.

LILLARD H S (1989), 'Incidence and recovery of salmonellae and other bacteria from commercially processed poultry carcasses at selected pre- and post-evisceration steps', *Journal of Food Protection*, **52**, 88–91.

LUBER P, SCHERER K and BARTELT E (2005), 'Contamination of retail chicken legs with *Campylobacter* spp', *Fleischwirtschaft*, **85**, 93–96.

McMEEKIN T A and THOMAS C J (1978), 'Retention of bacteria on chicken skin after immersion in bacterial suspensions', *Journal of Applied Bacteriology*, **45**, 383–387.

McNAB W B, FORSBERG C M and CLARKE R C (1991), 'Application of an automated hydrophobic grid membrane filter interpreter system at a poultry abattoir', *Journal of Food Protection*, **54**, 619–622.

McNAB W B, RENWICK S A, LOWMAN H R and CLARKE R C (1993), 'Variability in broiler carcass bacterial load at three abattoirs, as measured by a hydrophobic grid membrane filter interpreter', *Journal of Food Protection*, **56**, 700–705.

MEAD G C (2004), 'Microbial hazards in production and processing', in *Poultry Meat Processing and Quality*, ed. Mead G C, Cambridge, UK, Woodhead Publishing, 222–257.

MEAD G C and DODD C E R (1990), 'Incidence, origin and significance of staphylococci on processed poultry', *Journal of Applied Bacteriology Symposium Supplement* 81S–91S.

MERCURI A J and KOTULA A W (1964), 'Relation of "breast swab" to "drip" bacterial counts in tray-packed chicken fryers', *Journal of Food Science*, **29**, 854–858.

MULDER R W A W and BOLDER N M (1981), 'The effect of different bird washers on the microbiological quality of broiler carcasses', *Veterinary Quarterly*, **3**, 124–130.

NOTERMANS S and KAMPELMACHER E H (1974), 'Attachment of some bacterial strains to the skin of broiler chickens', *British Poultry Science*, **15**, 573–585.

NOTERMANS S and KAMPELMACHER E H (1975a), 'Further studies on the attachment of bacteria to skin', *British Poultry Science*, **16**, 487–496.

NOTERMANS S and KAMPELMACHER E H (1975b), 'Heat destruction of some bacterial strains attached to broiler skin', *British Poultry Science*, **16**, 351–361.

NOTERMANS S, HINDLE V and KAMPELMACHER E H (1976), 'Comparison of cotton swab versus alginate swab sampling method in the bacteriological examination of broiler chickens', *Journal of Hygiene, Cambridge*, **77**, 205–210.

OJENIYI B, WEGENER H C, JENSEN N E and BISGAARD M (1996), '*Listeria monocytogenes* in poultry and poultry products: epidemiological investigations in seven Danish abattoirs', *Journal of Applied Bacteriology*, **80**, 395–401.

PATTERSON J T (1971), 'Microbiological assessment of surfaces', *Journal of Food Technology*, **6**, 63–72.

PATTERSON J T and GIBBS P A (1975), 'Sampling methods for poultry meat', *Proceedings of the Second European Symposium on Poultry Meat Quality*, Oosterbeek, Netherlands, 12–15 May, Paper 22, 1–10.

RENWICK S A, McNAB W B, LOWMAN H R and CLARKE R C (1993), 'Variability and determinants of carcass bacterial load at a poultry abattoir', *Journal of Food Protection*, **56**, 694–699.

RUSSELL S M, COX N A and BAILEY J S (1997), 'Sampling poultry carcasses and parts to determine bacterial levels', *Journal of Applied Poultry Research*, **6**, 234–237.

SHARPE A N (2001), 'Microbe suspender having a vibrating beater for agitating the contents of a bag', *United States Patent*, **6**, 273, 600.

SHARPE A N, ISIGIDI BIN KINGOMBE C, WATNEY P, PARRINGTON L J, DUDAS I and DIOTTE M P (1996), 'Efficient non-destructive sampler for carcasses and other surfaces', *Journal of Food Protection*, **59**, 757–763.

SHERIDAN J J (2000), 'Monitoring CCPs in HACCP systems', in *HACCP in the Meat Industry*, ed. Brown M, Cambridge, UK, Woodhead Publishing, 203–230.

SIMMONS M, FLETCHER D L, CASON J A and BERRANG M E (2003), 'Recovery of *Salmonella* from retail broilers by a whole-carcass enrichment procedure', *Journal of Food Protection*, **66**, 446–450.

SIMONSEN B (1971), 'Methods for determining the microbial counts of ready-to-cook poultry', *World's Poultry Science Journal*, **27**, 368.

SKARNES R C and WATSON D W (1957), 'Antimicrobial factors of normal tissues and fluids', *Bacteriological Reviews*, **21**, 273–294.

STERN N J (1995), 'Influence of season and refrigerated storage on *Campylobacter* spp. contamination of broiler carcasses', *Journal of Applied Poultry Research*, **4**, 235–238.

SVEUM W H, MOBERG L J, RUDE R A and FRANK J F (1992), 'Microbiological monitoring of the food processing environment', in *Compendium of Methods for the Microbiological Examination of Foods*, eds Vanderzant C and Splittstoesser D F, Washington DC, USA, American Public Health Association, 62–76.

THOMAS N L (1978), 'Observations on the relationship between the surface area and weight of eviscerated chickens, ducks and turkeys', *Journal of Food Technology*, **13**, 81–86.

THOMAS C J and McMEEKIN T A (1980), 'Contamination of broiler carcass skin during commercial processing procedures: an electron microscopic study', *Applied and Environmental Microbiology*, **40**, 133–144.

THOMAS C J and McMEEKIN T A (1981), 'Attachment of *Salmonella* spp. to chicken muscle surfaces', *Applied and Environmental Microbiology*, **42**, 130–134.

THOMAS C J and McMEEKIN T A (1982), 'Effect of water immersion on the microtopography of the skin of chicken carcasses', *Journal of the Science of Food and Agriculture*, **33**, 549–554.

USDA-FSIS (1996), Part II, Department of Agriculture, Food Safety and Inspection Service, 9 CFR Part 304, *et al.* Pathogen Reduction; Hazard Analysis and Critical Control Point (HACCP) Systems; Final Rule, *Federal Register*, **61**(144), 38806–38989.

WALDROUP A L (1996), 'Contamination of raw poultry with pathogens', *World's Poultry Science Journal*, **52**, 7–25.

8

Microbiological monitoring of cleaning and disinfection in food plants

J. F. Rigarlsford, Consultant, UK

8.1 Introduction

In order to ensure the highest possible hygiene standards in red meat, poultry and egg factories, it is essential that each one is cleaned and disinfected regularly. In most factories, this is carried out at the end of the day's processing. Production is stopped, gross debris is removed and then the plant is cleaned with detergents, before being rinsed with water and disinfected. Many factories also clean during staff breaks and this is known as 'mid-shift cleaning'. The effectiveness of cleaning can be extremely variable. It will depend on the training and commitment of the personnel, both those on the production lines and others responsible for the cleaning process itself. In a well-run factory, sufficient time is set aside to ensure a thorough and effective clean, and the whole process is properly documented, using appropriate check lists; furthermore, the cleaning operatives are specially trained, thus providing maximum effectiveness. The importance of ensuring that the whole factory is thoroughly cleaned and disinfected cannot be over-emphasised. This is done for various reasons as discussed below.

8.1.1 Product safety

If surfaces are left dirty, or inadequately cleaned and disinfected, pathogenic microorganisms may be present and could even grow and reach numbers capable of contaminating any food that subsequently touches those surfaces. The food may then become a potential health risk. The finding of any pathogens in a food product can have a major impact on the company. Inevitably, there will be considerable costs for stopping production, withdrawing the product from the market place, calling in specialist personnel to decontaminate the factory and subsequent loss of production time. There could also be a prosecution and the costs of preparing a 'due diligence' defence. This may be followed by a significant fine and consequent adverse publicity that could affect subsequent sales. There are many examples of a single food contamination incident that resulted in the closure of a business. Therefore, the implications of inadequate cleaning and disinfection could be catastrophic for any processor of red meat, poultry or eggs.

8.1.2 Product shelf-life

As with pathogens, if surfaces are left dirty, or inadequately cleaned and disinfected, spoilage organisms are likely to survive and grow to numbers that could subsequently contaminate any food coming into contact with such surfaces. Spoilage of the product may then start to occur. In the case of meat, the changes involved can include 'off' odour, slime formation and discolouration, and the product may well be spoiled before it reaches the end of its expected shelf-life. With pressure on the food industry to produce food as cheaply as possible, there is an increasing demand for the longest possible shelf-life. Therefore, cleaning and disinfection should be thorough enough to ensure that the numbers of spoilage organisms remaining on cleaned surfaces are minimal. Thus, any contamination of the food will also be minimal and the product should then meet its anticipated shelf-life.

8.1.3 Production efficiency

Inadequate cleaning can lead to a build up of debris on production belts and other surfaces, so that clogging of the line reduces its speed and efficiency. Blockages occurring in pipelines can even stop the production process. Also, the author has seen many items in food plants that have become damaged, sometimes beyond repair, due to inadequate cleaning. Unfortunately, many pieces of equipment used in food factories are not designed for easy cleaning and the equipment supplier must be consulted for advice. Sometimes, it may be necessary to call in an experienced hygiene consultant to identify the most appropriate procedure. To ensure that the plant can work as effectively as possible and maintain its longevity (thus minimising capital costs), regular, effective cleaning is essential.

8.1.4 Operative safety

The build-up of food debris, particularly on floors, can causing slipping and other accidents to personnel in the production area. These may result in lost working time and increased production costs. The accidents can result in simple sprains, bone breakages, severe injury and even death. In the UK, any accident is also reportable to the Health and Safety Executive, and this may well lead to prosecution and adverse publicity. The injured person may resort to litigation and expect substantial compensation. All food premises should be cleaned thoroughly and with sufficient frequency to ensure the safety of the factory personnel.

8.1.5 Aesthetic considerations

A dirty plant not only looks unsightly, but it can become a very unpleasant and smelly working environment. In meat and poultry plants, the author has often seen, in otherwise well-run factories, walls and ceilings with visible fungal growth (usually of the black mould, *Aspergillus niger*), or the floors have a build-up of 'rotting' debris in the more inaccessible areas. This looks very unsightly and colours the observer's view of the whole factory. A poorly cleaned factory can lead to a lack of commitment from the employees and an inefficient operation. People work better in pleasant conditions and, if they feel that the company has done everything possible to ensure the plant looks and is clean, they are more likely to work well and experience job satisfaction.

8.1.6 As a selling tool

Customers, such as major retailers or their consultant auditors, are likely to visit the factory frequently to check that the supplier is meeting their stringent requirements. If a factory looks clean, particularly for the first visit of a prospective customer, that company is more likely to do business with the factory. Clearly, the first impression is extremely important.

The cleaning and disinfection process, often termed 'sanitisation', should be as thorough as possible. To ensure that the highest standards are always achieved, it is necessary to have an objective protocol for monitoring surfaces after the cleaning process. This will not only allow the effectiveness of cleaning and disinfection to be assessed, but, if it can be done in 'real time', i.e. an answer can be obtained in minutes, it will also give an early warning of any potential problems and allow remedial action to be taken (usually a thorough re-clean of problem sites). It is, therefore, a key requisite for 'due diligence'.

8.2 Sanitisation

The term 'sanitisation' is frequently misused. In the food industry, it is often said that a plant is cleaned and sanitised, but this is tautology, because sanitisation encompasses both cleaning *and* disinfection. The British Standard (BS 1986) defines sanitisation (in the British Standard it is spelt sanitization) as 'A term used mainly in the food and catering industry. A process of both cleaning and disinfecting utensils, equipment and surfaces.'

Sanitisation also has different meanings in other countries. In the USA, the term is mostly used to imply mild disinfection. The confusion continues when the term 'sanitiser' is used. Even a well-known British retailer, when advising their suppliers on cleaning schedules, tended to use the term 'terminal sanitiser' in referring to the final disinfection phase of the cleaning process. (They now refer correctly to 'terminal disinfection'.) In order to avoid any confusion, there has been a trend in the food and beverage industry to replace the word sanitiser by 'detergent disinfectant', i.e. a product that both cleans and disinfects.

8.3 Sources of product contamination

The main source of microbial contamination is the animal itself. Only clean animals should be sent for slaughter, since it is impossible to produce meat with low levels of contamination if the hide or fleece is heavily soiled. Therefore, conditions of transport and lairage at the abattoir are critical to maintaining livestock in a clean condition. The coops or crates in which live poultry are transported for slaughter tend to become heavily laden with faecal material, resulting in a major contamination risk.

Personnel in processing plants are also a major potential source of contamination. They should be properly trained so that they are fully aware of good hygiene practices and each individual should be supplied with the appropriate protective clothing and equipment. These should be used only in the work area, and should be removed when entering the toilets or rest areas, such as canteens. Heavily soiled clothing and equipment should be replaced with clean items. Dirty clothing can harbour many microorganisms and act as a vehicle for cross-contamination of the food being processed, or surfaces in the factory that subsequently transfer microbes to the food. All operatives must wash their hands frequently and thoroughly. In high-care areas of the factory, it is strongly recommended that personnel use an alcoholic hand rub, preferably a gel, after hands are washed and dried thoroughly.

The likelihood of a build-up of microorganisms, as debris accumulates on food surfaces, is not helped by the fact that much processing equipment used in red meat and poultry plants is not designed for easy cleaning. Often, there are areas that are virtually inaccessible. Evisceration machines in poultry processing plants are a specific example. Fortunately, most hygiene crews are aware of these problem areas and put extra effort into cleaning them. Sometimes, this is to the detriment of other areas that appear easier to clean at first sight. Shackles are also difficult to clean. This can be a major issue in a poultry plant, where large numbers of shackles may be used. Usually, they are not constructed of the best quality stainless steel and this may be a problem in relation to the choice of detergent, since the more aggressive ones can cause corrosion. The cleaning process itself can spread contamination. Aerosols are created that transmit microbes around the factory; indeed, the highest microbial levels in the atmosphere can occur during cleaning. It is therefore important that the cleaning programme includes a disinfectant, particularly in high-care areas of the factory, to combat these organisms, as well as the few left on surfaces after cleaning.

8.4 Development of a testing programme

It is essential that any testing programme is meaningful, practical and cost-effective. Therefore, sufficient sites in the factory must be tested to allow a realistic assessment of the standard of cleaning and disinfection. In a large processing plant, at least 30 sites should be checked, but not too many, because the data thus obtained can then become confusing and difficult to assess. Therefore the sites chosen must be typical and meaningful. It should be realised that, if a chosen sampling site is visibly dirty, it is pointless to check that area microbiologically because, inevitably, it will yield high counts. Therefore, any operative taking swab samples needs to be trained first to make visual assessments and check each site prior to sampling. This also applies when the operative is using a rapid method, such as adenosine triphosphate (ATP) analysis (see below). In addition, the testing programme should be dynamic. With experience, it may be realised that the chosen sampling sites are not necessarily the most appropriate. Thus, it may be necessary to alter the programme, as trends in the data indicate more pertinent sampling sites. In well-run plants, sampling is linked to the Hazard Analysis Critical Control Point (HACCP) programme, which will have identified the Critical Control Points (CCPs). In most factories, cleaning and disinfection will be a collective CCP. As such, it will be monitored, and therefore requires a fully appropriate testing programme.

The responsibility for setting up the sanitisation monitoring programme usually lies with one or more of the following: the technical manager, microbiologist, veterinarian, quality manager and hygiene manager. However, the technicians taking the samples and the cleaning personnel may also need to be consulted. Unfortunately, the contribution made by the cleaning crew to the overall safety of the end product is often overlooked. The job of cleaning operatives is relatively unpleasant, monotonous and under-valued. Moreover, they often include immigrant labour, meaning that language issues can arise. However, the technicians and cleaning operatives are in close contact with the processing operation and soon learn to identify the areas that are difficult to clean. Thus, they may well have a contribution to make in identifying areas that need to be monitored. The author has found their contribution invaluable when trying to identify the source of a microbial problem.

It is important to randomise the areas that are to be sampled. Operatives soon become aware of established sampling sites and will put extra effort into the cleaning of those areas. As a consequence, they make less effort in other areas. This may well lead to some food-contact areas being inadequately cleaned. In contrast, if a cleaning operative is aware that virtually any area could be sampled, it is likely that all parts of the plant will be cleaned thoroughly.

8.5 Testing plant surfaces

For many years, the traditional way to check the effectiveness of the cleaning and disinfection process has been to sample plant surfaces for microbiological examination. Swabs are taken from various, representative areas of the plant and then used to ascertain their microbial content. This usually includes a total viable count and presumptive counts of one or more of the following: Enterobacteriaceae, coliforms, *Staphylococcus aureus*, *Pseudomonas* spp. (in relation to meat spoilage), *Listeria monocytogenes*. yeasts and moulds. The choice of organisms to be tested for will depend on the type of food and the assurances needed to satisfy the good manufacturing standards of the factory. However, the customer, particularly a major retailer, will often stipulate the test organisms as part of the buying contract. If done correctly, microbiological testing can provide a good assessment of the numbers and types of organisms left on surfaces after cleaning and disinfection. However, a difficulty with microbial counts is that they rely on multiplication of cells to yield countable colonies. Incubating the samples is time consuming, often requiring at least 48 hours. Therefore, any data obtained are retrospective, since a factory cannot wait for the test results to become available before re-starting the production process. This means that microbiological monitoring can only be used for trend analysis or in attempting to identify the source of a microbial problem. It cannot be used as a means of evaluating the effectiveness of cleaning in 'real time'.

For many years, attempts were made to develop rapid methods, so that surfaces could be checked for cleanliness and microbial content in sufficient time to make an assessment of the need for re-cleaning. The breakthrough came in the 1980s, with the development of ATP technology, and this will be discussed in more detail below.

8.5.1 Swab sampling

Swabs

The swabs selected should be sterile and stored in such a way that they remain in good condition. A variety of swabs made of various materials and wrapped in protective sleeves (mainly sterilised by gamma irradiation) is available commercially. For sampling in food factories, the swab-stick is usually made of shatter-proof plastic. Wooden swabs should not be used, because the wood could splinter and leave small fragments on plant surfaces, creating the risk of 'foreign body' contamination for the food product. The tip of the swab, or bud, is usually made of cotton or a similar material. Alginate swabs are also available and, in this case, the alginate dissolves when the diluent contains sodium hexametaphosphate. In theory, more microorganisms should be released into the diluent than is the case with cotton-wool swabs. However, in the author's experience, microbial counts can be reduced, when these swabs are used, because the alginate–sodium hexametaphosphate mixture may be inhibitory to organisms already stressed by the cleaning process.

Neutraliser

The organisms on the swab will have come from a surface that has been both cleaned and disinfected. Residuals of the chemicals involved may still be present on the test surface and will be taken up by the swab. These chemicals may have the potential to inhibit subsequent growth of the organisms present, and therefore require neutralisation. Since a variety of chemicals may occur on the surface, the author recommends a universal neutraliser. Over 30 years' experience has shown that the following neutraliser is probably the best compromise. The neutraliser in question contains a combination of 15 ml Tween 80 ™(CIC1 American Inco, Wilmington, DE, USA), 3 g lecithin and two sodium thiosulphate–Ringer tablets (supplied by, e.g. Oxoid Ltd, Basingstoke, UK), all of which are made up to 1 litre and sterilised at 121 °C for 15 min. The neutraliser is prepared and then dispensed into plastic Universal containers in 10 ml amounts. (Glass is prohibited in food factories, because of the risk of 'foreign body' contamination.) Other neutralisers, such as that used in the BS EN Standard (BSI, 2005), may also be used. Since organisms left on the surface are probably damaged or at least under stress from the cleaning and disinfection procedures, it is important that the neutraliser is not in itself inhibitory, either chemically or physically. For instance, the neutraliser must be isotonic, so that osmotic effects on the organisms are minimal.

Swabbing the surface

The procedures involved are summarised in Fig. 8.1. The sterile swab should be removed carefully from its sleeve, as should the lid of the neutraliser bottle by using the crook of the little finger ('microbiologist's finger'). The swab is then dipped briefly in the neutraliser and the tip wiped over the entire area to be tested, with the swab being rotated slowly as it does so. The size of the area being sampled will vary according to the type of surface or equipment being swabbed, but, as a general guideline, an area of 25 cm^2 (i.e. 5×5 cm) is suitable for most situations. Templates can be made to ensure that the size of the area being swabbed is always the same. However, these must be sterilised before use. This means that, in a factory where at least 30 swabs are taken, it would be necessary to have an

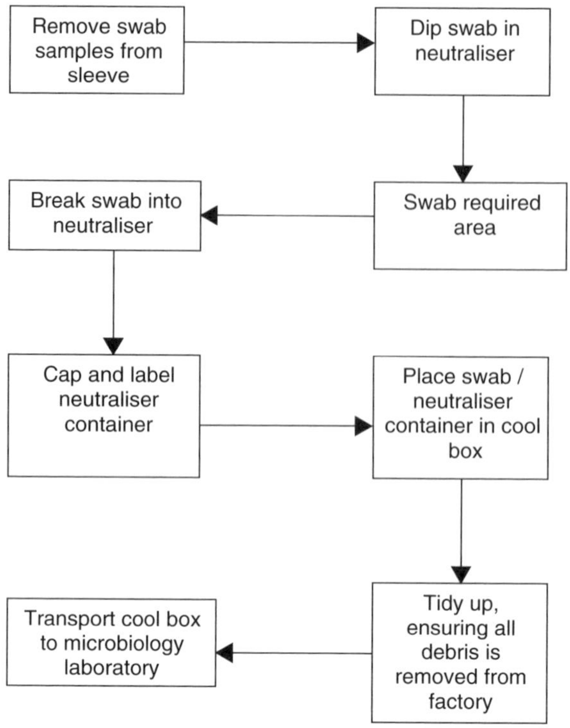

Fig. 8.1 Flow diagram of swab sampling and handling procedures.

equivalent number of pre-sterilised templates. The author has found that, with experience, the size of the area can be judged by eye with reasonable accuracy, rendering the use of templates unnecessary. For all sampling, care must be taken not to touch the tip of the swab or the area being swabbed, or let the swab touch any other surfaces, because this could give misleading results. Immediately after sampling, the swab is returned to the bottle of neutraliser and, by pressing it against the side of the bottle, the swab-stick is broken off, ensuring that the part going into the bottle is untouched, while the upper part is rejected. The cap of the bottle is then replaced. The bottle should be labelled so that the swab can be related to the site that was sampled. In some factories, there is a preference to use larger swabs for large, flat surfaces. The procedure is then a scaled-up version of the process already described above.

Transport

As each sampling session is completed, the bottles containing the swabs are transferred to a cool box containing ice packs, so that the samples can be kept at 2–6 °C. A cool box may be unnecessary, if the samples can be tested in less than an hour from the time of sampling. Samples must not be frozen. In all cases, samples should be processed as soon as possible, and no longer than 24 hours after the swabs are taken.

Tidying up

When all sites have been sampled, it is essential to ensure that the sampling equipment is carefully packed away, ensuring that nothing, including debris, such as swab sleeves and swab-sticks, is left in the production area. It may be advisable to wipe down the swabbed areas with alcoholic wipes and then the spent wipes must also be removed from the production area.

Dilution of swab neutraliser solution

The steps involved in processing swab samples in the laboratory are shown in Fig. 8.2. Firstly, each of the bottles containing the swabs should be mixed by means of a vortex mixer. Then, the samples are serially diluted, making ten-fold dilutions, as required. For this purpose, a 1 ml aliquot is pipetted into 9 ml of sterile, isotonic diluent (usually 0.1 % peptone water, Maximum Recovery Diluent (Oxoid) or quarter-strength Ringer solution). This gives a dilution of one in 100. The diluent is then shaken and a 1 ml aliquot transferred to a fresh bottle of diluent, giving a dilution of one in 1000.

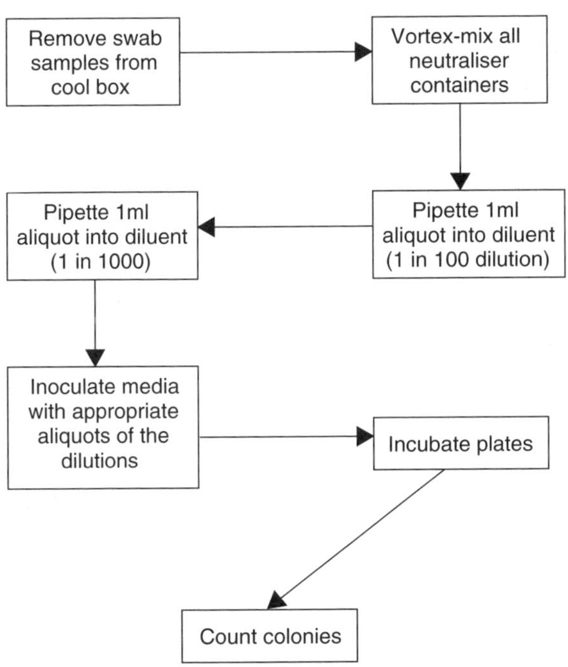

Fig. 8.2 Processing swabs in the laboratory after sampling.

Controls
It is good practice to carry out sterility checks on an unused swab and one bottle each of the neutraliser and diluent, to help rule out false positives from contaminated testing materials.

Plating out
Suitable aliquots of the neutraliser solution and the two further dilutions are used to prepare pour-plates or are used for surface inoculation of agar media, depending on the organisms being sought. Ideally, all samples should be tested in duplicate.

Incubation
The inoculated plates are placed in the appropriate incubator for the required incubation period. Ideally, stacking of plates must be avoided to ensure uniform incubation conditions.

Colony counting and calculation of results
At the end of the incubation period, the colonies on each plate are counted. Only those plates with counts of 30–300 colonies should be used to calculate the number of organisms present in the sample. The calculation must take account of the sample dilution factor.

8.5.2 Contact plates and slides

A commercially-available contact plate comprises a ready-prepared, small petri dish that is filled above the lip with an agar medium and can be applied directly to the test surface. Thus, microbes are transferred to the surface of the medium. The plate can then be incubated and a colony count obtained. A major deficiency is that not all the organisms on the test surface are transferred by this means, while those occurring close together may grow as only one colony. Also, the medium rarely contains any specific neutraliser. Therefore, some organisms may be inhibited by chemical residues present and fail to grow. If the plate contains a selective medium, organisms that are stressed or sublethally injured after contact with the cleaning chemicals may also fail to grow.

Contact slides are similar to contact plates, but the agar medium is carried on a plastic slide. The slide may contain a different medium on each side or the medium may be the same on both sides. Each version is available commercially as a 'dip slide'. Glass contact slides are also available, but should not be used in food factories. The slides have the same limitations as contact plates.

8.5.3 Surface rinsing

Surface rinsing can be used to assess the cleanliness of sites that cannot be reached by swabs or contact plates. It is particularly suitable for cleaning-in-place systems. Sterile neutraliser is passed through the site to be tested, ensuring that it reaches as much of the surface as possible. This may well involve manipulating the equipment, if possible. After a specified contact time (usually 20–30 minutes), the neutraliser is recovered aseptically and examined microbiologically.

8.5.4 Choice of media for detection and enumeration of microbes

Total viable count

To obtain the best estimate of the total number of viable bacteria on a surface, a general-purpose medium, such as tryptone-soya agar (TSA) or plate count agar (PCA), should be used. The author has found that, for most factories, TSA is the preferred medium, because it allows better growth of *Pseudomonas* spp. than PCA. The plates should be incubated at 30 °C for 48 hours. If *Pseudomonas* spp. are suspected, it may be necessary to incubate the plates for up to a week, when a further count can be obtained. Some *Pseudomonas* spp. are very slow growers and need at least five days for colonies to become visible.

Enterobacteriaceae and coliform bacteria

Counts of Enterobacteriaceae as a whole can be obtained on violet–red–bile–glucose agar (Chapter 4). Coliform bacteria are a sub-set of this family and their presence is usually taken as an indication of faecal contamination and the possible presence of enteric pathogens. They are enumerated on violet-red-bile agar, in which glucose is replaced by lactose. For both groups of organisms, the pour-plate method is preferred, with an overlay of sterile medium. The plates are normally incubated at 37 °C for 48 hours before the typical red colonies are counted, but see also Chapter 4.

Coagulase-positive staphylococci (e.g. *Staphylococcus aureus)*

Coagulase-positive staphylococci bacteria are potential pathogens that are carried by 30–40 % of the healthy population. Baird–Parker agar is often used for enumeration, with incubation at 37 °C for 48 hours, but atypical colonies can cause confusion. The medium contains a supplement of egg yolk and tellurite. Typical, coagulase-positive *Staph. aureus* produces colonies that are grey–black and shiny, with a narrow, white margin surrounded by a zone of clearing (often described as 'rabbit eye' colonies). It should be noted, however, that bovine strains, in particular, do not always produce this zone and confirmatory testing is needed.

Pseudomonas spp.

Pseudomonas spp. are major spoilage bacteria. Pseudomonas agar containing the appropriate CFC (10 mg/l cetrimide, 10 mg/l fucidin, 50 mg/l cephalosporin) supplement (Oxoid Ltd) is the medium of choice and plates are usually incubated at 30 °C for 24 hours. However, pseudomonads exposed to chemicals used in plant cleaning and disinfection may become sublethally injured and take longer to form colonies on the selective medium. In the author's experience, it is necessary to count the plates after 48 hours and again at seven days.

Listeria spp.
Certain *Listeria* spp. (particularly *L. monocytogenes*) are pathogens and even low numbers of *L. monocytogenes* can cause severe illness. Traditionally, the preferred medium is Oxford agar, with varying incubation times, such as 24 hours, 48 hours and seven days at 37 °C.

Escherichia coli
For the *Escherichia coli* organism, an aliquot of diluted sample suspension is added to a membrane filter placed on tryptone-bile agar. The isolation and enumeration of *E. coli* is described more fully in Chapter 4.

Yeasts and moulds
Yeasts and moulds are potential spoilage organisms and are grown on Sabouraud dextrose agar incubated at 25 °C, colony counts are made after 5–7 days.

Other bacteria
Some plants may wish to look for specific bacteria other than those described above, e.g. *Clostridium estertheticum* (Chapter 5) or *Salmonella* (Chapter 10).

8.5.5 Identification of isolates

The traditional method of preparing a series of tubes containing different substrates has been superseded by incorporating the necessary reagents in test strips that are produced commercially, such as the API® system (biomérieux marcy l'Étoile, France) (see Chapter 11). The author has used the various types of API strip with some success. A pure culture of the isolate is added to the cupules on the strip, which is then incubated. After incubation, the appropriate reagents are added and the reactions observed. The results are recorded as a set of code numbers that can be inserted into a computer programme. This will then give the most likely identification, together with the probability of its accuracy.

8.6 Rapid methods

For many years, attempts have been made to find a system for assessing the effectiveness of cleaning and disinfection within a time-frame that would allow almost immediate remedial action. Any surface that still has visible debris on it after cleaning obviously needs re-cleaning. However, residual dirt can be invisible to the naked eye and this could well include microorganisms. The disinfection stage of the cleaning process may kill any remaining bacteria, but it will have a minimal effect on residual soil. This material could be a significant source of nutrients for any organisms that subsequently contaminate the surface. It is important, therefore, that all surfaces are cleaned as thoroughly as possible. After cleaning, the surface is rinsed with potable water and then the disinfectant is applied to combat any remaining microbial contamination. Traditional microbiological methods can be used to determine the number of organisms that survive, and even the types remaining on freshly-cleaned surfaces. However, the traditional methods are time consuming, because they require an incubation step, usually of 48 hours or more. Obviously, it is impractical to wait until the results are available in order to assess whether it is safe to re-start production. Usually, the factory requires cleaning and disinfection to be done as quickly as possible. Therefore, the value of conventional testing is limited. Nevertheless, it can provide information to explain how product contamination has occurred, data that can be used for trend analysis and a means of identifying sites that may be overlooked or are difficult to clean. Re-starting production without knowing whether all surfaces, particularly those at CCPs, have been cleaned and disinfected adequately has inherent risks. At worst, it could mean that all product produced during the microbiological incubation period has unacceptable levels of contamination. In such a condition, the product may even represent a potential health risk or spoil before the end of shelf-life. In consequence, the factory could incur significant costs.

8.6.1 Counting of micro-colonies

In earlier attempts to find a method that would give an immediate indication of the presence of microorganisms, consideration was given to the counting of micro-colonies. An agar culture of a swab sample was incubated for four hours. Then, a microscope slide was placed on the agar surface, carefully lifted off, thereby removing part of each developing colony, and the material 'fixed' with picric acid. Finally, the slide was dried and colony counts obtained. This method would clearly indicate a heavily contaminated site; however, it was qualitative rather than quantitative, because, in practice, the slide would only touch the larger colonies. Also, the method was very labour-intensive and it still took more time than would be practical for production to cease until results became available. Although the method was of limited value, it was a move in the right direction. The real breakthrough in rapid methods came in the 1980s.

8.6.2 Other methods

At around the same time, other methods were beginning to emerge. These included measurement of electrical impedance and conductance, micro-calorimetric assessment and use of gene probes and bioluminescence. Most could give a rapid assessment of microbial numbers, but were expensive and more suitable as research tools than for routine use in food factories. They were also limited to the detection of microbes and were not able to detect the presence of any residual soil.

8.6.3 Adenosine Triphosphate (ATP) technology

ATP technology has been available for some years. It is based on the fact that all living cells contain adenosine triphosphate, which is the high-energy compound is involved in cell metabolism. A method was developed based on bioluminescence. The phenomenon was discovered in fireflies, which have tails that glow in the dark. This is because the firefly contains luciferin and the enzyme luciferase. In the presence of ATP and metal ions (particularly magnesium), light is produced immediately. The luciferin and luciferase were extracted from the firefly and used to assess the quantity of ATP present by measuring the amount of light produced with a luminometer (Fig. 8.3). Extractants based on cationic quaternary ammonium compounds were also developed to remove the ATP from microbial cells. Although, in theory, the larger the number of microorganisms, the more ATP is available and the more light is produced when the ATP is added to luciferin, it was soon found difficult to correlate the number of organisms present with the quantity of ATP. This is because the amount of ATP in each cell is not only dependent on the microbial species, but also on its growth phase. Stationary phase organisms contain very little ATP whilst, in the log phase, there is much more. Similarly, bacterial spores contain minimal amounts of ATP, because this is a resting stage for the bacterium. Another problem was that luminometers were large machines that could not be readily carried around production areas. At this point, it was decided that ATP technology had limited application for assessing microbial numbers.

A change in attitude occurred, when the author and others realised that dirty surfaces contain ATP. This ATP could be in microorganisms present on the surface, in residual food soil, or even in contamination from people touching the test surface. Therefore, the dirtier the surface, the greater the amount of ATP likely to be present. It was soon shown that the ATP technique could be the first rapid method to give an assessment of hygiene in 'real time'. After cleaning, surfaces were swabbed, the swabs added to

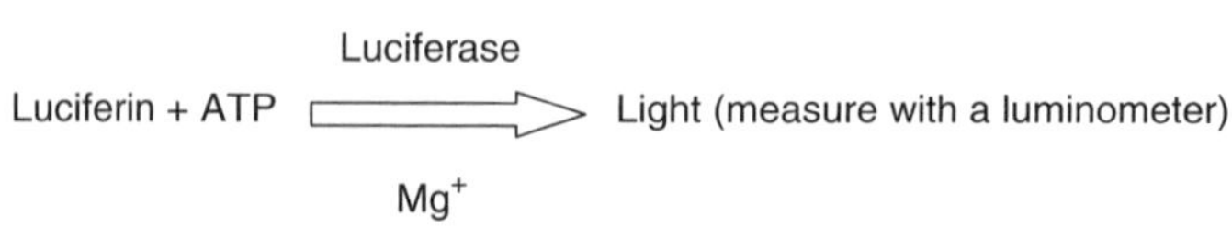

Fig. 8.3 Generation of light from ATP with luciferase.

extractant and the extract examined in a luminometer. It was also apparent that the method would be even more practical, if the luminometer could be carried to the production area. Accordingly, portable luminometers were developed. Initially, these were still heavy and one food microbiologist described them as 'luggable' rather than portable. However, they rapidly became smaller and lighter, particularly as battery technology developed. The problem of taking chemicals into the production area was also overcome by the development of 'pens' that contained the chemicals and the swabs. The cost of the technique fell, as the technology advanced, particularly when a method of producing luciferin and luciferase synthetically was discovered.

Now, it is commonplace to use ATP luminometers to assess the effectiveness of cleaning. Results are available almost immediately and the technique can be easily taught to the cleaning crew, since it no longer requires a specialist to operate the equipment. The luminometers can be programmed so that the operator knows which sites to test, and the data can be downloaded to a computer, enabling rapid communication of the results. Trend analysis can be done automatically.

Thus, ATP technology is a very valuable tool for red meat, poultry and egg plants and, in some respects, is comparable to conventional swab sampling. However, it should be stressed that the technique is actually used to provide 'a dirt meter', which estimates how much residual soil is present on the surface. It cannot quantify the number of organisms on the surface or indicate the species present. To obtain these data, it is still necessary to use traditional microbiological methods. However, ATP technology does allow an almost immediate assessment of whether a surface has been cleaned adequately. Accordingly, the test swabs should be taken after the surface has been cleaned and rinsed, i.e. before the disinfection stage. This is also important in view of the fact that some disinfectants can interfere with the luciferin/luciferase reaction. Suppliers of ATP equipment have tried to overcome these problems by modifying the extractant system but, to be certain of obtaining the most reliable results, it is still best to test the surfaces before the disinfectant is applied.

It should be realised that the technology can be abused. The author remembers a situation where the introduction of ATP testing into a factory suddenly appeared to improve the hygiene, even though there had been no change in the cleaning process. The author was suspicious and decided to spend a night in the factory over the cleaning period. During a meal break, the author gained the confidence of the cleaning crew and asked the supervisor what he thought of the ATP system. The supervisor replied that it had given them considerably more work, because the values obtained were high and they were having to re-clean several times. This meant that they had little spare time and were exhausted at the end of the shift! The supervisor then said that he had 'worked it out'. If the swab did not actually touch the surface, very low values were always obtained! The obvious lesson from this is that management needs to check that all procedures are being carried out properly.

8.6.4 Protein assessment

Test kits are now available that estimate the amount of protein left on surfaces after cleaning. The test is based on the traditional biuret reaction, which detects peptide bonds. In this test, treatment with copper sulphate under alkaline conditions and addition of an indicator (bicinchoninic acid) yields a purple colour when protein is present. The strength of the colour depends on the amount of protein. The test has found little favour in the meat and poultry industries and tends to be used only in certain catering establishments that are too small to justify the use of ATP technology.

8.6.5 Nicotinamide adenine dinucleotide (NAD) assessment

NAD can be found in all living matter and therefore is present in most foods and all microorganisms. Various test kits are available to detect NAD. These contain a cationic detergent to lyse microbial cells and release the compound. They make use of an enzyme reaction that produces a bluish colour in the presence of NAD. Like the protein assessment technique, use of this test is mostly confined to small food premises.

8.7 Air testing

Many poultry plants and some red-meat processing factories have areas where the quality of the air is critical to the hygienic production of certain food items. This is particularly so in high-care areas, where the air is often refrigerated to give a working temperature of 10–12 °C, and sometimes even cooler. The necessary conditions depend on the use of chiller units, often attached to wind socks to ensure an even temperature distribution in the room. The air may be filtered to remove gross debris. After cleaning, it may well be considered necessary to check the microbial content of the air. The procedure simply involves exposing agar plates (settle plates) at various points in the production area for at least 30 minutes. The lids of the petri dishes are then closed and the plates incubated so that counts can be obtained. There is also a variety of commercially-available air samplers. With these, agar plates are inserted into the sampler and a known volume of air is drawn through. The plates can then be incubated as usual. Some air samplers, such as the Anderson sampler (Anderson, 1958), contain a series of stainless steel plates with pore sizes down to micron level. At the end of the sampling period, the plates are weighed and the quantity of captured material of a particular size can then be calculated.

8.8 Data utilisation and limit values

It is important that all the data obtained should be meaningful and usable for control purposes. Various computer programmes are available that allow data to be stored and used for subsequent trend analysis. It is vital that all samples are logged within minutes of their being taken. A simple numbering system on the sample bottle may suffice, as long as it can be traced readily to the site tested. The author has experienced the frustration, on returning to the laboratory, of finding that the number of samples and the logged data do not tally. It may then be impossible to repeat the sampling, because production may have re-started. The importance of checking that the samples tally with the documented site information, as each sample is taken, cannot be over-emphasised.

Acceptable microbial limits for contamination of cleaned and disinfected surfaces can be established. Ideally, the limit would be a total absence of viable bacteria. However, it would rarely be achieved and a total viable count of 100 cfu/cm^2 is often used. This can be supplemented by the absence of specific pathogens or a limit for coliforms, such as $<10/cm^2$. The factory often has a problem in setting its action limits, because each customer has their own view on what is achievable. When a rapid method, such as ATP measurement, is used, it is even more difficult to set action limits. This is because there is no standard means of calibrating the luminometer. Thus, the calibration method will vary from one supplier's machine to another. In this case, the food factory should be guided by the equipment supplier to determine what is achievable and meaningful in a given situation. However, it is always advisable to set the strictest limits possible. This will then ensure that hygiene standards for food plant and equipment are maintained at the highest level, so that the chances of subsequent microbial problems are minimal.

8.9 Future trends

The ideal monitoring system has not yet been achieved and there is considerable scope for improvement. The ideal would be a device that could be passed over a cleaned surface and immediately indicate the degree of residual soiling, together with the numbers and types of specific-risk microorganisms. At present, the best technique seems to be ATP analysis, since it enables an assessment to be made within minutes, showing whether or not a particular piece of plant or equipment needs re-cleaning. It is also convenient for repeat analyses, when further re-cleaning is needed. However, the ATP system cannot be used to determine microbial numbers or the types of organisms present. Developments in gene probes look very promising for detecting specific organisms, but are still some way from being applicable to the routine assessment of plant cleaning and disinfection.

8.10 Sources of further information

The terms used in relation to disinfectants are defined in BSI (1986), while a standard method for evaluating the bactericidal activity of disinfectants used in the food industry is described by BSI (1997, 2005). Appropriate hygiene control methods are given in a manual (CCFRA, 2003) and hygiene monitoring is reviewed by Easter (2003).

8.11 References

ANDERSON A A (1958), 'New sampler for the collection, sizing and enumeration of viable airborne particles', *Journal of Bacteriology*, **76**, 471–484.

BSI (1986), 'BS5283:1986 Glossary of terms relating to disinfectants', UDC 001. 4:614.484:648.6, London, UK, British Standards Institution.

BSI (1997), 'BS EN 1276:1997 Chemical disinfectants and antiseptics – Quantitative suspension test for the evaluation of bactericidal activity of chemical disinfectants used in food, industrial, domestic and institutional areas – Test method and requirements (phase 2, step1)', London, UK, British Standards Institution.

BSI (2005), 'BS EN 1040:2005 Chemical disinfectants and antiseptics – Quantitative suspension test for the evaluation of basic bactericidal activity of chemical disinfectants used in food, industrial, domestic and institutional areas – Test method and requirements (phase 1)', ICS 11.080.20: 71.100.35, London UK, British Standards Institution.

CCFRA (2003), *Manual of Hygiene Methods for the Food and Drink Industry*, Guideline No.45 ISBN: 0 905942 70 1, Chipping Campden, UK, Campden & Chorleywood Food Research Association.

EASTER M (2003), 'Hygiene monitoring in support of food safety – a review of methods and industry trends', *Society for Food Hygiene and Technology Focus*, http://www.sofht.co.uk/technical/hygiene_monitoring.htm.

9

Microbiological analysis of eggs and egg products

J. E. L. Corry, University of Bristol, UK

9.1 Introduction

Eggs of the domestic hen (*Gallus gallus*) are an important part of the human diet in many countries and are traded internationally. In the European Union (EU), production increased by 12.2 % between 1993 and 2003, reaching an annual total of 96.4 billion eggs (Windhorst, 2005). Current demand is such that, in the UK alone, approximately 27 million eggs are consumed every day. Despite their water content (74 %), eggs provide high-quality protein and are a source of unsaturated fatty acids, iron, phosphorus, minerals and vitamins A, B, D, E and K (ICMSF, 1998). Because of their functional properties, eggs have many uses in the food industry and are key ingredients in a variety of foods. They are widely used, for example, in the manufacture of noodles, salad dressings, ice cream, confectionery and bakery goods. While many are retailed as shell eggs, more are used as bulk materials for manufacturing purposes, including pasteurised liquid whole egg, albumen and yolk, and can be obtained in fresh, frozen or dried form, according to requirement.

The majority of newly-laid eggs are sterile internally; however, contamination can occur with organisms that are potentially pathogenic for humans, especially *Salmonella*. The hazard posed by egg products that contain these bacteria has been recognised for many years (e.g. Scott, 1930; Thatcher and Montford, 1962) and, for this reason, liquid egg products were required to be pasteurised (Anonymous, 1963). However, in the mid to late 1980s, an unprecedented pandemic of human salmonellosis became evident that was due to *S.* Enteritidis (SE). This was attributed to a panzootic of the serotype, affecting both broiler and layer flocks (ACMSF, 2001). It led to an increase in the number of *Salmonella*-contaminated eggs being produced and an enhanced risk of human infection from in-shell eggs. Subsequently, the situation improved considerably as control measures, such as flock vaccination, were introduced, but the potential for this and other *Salmonella* serotypes to cause similar problems in the future has been recognised. Because of the public health risks associated with egg contamination, microbiological monitoring of eggs and egg products continues to be an important requirement.

In addition to foodborne pathogens, the contents of in-shell eggs can also become contaminated with bacteria capable of causing spoilage, especially if the shell of the egg has been cracked. The organisms responsible for the rotting of eggs are of various kinds, but include especially *Pseudomonas* spp. and certain Enterobacteriaceae. Further microbial contamination occurs during the egg breaking-out process, hence the need for pasteurisation of raw egg products to ensure product safety.

This chapter describes briefly the main factors affecting microbial contamination of eggs and egg products with pathogens and spoilage organisms. Methods for product sampling are described and discussed, as well as those for the detection/isolation of the principal organisms of concern.

9.2 Shell eggs

9.2.1 Microbiological aspects

Although the biological role of the egg is to protect and nourish the developing embryo, many of the features that allow it to do so have relevance to the use of eggs as a human food, especially in relation to the control of microbial invasion. The physical and chemical defences of the egg against microbial contamination have been reviewed extensively by Board and Fuller (1994), and will only be summarised here.

The shell of the egg is a rigid structure made largely of calcium carbonate on an organic matrix. It contains a total of 7000–17000 pores, ranging in diameter from nine to 35 μm. The surface of the shell is covered by a glycoprotein cuticle which, immediately after laying, plugs most, but not all, of the pores. Internally, there are two shell membranes and the inner one, in particular, acts as a bacterial filter when an organism penetrates the shell; however, some organisms can breach this barrier and reach the contents. Within the egg, the yolk (an excellent microbial growth medium) is isolated from the surrounding albumen and held in place centrally by the chalazae. Any microbial movement in the albumen appears to be impeded by the viscous nature of the albuminous sac. After lay, the egg cools and the pH of the albumen rises, due to loss of CO_2, from 7.4 to about 9.2. This high pH value combines with other factors, including a low level of available nitrogen, to impede or prevent microbial growth. The factors involved tend to act synergistically and the more important ones include ovotransferrin (conalbumin), which binds iron, and lysozyme, an enzyme that attacks cell-wall peptidoglycan in Gram-positive bacteria. Thus, fresh albumen is an unfavourable environment for microbial persistence and growth.

During storage of eggs, certain changes occur that make the egg more susceptible to microbial invasion. These include a reduction in the viscosity of the albuminous sac and weakening of the chalazae. In consequence, the yolk, which is contained in the vitelline membrane, gradually drifts towards the shell membranes. On touching the inner membrane at a particular point, any penetrating bacteria will be able to reach the yolk. Also, there may have been diffusion of nutrient materials from the yolk to the albumen that would make the latter more susceptible to microbial growth (Humphrey *et al.*, 1991).

With regard to *Salmonella*, there are three possible routes by which the organism could gain access to the contents of whole eggs. One occurs as a result of infection of reproductive tissue in the hen; certain strains of SE possess the unusual, but not unique, property of being able to invade the reproductive tract of the chicken (ACMSF, 2001). The second route follows invasion of the oviduct from the intestinal tract (cloaca), which could occur during the formation of the egg as it moves down the oviduct, where first the albumen, then the membranes and finally the shell are laid down. Thirdly, penetration of the egg could occur via passage through the shell, the organisms being derived from either the intestinal tract of the hen or the environment in which the egg is laid.

In the case of SE, any site within the reproductive tract appears capable of becoming infected with the organism, and it seems likely that the specific site of the infection will influence the location of contaminating bacteria in the egg contents. In practice, they are usually located in the albumen or on the outside of the vitelline membrane. Infected hens tend to produce contaminated eggs only sporadically and with no predictable pattern. Thus, the incidence of contents-positive eggs is generally very low. In routine surveys, the proportion of positives varied from 0 to 0.05 % (ACMSF, 2001) and the number of *Salmonella* cells in positive eggs was also low, being <50 per egg, but could multiply to very high numbers during storage at >8 °C (Humphrey, 1994). It should be noted that the shell is more frequently contaminated with salmonellas than the contents, due mainly to infection of the reproductive tract or faecal contamination of the egg. The serotypes found have included not only SE but others, such as *S.* Braenderup, *S.* Infantis, *S.* Livingstone and *S.* Typhimurium (de Louvois, 1993). Salmonellas on the shell can contaminate the contents when the egg is used in the kitchen (or broken out to make egg products), and can also cross-contaminate other foods, where they may multiply and cause illness.

9.2.2 Spoilage of shell eggs

When laid, the egg passes through the cloaca of the hen into an environment that is contaminated with a variety of microorganisms derived from faeces, dust, nesting material, feed, etc. Those organisms that predominate on the shell surface are not usually the types most often associated with egg spoilage (Mayes and Takeballi, 1983). The main organisms on the shell are Gram-positive and include, particularly, *Micrococcus* spp., although the microflora varies qualitatively and quantitatively from one geographical region to another. The organisms causing spoilage, however, are a relatively restricted number of Gram-negative types, as indicated in Table 9.1, and their occurrence in egg rots is not influenced by either geographical area or husbandry methods. This suggests that the intrinsic defence mechanisms of the egg have an overriding influence on the selection of organisms that are capable of spoilage (Mayes and Takeballi, 1983).

In order to cause spoilage of the intact egg, an organism must be able to penetrate the shell and overcome the various antimicrobial barriers of the membranes and albumen. The extent to which penetration occurs increases with the age of the egg and the duration of contact with contaminated material, especially at high relative humidity. The process is influenced by the temperature of storage, and is similar for both salmonellas and spoilage bacteria (Simmons *et al.*, 1970), although spoilage bacteria are capable of multiplying at lower temperatures. Penetration is facilitated by the presence of wet and dirty shells and by contraction of the contents during cooling, which has a tendency to draw in any liquid on the shell surface, including any microbes therein. Fluorescent *Pseudomonas* spp. are a major cause of egg spoilage: they are motile, produce fluorescent pigment (pyoverdine), that competes with ovotransferrin for metal ions needed for microbial growth, and are resistant to other protective factors in the albumen. With this type of organism, penetration of the inner shell membrane appears to be enzyme-mediated, rather than being due to movement of the bacterium through the system of overlapping fibres (Stokes *et al.*, 1956; Brown *et al.*, 1965). After penetrating the shell, the organism is

Table 9.1 Examples of bacterial genera associated with the spoilage of shell eggs

Genus	Rot colour
Pseudomonas	Green
Acinetobacter–Moraxella	Colourless
Pseudomonas	Black
Proteus	
Aeromonas	
Alcaligenes	
Enterobacter	
Pseudomonas	Pink
Pseudomonas	Red
Serratia	

Source: adapted from ICMSF (1998).

relatively uninhibited and grows rapidly in the contents. Other primary invaders of the egg include *Alcaligenes*, *Citrobacter*, *Flavobacterium* and *Proteus* spp. Further types can also grow once the defences of the egg have been breached. Examples are species or other strains of *Acinetobacter*, *Alcaligenes*, *Enterobacter*, *Escherichia*, *Flavobacterium* and *Proteus*. The rots that develop often have a characteristic colour (Table 9.1). That due to the non-proteolytic *Ps. putida* leads to a fluorescence in the albumen, when viewed under ultraviolet light, while *Ps. fluorescens*, which produces lecithinase, causes a pink discolouration. Organisms that are strongly proteolytic, such as *Aeromonas*, *Alcaligenes*, *Escherichia* and *Proteus*, digest the albumen and turn the yolk black. Some organisms, including many salmonellas, produce no obvious changes in the egg, despite their ability to multiply extensively under appropriate conditions (Board, 1965). If egg collection on the farm is delayed, or the humidity is too high during refrigerated storage, mould growth may occur. This is due usually to *Cladosporium herbarum* and is termed 'whiskers' in the trade (Board *et al.*, 1994). The fungal hyphae can penetrate the shell and spread throughout the egg.

9.2.3 Washing of eggs

Within the EU, shell eggs are separated into two categories. Those in Grade A have a normal, clean, undamaged shell and cuticle. Currently, they must not be washed or cleaned, either before or after grading, and cannot be chilled below 8 °C or otherwise treated for preservation. Grade B eggs do not meet the basic requirements and can only be used by the food industry or for non-food purposes, such as shampoo manufacture. Such eggs can be subjected to washing in order to clean them and reduce the microbial load at the surface prior to use. The situation is different in the USA, where almost all commercial eggs are washed (Baker and Bruce, 1994). Other countries, too, including Australia and Japan, allow washing of table eggs. A recent report from the European Food Safety Authority (EFSA, 2005) shows that washing of Grade A eggs is now being considered in the EU.

Egg washing is usually carried out on-line and involves three distinct stages: (i) a pre-wash or wetting stage to soften any adherent material; (ii) spray-washing/brushing in potable water containing added chemicals to clean and sanitise the egg surface and (iii) rinsing in plain water to remove loose debris, chemical residues, etc. Subsequently, the eggs are dried and may be oiled and/or cooled to maintain quality. For eggs to be sold in shell at retail, microbial contamination on the surface can be reduced by 1–6 $\log_{10}$ units, thereby diminishing the potential for cross-contamination of foods during preparation (EFSA, 2005). However, washing of eggs has the disadvantage of possible damage to the cuticle and shell. Also, it has become clear that, unless strictly controlled, the process can lead to contamination of egg contents with pathogens, the organisms being drawn through the physical barriers of the egg *via* contaminated water. Similarly, early studies (Brooks, 1951) showed higher rates of spoilage for eggs washed under sub-optimal conditions. The key requirements are that eggs should be washed as soon as possible after lay and the water must be at least 11 °C higher than the temperature of the eggs (Brant and Starr, 1962). The wash-water should be treated with an appropriate detergent–disinfectant, and alkalinity maintained at pH 10–11 to avoid microbial growth in the water (Kinner and Moats, 1981). Furthermore, the iron content of the water should be <2 parts per million to avoid interference with the protective effect of ovo-transferrin in the egg (Garibaldi and Bayne, 1962). In a recent study, Hutchison *et al.* (2004) showed that washing eggs under controlled conditions improved their microbiological condition, reduced *Salmonella* levels on pre-inoculated eggs by 5–6 $\log_{10}$ units and avoided any detectable contamination of egg contents with these organisms. The most important factor in the study was the temperature of the water used for washing and rinsing the eggs: only when it was allowed to fall below 34 °C did internal contamination of eggs occur.

9.2.4 Sampling of shell eggs

Since the mid-1980s, there have been numerous surveys of *Salmonella* contamination in shell eggs in different countries (ACMSF, 2001). These have centred around outbreaks of human salmonellosis, specific cases of illness, eggs from infected flocks or egg production more generally. The work has demonstrated unequivocally that SE can be present in the contents of clean, dry, intact eggs.

Methods for sampling and testing of shell eggs have been described by Roberts and Greenwood (2003) and appear to be those favoured by the UK Public Health Laboratory Service. Using these methods, contamination of the shell may be distinguished from that of the yolk and albumen, whether examined together or separately. Also, it is sometimes appropriate to pool eggs in batches of six or more for survey purposes, e.g. Elson *et al.* (2005), despite the loss of sensitivity. With all methods, it is recommended that laboratory staff wear disposable gloves when handling eggs (Roberts and Greenwood, 2003).

Method 1, as described below, does not involve any disinfection of the shell. The steps involved in egg sampling are as follows.

A. The egg is cracked against the top of a sterile vessel containing 180 ml of buffered peptone water (BPW) and the contents of the egg are dropped into it.
B. The resultant mixture is homogenised by shaking and incubated at 37 °C for 18 ± 2 hours, prior to detection or selective isolation of *Salmonella*.
C. The shell is dropped into a further 180 ml of BPW in a separate container and incubated as above.

D. If required, the albumen can be removed by pipette into a separate vessel, together with an amount of BPW that is nine times its weight. The material is mixed to form a 1 : 10 homogenate for pre-enrichment, as before.

The initial pre-enrichment culture is subcultured into an enrichment medium and then plated, or plated directly onto a selective agar (see Section 9.2.5).

The yolk can be treated in a similar manner, but complete separation of the egg components is unlikely to be achieved, because some albumen will adhere to the yolk, as it does to the shell.

Method 2 involves sampling the shell surface by swabbing, followed by a disinfection step. On breaking out the egg, the shell is then discarded. Therefore, the method is less sensitive than Method 1, but aims to avoid the risk of contaminating the egg contents from the shell.

A. The shell is first wiped with a large, sterile cotton-wool swab that has been moistened with BPW. The swab is then transferred to 180 ml of BPW.
B. The shell is wiped again with cotton wool soaked in 70 % industrial methylated spirit (IMS) or the egg is immersed completely in IMS. It is then removed and allowed to dry. Another alternative is to disinfect the shell surface using a 'wipe' impregnated with isopropyl alcohol. (A method described by Ricke *et al.* (2001) involves immersion of the egg in 70 % ethanol for 10 minutes, draining and then flaming before breaking out the egg.)
C. The egg is broken out and the contents collected and homogenised, as described for Method 1. The shell is discarded.
D. The swab suspension from the shell and the homogenate of egg contents are incubated, as before.
E. The albumen and yolk can be examined separately, if necessary (see Method 1).

A modification of Method 2 for examining 20 eggs at a time is described by the British Egg Industry Council (COP, 2004). The eggs are swabbed with a cotton-wool pad, which is then placed in 250 ml BPW and incubated at 41.5 °C for 24 hours. After disinfecting and drying the eggs, the contents of all 20 are added to 100 ml BPW in a large stomacher bag and homogenised before incubating at 41.5 °C for 48 hours.

9.2.5 Detection/isolation of *Salmonella*

The topic is considered more generally in Chapter 10. In principle, eggs and egg products are no different in this respect from many other foods, and a variety of different methods would be applicable. The cultural method described by Roberts and Greenwood (2003) involves selective enrichment from shell cultures in BPW, using Rappaport–Vassiliadis soya–peptone broth (RVS), incubated at 41.5 ± 1 °C, with subsequent plating on xylose–lysine desoxycholate (XLD) agar and a second medium, such as modified brilliant green agar or mannotol–lysine–crystal violet–bile agar. It is noteworthy that no selective enrichment step is used for egg contents and plating is carried out directly from BPW cultures. This is because, in the case of the contents, no interference would be expected from other organisms. RVS cultures of shells are incubated for 20–24 hours; however, for both artificially- and naturally-infected hens, the *Salmonella* isolation rate from pooled eggs was increased significantly by extending the incubation period to 48 hours or by adding a source of iron to the medium to negate the inhibitory effect of albumen (Gast and Beard, 1992; Humphrey and Whitehead, 1992). This is especially useful when albumen alone is being examined.

Where required, *Salmonella* in egg contents can be quantified by means of a cultural method and use of the multiple-tube technique to determine the Most Probable Number, as described by Roberts and Greenwood (2003).

Various studies have sought to identify suitable methods for *Salmonella* detection and isolation. Using artificially-inoculated shell eggs and naturally-contaminated liquid egg, Hara-Kudo *et al.* (2001) compared a number of enrichment and plating media for recovering SE. The combination of pre-enrichment in BPW + cysteine and selective enrichment in tetrathionate broth gave the best results, and six out of seven selective plating media used in conjunction with this enrichment system allowed detection of the organism in all samples tested. More rapid methods have also been studied. For example, De Medici *et al.* (1998) recovered SE effectively from inoculated shell eggs with the Vidas® Immuno-Concentration Salmonella (ICS) System together with VITEK® (BioMérieux, Marcy-L'Étoile, France) and modified semi-solid Rappaport–Vassiliadis medium. The latter was favoured in practice on the grounds of low cost, simplicity and labour-saving, with results being available within 48 hours. Other rapid methods that have been studied in relation to shell eggs have included a polymerase chain reaction method for detecting SE (McElroy *et al.*, 1996) and the use of *Lux*$^+$ bacteriophage (Chen and Griffith, 1996). Also, Gast *et al.* (2003) evaluated two different assays, one involving fluorescence polarisation and the other lateral-flow immunodiffusion. Although significantly less sensitive than culturing methods, these assays detected SE consistently when pools of 10 eggs were inoculated with approximately 10 cfu of the test organism and incubated at 25 °C for 72 hours.

9.3 Bulk liquid egg

9.3.1 Microbial contamination

Commercial preparation of bulk liquid egg, whether whole egg, albumen or yolk, usually involves both normal, intact eggs and those that are misshapen and/or damaged due to fine cracks. Within the EU and USA, any 'leakers', in which the shell and its membranes have been breached, cannot be included in any product intended for human consumption. Also excluded are eggs showing evidence of spoilage and those known as 'incubator clears', which are incubated eggs that have failed to hatch and were found not to contain an embryo, when candled a few days before the expected day of hatch. Incubated eggs are considered unsuitable, because of the risk of microbial growth during incubation and hence contamination of the bulk product. A test for them was developed by Robinson *et al.* (1975) and is based on the detection of 3-hydroxybutyric acid that forms when embryonic growth is inhibited. Incubator clears sometimes contain $>10^8$ cfu per gram of bacteria, mostly *Staphylococcus/Micrococcus* spp. or Enterobacteriaceae, without appearing obviously spoiled (Corry, unpublished observations).

For liquid-egg production, shell eggs that have been candled, washed and inspected visually are processed on-line, using automatic machines capable of breaking out many thousands of eggs during a working shift and, where necessary, separating yolk and albumen. Subsequent homogenisation of the bulk product distributes microbial contaminants throughout the batch in a more uniform manner. The degree of such contamination is influenced by a number of factors, including the condition of the shell eggs used. The initial flora is a diverse mixture of organisms derived mainly from the shell, but also including those from occasionally-contaminated egg contents, processing equipment and its environment and plant operatives (ICMSF, 1998). A process no longer permitted in the EU for liquid egg used in human food is crushing of shell eggs, followed by centrifugation to remove the broken shells. This causes extensive microbial contamination of the resultant 'melange'.

9.3.2 Effects of pasteurisation

Prior to pasteurisation, liquid egg is likely to yield the following counts per g (Ricke *et al.*, 2001): aerobic plate count 10^3–10^6; coliform bacteria 10^2–10^5; yeasts and moulds <10; *Salmonella* <1. Despite the low level of *Salmonella* contamination pasteurisation is essential to minimise the risk to human health, although according to Ricke *et al.* (2001), yolk containing 10 % salt does not require pasteurisation, if used in mayonnaise or salad dressing with a pH value of 4.1 or less. For most products, the heating regime needed to destroy *Salmonella* is close to that having an adverse effect on the physical and functional properties of the material. Heating regimes vary between countries, but aim to reduce *Salmonella* contamination by 1000–10 000 fold, so that the organism is below the normal level of detection (at least <1 per kilogram of product). In the USA, for example, liquid whole egg is heated at 60 °C for 210 seconds, while the requirement in Australia is 62 °C for 150 seconds (Cunningham, 1986).

According to Ricke *et al.* (2001), all pasteurised egg products, including liquid, frozen and dried products, should meet the following specifications per gram: aerobic plate count <25 000, coliform bacteria <10, yeasts and moulds <10 and *Salmonella* negative by prescribed sampling and testing procedures. Reasons cited for failure to meet these criteria include excessive microbial contamination of the initial material, inadequate pasteurisation or re-contamination of the pasteurised product. Temperature abuse of liquid and frozen egg may also be a significant factor.

EU legislation (EU, 2005) requires the use of a sampling plan and testing for Enterobacteriaceae according to ISO 21528-2 (ISO, 2004), which describes the use of violet–red–bile–glucose agar in a pour plate method, with incubation at 37 °C. For this purpose, $n = 5$, $c = 2$, $m = 10$ cfu per gram or ml and $M = 100$ cfu per gram or ml. The criterion applies at the end of the manufacturing process and, in the case of unsatisfactory results, requires checks to be made on the efficiency of the heat treatment and prevention of re-contamination. In addition, egg products other than those for which the *Salmonella* risk has been eliminated are required to meet a criterion for *Salmonella*. The sampling plan is $n = 5$, $c = 0$ and the organism must be absent from 25 g of the sample, when tested in accordance with EN/ISO 6579 (ISO, 2002). The criterion applies to products placed on the market during their shelf-life. The same criterion also applies to ready-to-eat foods containing raw egg, but excluding those for which the manufacturing process or the composition of the product will eliminate the *Salmonella* risk.

9.3.3 Shelf-life and spoilage of pasteurised egg

Egg products are perishable, even when pasteurised and stored under refrigeration. Their shelf-life is influenced by the microbiological status of the original eggs and subsequent conditions of processing and packaging. While pasteurised material obtained from clean eggs can be expected to remain edible under chill conditions for about three weeks, the use of ultra-pasteurisation combined with aseptic packaging extended shelf-life at 4 °C to 3–6 months (Ball *et al.*, 1987). When spoilage does occur, 'off' odours are usually much more intense in the case of yolk and whole egg than they are with albumen. Pasteurisation destroys most of the Gram-negative organisms that can multiply in the raw product and leaves mainly the more resistant Gram-positive types, including species of *Bacillus*, *Enterococcus*, *Micrococcus* and *Staphylococcus*. The principal survivors in whole egg heated at 65 °C for three minutes were *Microbacterium lacticum* and *Bacillus* spp., which grew readily at 10 °C and 15 °C (Payne *et al.*, 1979). With pasteurised albumen, Barnes and Corry (1969) showed that the surviving flora comprised mainly 'faecal streptococci' (enterococci), *M. lacticum*, strains resembling *Aerococcus viridans* and two groups of 'atypical lactobacilli' recognised by Thornley and Sharpe (1959). All of the organisms were isolated on heart infusion agar, and tests for their characterisation and identification are described by Barnes and Corry (1969). It should be noted, however, that the naming of these and other egg spoilage bacteria in the earlier literature does not always correspond to the modern classification.

9.3.4 Contamination with pathogens other than *Salmonella*

Despite some evidence that *Campylobacter jejuni* can be transmitted vertically *via* the egg (Pearson *et al.*, 1996; Cox *et al.*, 2002), this appears to be uncommon and survival of the organism on egg shells and in albumen is poor (Doyle, 1984; Kollowa and Kollowa, 1989). Even when present on the egg surface, the organism has little capability for penetrating the shell membranes and contaminating the contents (Doyle, 1984; Shane *et al.*, 1986; Sahin *et al.*, 2003). In a study of egg products (Izat and Gardner, 1988), *C. jejuni* could not be detected in pasteurised whole egg, yolk or albumen, scrambled egg mix, chopped whole eggs or egg-and-cheese omelettes. For these reasons, egg products are unlikely to be a significant vehicle for human campylobacteriosis and only a single outbreak in which uncooked eggs were implicated has been recorded (Finch and Blake, 1985).

The psychrotrophic pathogens, *Listeria monocytogenes* and *Yersinia enterocolitica*, are potentially of greater concern. *L. monocytogenes* is a common contaminant of raw egg products and can multiply at 5 °C and above in raw and pasteurised liquid whole egg and yolk. Isolation and enumeration of this organism is described in Chapter 10. Its inability to survive for long periods in raw albumen has been attributed to the presence of lysozyme (Khan *et al.*, 1975; Foegeding and Leasor, 1990, Sionkowski and Shelef, 1990). The organism is considered to be a potential hazard in some egg products, because it is marginally more heat resistant than the vegetative cells of other pathogens and may survive minimal pasteurisation treatments, if present in unusually high numbers (Ricke *et al.* 2001).

However, Moore and Madden (1993) were unable to detect *Listeria* spp. in 500 egg-product samples examined after pasteurisation and, so far, no human cases of listeriosis have been associated with egg products. Strains of *Y. enterocolitica* are also common in poultry and poultry products, and survive well under the alkaline conditions occurring in egg albumen. The organism has the ability to penetrate the shell of whole eggs and contaminate the contents (Amin and Draughon, 1990), but the serotypes occurring in poultry are usually different from those causing human illness (Cox *et al.*, 2005).

Another relevant foodborne pathogen is *Staphylococcus aureus*, which is carried asymptomatically by many humans and may also be present on egg shells. The organism can grow well in liquid whole egg, but is unable to multiply under chill conditions. It is a potential hazard in salted yolk, being tolerant of the low water activity (0.90) of this product, but would need to multiply to *ca* 10^5 cfu per ml in order to form the enterotoxin responsible for food poisoning. Such a level of growth would involve considerable temperature abuse of the product (Ijichi *et al.*, 1973). In the EU, recently repealed legislation (enacted by the UK Egg Products Regulations (Anonymous, 1993)) included a criterion for *Staph. aureus*, which was absence from one gram of sample when examined by a colony-count technique described in ISO 6888 (latest version: ISO, 1999). The required isolation medium was Baird–Parker agar, with confirmation of colonies by positive results in tests for coagulase and DNAse. In a study of egg products, Scotter *et al.* (1994) reported a total of 9 % false-positive and 16 % false-negative results, when Baird–Parker agar was used. The presence of atypical colonies appears to have contributed to this problem.

Bacillus cereus can be found sometimes in liquid egg products, especially in the summer, although no outbreaks of disease from egg containing this organism have been reported. Large numbers ($>10^6$ per gram) of *Bacillus* spp. are normally needed to cause illness. However, spores of this and other *Bacillus* spp. are likely to survive pasteurisation and persist in the pasteurised product.

Upper microbial limits currently used by industry in the UK for pasteurised liquid egg products and similar (e.g. scrambled egg) are as follows (per gram): Enterobacteriaceae 10 or 100; *E. coli* < 10; *B. cereus* < 100 or < 200; *Staph. aureus* < 20. *Listeria* spp. should be absent in 25 g and *Salmonella* spp. absent in 2 × 25 g.

9.3.5 Sampling of egg products

The methods described below are based on those of Roberts and Greenwood (2003) and Ricke *et al.* (2001). With liquid products, it is important that the material being sampled is mixed thoroughly to ensure an even distribution of any contaminants. Samples should be kept below 4 °C for no more than four hours prior to examination, and raw and pasteurised products should always be tested separately to avoid possible cross-contamination.

- *Raw liquid egg.* The unpasteurised material is sampled from the raw egg balance tank. If there is a sampling tap, this should be run to waste briefly before collecting the sample in a sterile container. Otherwise, a sterile dipper is needed. Samples of at least 25 g are required.
- *Pasteurised liquid egg.* Samples taken as above are collected from the holding tank, as close as possible to the pasteuriser. For each tank being sampled, the temperature of the contents should be recorded, in case of any temperature abuse.
- *Frozen egg.* Cans should be selected that are representative of the lot being sampled. Each can is opened aseptically and any surface frost removed from the frozen egg with a sterile spoon. Then, samples are taken with an electric drill, passing through the centre of the block. At least 25 g of the resultant shavings are collected and should be maintained in the frozen state until required for testing. The material is thawed at 0–4 °C or by holding at room temperature for 2–3 hours.

- *Dried egg.* Representative packs are opened carefully and sampled aseptically, using a sterile spoon. The material thus obtained is mixed thoroughly and 25 g transferred to a sterile container. While BPW is being added gradually, the mixture is stirred with a sterile spatula to obtain an homogeneous suspension. To facilitate rehydration, this is allowed to stand at room temperature for 60 minutes before proceeding further.

9.3.6 Testing for *Salmonella*

In the EU, Commission Regulation 2073 (EU, 2005) requires testing to be carried out according to ISO 6579, a method that is applicable to foods in general (ISO, 2002). The media for sample pre-enrichment and enrichment are respectively BPW, RVS and Muller–Kauffmann tetrathionate–novobiocin broth. Enrichment cultures are plated on XLD agar and a second medium of choice. Typical colonies are then confirmed by biochemical and serological tests. When examining albumen (and possibly other egg products), improved detection of salmonella may be obtained by supplementing the pre-enrichment and/or enrichment medium with ferrioxamine E in order to counter the iron-sequestering properties of albumen (Reissbrodt and Rabsch, 1993; Reissbrodt *et al.*, 1996; Thammasuvimol *et al.*, 2006).

As with shell eggs, other methods can be used to detect and/or isolate *Salmonella* from egg products, including rapid methods (see p. 189 and Chapter 10). For example, a study involving 17 participating laboratories (Gibson *et al.*, 1992) compared an automated conductance method for *Salmonella* with a standard cultural method. Liquid whole egg was among the foods examined and, for various serotypes and inoculum levels of 1–5 or 10–50 cells per 25 g, there was no significant difference between the two methods.

9.4 Future trends

Salmonella is, and will continue to be, the most important foodborne pathogen associated with shell eggs and raw egg products. Prior to the late 1980s and emergence of the SE pandemic, shell eggs were regarded as a relatively safe food item that could be eaten raw or under-cooked by all members of society, including the most vulnerable. However, with the advent of egg-associated SE infections, the situation changed and the risk of contracting salmonellosis by this means became clear. A consequence of the problem has been the need to develop national surveillance programmes for shell eggs and, because of the low contamination rate, to examine many thousands of eggs for *Salmonella*. Since eggs are involved in international trade, there is also a need for international harmonisation of egg surveillance (ACMSF, 2001). Not surprisingly, the epidemiological picture continues to change. While the SE strains responsible for eggborne illness are known to vary between countries, they can equally vary over time in individual countries. For example, the predominant strain-type in Europe has been 'phage type (PT) 4 but, in the USA, PTs 8 and 13 have predominated. More recently, the UK has experienced an increase in PT14b, which has risen from <200 laboratory-confirmed cases of illness per year prior to 2001 to 922 cases in 2003. The organism in question was also isolated from shell eggs imported from Spain (Anonymous, 2004). The changes occurring are not necessarily confined to SE, because other serotypes, including *S.* Heidelberg, *S.* Typhimurium and *S.* Typhimurium var. Copenhagen, are invasive in poultry and have been shown to colonise the reproductive tract in infected hens (Snoeyenbos *et al.*, 1969). Any of these, as well as 'new' strains of SE, may be capable of direct contamination of egg contents. Also, predicted changes in global export/import patterns for eggs (Ernst, 2002) may lead to new *Salmonella* problems in the future.

In this context, the value of microbiological testing is to pinpoint sources of contaminated eggs that are entering the country and to indicate those eggs that need to be re-directed for pasteurisation rather than retail sale. To minimise any delay in releasing eggs that are *Salmonella*-negative, there will be continuing interest in rapid methods of *Salmonella* detection, especially those that avoid the need for pre-enrichment of samples.

For raw egg products, heat pasteurisation is an important control measure, the efficacy of which depends partly on hygiene control throughout the supply chain. This is because there is an inevitable compromise between the heat treatment needed to make the product safe and the adverse effects of heat on the functional properties of the material. Ways of optimising the former, while reducing the latter, will continue to be sought. Egg albumen is particularly vulnerable in this respect. For example, Baron *et al.* (2003) compared two different heating regimes for spray-dried albumen, but found that neither was entirely satisfactory for maintaining functional quality. It is likely that this problem will be solved ultimately by the application of new technology. Present possibilities include the use of ultrasonic waves and pulsed electric fields, as discussed by Sheldon (2005). More effective treatments for raw egg products would reduce the risks, not only from *Salmonella* but also *Listeria* and other pathogens that can be present.

9.5 Sources of further information

The book produced by the International Commission on Microbiological Specifications for Foods (ICMSF, 1998) includes a chapter on eggs and egg products that gives an authoritative account of factors affecting microbial contamination and survival. A more recent review of bacterial infection of eggs is provided by Gast (2005). For further information on analytical aspects, the reader is recommended to more general books on food analysis, such as Downs and Ito (2001) and Roberts and Greenwood (2003), which cover methods favoured in the USA and UK, respectively. In relation to egg products, a useful text is that of Stadelman and Cotterill (1995), while the latest techniques for reducing pathogen contamination, including possible alternatives to heat treatment and the pasteurisation of shell eggs, are covered by the US Institute of Food Technologists (IFT, 2000) and Sheldon (2005).

9.6 Acknowledgement

The author is indebted to Professor Geoff Mead for assistance in drafting the manuscript.

9.7 References

ACMSF (2001), *Second Report on* Salmonella *in Eggs*, Advisory Committee on the Microbiological Safety of Food, Norwich, UK, The Stationery Office.

AMIN M K and DRAUGHON F A (1990), 'Infection of shell eggs with *Yersinia enterocolitica*', *Journal of Food Protection*, **56**, 826–830.

ANONYMOUS (1963), 'The Liquid Egg (Pasteurisation) Regulations, 1963', *Statutory Instruments (Food and Drugs) No. 1503*, London, UK, HMSO.

ANONYMOUS (1993), 'The Egg Products Regulations 1993', *Statutory Instrument No. 1520*, London, UK, HMSO.

ANONYMOUS (2004), 'National case-control study of *Salmonella* Enteritidis Phage Type (PT) 14b infection implicates eggs used in the catering trade, *CDR Weekly*, **14**(8): News.

BAKER R C and BRUCE C (1994), 'Effects of processing on the microbiology of eggs', in *Microbiology of the Avian Egg*, eds Board R G and Fuller R, London, UK, Chapman & Hall, 153–173.

BALL H R Jr, HAMID-SAMIMI M, FOEGEDING P M and SWARTZEL K R (1987), 'Functionality and microbial stability of ultrapasteurized, aseptically packaged refrigerated whole egg', *Journal of Food Science*, **52**, 1212–1218.

BARNES E M and CORRY J E L (1969), 'Microbial flora of raw and pasteurised egg albumen', *Journal of Applied Bacteriology*, **32**, 193–205.

BARON F, NAU F, GUÉRIN-DUBIARD C, GONNET F, DUBOIS J-J and GAUTIER M (2003), 'Effect of dry heating on the microbiological quality, functional properties, and natural bacteriostatic ability of egg white after reconstitution', *Journal of Food Protection*, **66**, 825–832.

BOARD R G (1965), 'The properties and classification of the predominant bacteria in rotten eggs', *Journal of Applied Bacteriology*, **28**, 437–453.

BOARD R G and FULLER R, eds (1994), *Microbiology of the Avian Egg*, London, UK, Chapman & Hall.

BOARD R G, CLAY C, LOCK J and DOLMAN J (1994), 'The egg: a compartmentalized, aseptically packaged food', in *Microbiology of the Avian Egg*, eds Board R G and Fuller R, London, UK, Chapman & Hall, 43–61.

BRANT A W and STARR P B (1962), 'Some physical factors related to egg spoilage', *Poultry Science*, **41**, 1468–1473.

BROOKS J (1951), 'The washing of eggs', *Food Science Abstracts*, **23**, 545–554.

BROWN W E, BAKER R C and NAYLOR H B (1965), 'The role of the inner shell membrane in bacterial penetration of chicken eggs', *Poultry Science*, **44**, 1323–1327.
CHEN J and GRIFFITHS M W (1996) '*Salmonella* detection in eggs using *Lux*$^+$ bacteriophages', *Journal of Food Protection*, **59**, 908–914.
COP (2004), 'Code of Practice for Lion Eggs. 5th Version, July 2004', London, UK, British Egg Industry Council, www.stonegate.co.uk/pdfs/lioncodeofpractice.pdf.
COX N A, STERN N J, HIETT K L and BERRANG M E (2002), 'Identification of a new source of *Campylobacter* contamination in poultry: transmission from breeder hens to broiler chickens', *Avian Diseases*, **46**, 535–541.
COX N A, RICHARDSON L J, BAILEY J S, COSBY D E, CASON J A and MUSGROVE M T (2005), 'Bacterial contamination of poultry as a risk to human health', in *Food Safety Control in the Poultry Industry*, ed. Mead G C, Cambridge, UK, Woodhead Publishing, 21–43.
CUNNINGHAM F E (1986), 'Egg product pasteurization', in *Egg Science and Technology*, 3rd edn, eds Stadelman W G and Cotterill O J, Bringhamton, NY, USA, Haworth Press, 243–272.
DE LOUVOIS J (1993), '*Salmonella* contamination of eggs: a potential source of human salmonellosis', *PHLS Microbiology Digest*, **10**(3), 158–162.
DE MEDICI D, DI PASQUALE S, CICCAGLIONI G, FALZI G and OREFICE L (1998), 'Comparison between two rapid methods and the conventional cultural methods for the detection of Salmonella in eggs', in *Prevention of Contamination of Poultry Meat, Eggs and Egg Products*, eds Franchini A and Mulder R W A W, Luxembourg, Office for Official Publications of the European Communities, 81–84.
DOWNS F P and ITO K, eds (2001), *Compendium of Methods for the Microbiological Examination of Foods*, Washington DC, USA, American Public Health Association.
DOYLE M P (1984), 'Association of *Campylobacter jejuni* with laying hens and eggs', *Applied and Environmental Microbiology*, **47**, 533–536.
EFSA (2005), 'Opinion of the Scientific Panel on Biological Hazards on the request from the Commission related to the Microbiological Risks on Washing of Table Eggs', *EFSA Journal*, **269**, 1–39. http://www.efsa.eu.int/science/biohaz/biohaz_opinions/1196_en.html
ELSON R, LITTLE C L and MITCHELL R T (2005), '*Salmonella* and raw shell eggs: results of a cross-sectional study of contamination rates and egg safety practices in the United Kingdom catering sector in 2003', *Journal of Food Protection*, **68**, 256–264.
ERNST M (2002), 'Egg processing revolution ahead', *Poultry International*, November, 12–16.
EU (2005), 'Commission Regulation (EC) No. 2073/2005 of 15 November 2005 on microbiological criteria for foodstuffs', *Official Journal of the European Union*, **L338**, 1–26.
FINCH M J and BLAKE P A (1985), 'Foodborne outbreaks of campylobacteriosis: the US experience, 1980–1982', *American Journal of Epidemiology*, **122**, 262–268.
FOEGEDING P M and LEASOR S B (1990), 'Heat resistance and growth of *Listeria monocytogenes* in liquid whole egg', *Journal of Food Protection*, **53**, 9–14.
GARIBALDI J and BAYNE H G (1962), 'The effect of iron on the *Pseudomonas spoilage* of farm washed eggs', *Poultry Science*, **41**, 850–853.
GAST R K (2005), 'Bacterial infection of eggs', in *Food Safety Control in the Poultry Industry*, ed. Mead G C, Cambridge, UK, Woodhead Publishing, 1–20.

GAST R K and BEARD C W (1992), 'Detection and enumeration of *Salmonella enteritidis* in fresh and stored eggs laid by experimentally-infected hens', *Journal of Food Protection*, **55**, 152–156.

GAST R K, HOLT P S, NASIR M S, JOLLEY M E and STONE H D (2003), 'Detection of *Salmonella enteritidis* in incubated pools of egg contents by fluorescence polarization and lateral flow immunodiffusion', *Poultry Science*, **82**, 687–690.

GIBSON D M, COOMBS P and PIMBLEY D W (1992), 'Automated conductance method for the detection of *Salmonella* in foods', *Journal of AOAC International*, **75**, 293–302.

HARA-KUDO Y, KUMAGAI S, MASUDA T, GOTO K, OHTSUKA K, MASAKI H, TANAKA H, TANNO K, MIYAHARA M and KONUMA H (2001), 'Detection of *Salmonella enteritidis* in shell and liquid eggs using enrichment and plating', *International Journal of Food Microbiology*, **64**, 395–399.

HUMPHREY T J (1994), 'Contamination of egg shell and contents with *Salmonella enteritidis*', *International Journal of Food Microbiology*, **21**, 31–40.

HUMPHREY T J and WHITEHEAD A (1992), 'Techniques for the isolation of *Salmonella* from eggs', *British Poultry Science*, **33**, 761–768.

HUMPHREY T J, WHITEHEAD A, GAWLER A H L and ROWE B (1991), 'Numbers of *Salmonella enteritidis* in the contents of naturally contaminated hens' eggs', *Epidemiology and Infection*, **106**, 489–496.

HUTCHISON M L, GITTINS J, WALKER A, SPARKS N, HUMPHREY T J, BURTON C and MOORE A (2004), 'An assessment of the microbiological risks involved with egg washing under commercial conditions', *Journal of Food Protection*, **67**, 4–11.

ICMSF (2005), 'Eggs and egg products', in *Micro-organisms in Foods 6. Microbial Ecology of Food Commodities*, 2nd edition, International Commission on Microbiological Specifications for Foods, New York, USA, Kluwer Academic/Plenum Publishers, 597–642.

IFT (2000), Institute of Food Technologists, 'Kinetics of microbial inactivation for alternative food processing technologies', *Journal of Food Science Special Supplement*, Nov./Dec., 1–8.

IJICHI K, GARIBALDI J A, KAUFMAN V F, HUDSON C A and LINEWEAVER H (1973), 'Microbiology of a modified procedure for cooling pasteurised salt yolk', *Journal of Food Science*, **38**, 1241–1243.

ISO (1999), 'ISO 6888-3:1999 – Microbiology of food and animal feeding stuffs – horizontal method for the enumeration of coagulase-positive staphylococci (*Staphylococcus aureus* and other species). Part 1: Technique using Baird–Parker agar medium', Geneva, Switzerland, International Organization for Standardization.

ISO (2002), 'ISO 6579:2002 (E) – Microbiology of food and animal feeding stuffs – horizontal method for the detection of *Salmonella* spp.', Geneva, Switzerland, International Organization for Standardization.

ISO (2004), 'ISO 21528-2:2004(E) – Microbiology of food and animal feeding stuffs – horizontal methods for the detection and enumeration of Enterobacteriaceae. Part 2: Colony-count method', Geneva, Switzerland, International Organization for Standardization.

IZAT A L and GARDNER F A (1988), 'The incidence of *Campylobacter jejuni* in processed egg products, *Poultry Science*, **67**, 1431–1435.

KHAN M A, NEWTON J A, SEAMAN A and WOODBINE M (1975), 'The survival of *Listeria monocytogenes* inside and outside its host', in *Problems of Listeriosis*, ed. Woodbine M, Leicester, UK, Leicester University Press, 75–83.

KINNER J A and MOATS W A (1981), 'Effect of temperature, pH, and detergent on survival of bacteria associated with shell eggs', *Poultry Science*, **60**, 761–767.

KOLLOWA J and KOLLOWA C (1989), 'Occurrence and survival of *Campylobacter jejuni* on the shell surface of hen eggs', *Monatshefte Veterinarmedizin*, **44**, 63–65.

MAYES F J and TAKEBALLI M A (1983), 'Microbial contamination of the hen's egg: a review', *Journal of Food Protection*, **46**, 1092–1098.

McELROY A P, COHEN N D and HARGIS B M (1996), 'Evaluation of polymerase chain reaction for the detection of *Salmonella enteritidis* in experimentally inoculated eggs and eggs from experimentally challenged hens', *Journal of Food Protection*, **59**, 1273–1278.

MOORE J and MADDEN R H (1993), 'Detection and incidence of *Listeria* species in blended raw egg', *Journal of Food Protection*, **56**, 652–654, 660.

PAYNE J, GOOCH J E T and BARNES E M (1979), 'Heat-resistant bacteria in pasteurized whole egg', *Journal of Applied Bacteriology*, **46**, 601–623.

PEARSON A D, GREENWOOD M H, FELTHAM R K A, HEALING T D, DONALDSON J, JONES D M and COLWELL R R (1996), 'Microbial ecology of *Campylobacter jejuni* in a United Kingdom chicken supply chain: intermittent common source, vertical transmission and amplification by flock propagation', *Applied and Environmental Microbiology*, **62**, 4614–4620.

REISSBRODT R and RABSCH W (1993), 'Selective pre-enrichment of salmonella from eggs by siderophore supplements', *Zentralblatt für Bakteriologie*, **279**, 344–353.

REISSBRODT R, VIELITZ E, KORMANN E, RABSCH W and KUHN H (1996), 'Ferrioxamine E-supplemented pre-enrichment and enrichment media improve various isolation methods for *Salmonella*', *International Journal of Food Microbiology*, **29**, 81–91.

RICKE S C, BIRKHOLD S G and GAST R K (2001), 'Eggs and egg products', in *Compendium of Methods for the Microbiological Examination of Foods*, eds Downs F P and Ito K, Washington DC, USA, American Public Health Association, 473–481.

ROBERTS D and GREENWOOD M (2003), *Practical Food Microbiology*, 3rd edn, Oxford, UK, Blackwell.

ROBINSON D S, BARNES E M and TAYLOR J (1975), 'Occurrence of β-hydroxybutyric acid in incubator reject eggs', *Journal of the Science of Food and Agriculture*, **26**, 91–98.

SAHIN O, KOBALKA P and ZHANG Q (2003), 'Detection and survival of *Campylobacter* in chicken eggs', *Journal of Applied Microbiology*, **95**, 1070–1079.

SCOTT W M (1930), 'Food poisoning due to eggs', *British Medical Journal*, **11**, 56–48.

SCOTTER S, ALDRIDGE M and WOOD R (1994), 'Methods of examination for eggs and egg products: collaborative trial', *Journal of the Association of Public Analysts*, **30**, 155–176.

SHANE S M, GIFFORD D H and YOGASUNDRAM K (1986), '*Campylobacter jejuni* contamination of eggs', *Veterinary Research Communications*, **10**, 487–492.

SHELDON B W (2005), 'Techniques for reducing pathogens in eggs', in *Food Safety Control in the Poultry Industry*, ed. Mead G C, Cambridge, UK, Woodhead Publishing, 273–309.

SIMMONS E R, AYRES J A and KRAFT A A (1970), 'Effect of moisture and temperature on ability of salmonellae to infect shell eggs', *Poultry Science*, **49**, 761–768.

SIONKOWSKI P J and SHELEF L A (1990), 'Viability of *Listeria monocytogenes* in egg wash water', *Journal of Food Protection*, **50**, 103–107.

SNOEYENBOS G H, SMYSER C F and VAN ROEKEL H (1969), '*Salmonella* infections of the ovary and peritoneum of chickens', *Avian Diseases*, **13**, 668–670.

STADELMAN W J and COTTERILL O J, eds (1995), *Egg Science and Technology*, 4th edition, New York, USA, Haworth Food Products Press.

STOKES J L, OSBORNE W W and BAYNE H G (1956), 'Penetration and growth of *Salmonella* in shell eggs', *Food Research*, **21**, 510–518.

THAMMASUVIMOL G, SEO K H, SONG K Y, HOLT P S and BRACKETT R E (2006), 'Optimization of ferrioxamine E concentration as effective supplementation for selective isolation of *Salmonella* Enteritidis in egg white', *Journal of Food Protection*, **69**, 634–638.

THATCHER F S and MONTFORD J (1962), 'Egg-products as a source of salmonellae in processed foods', *Canadian Journal of Public Health*, **53**, 61–69.

THORNLEY M J and SHARPE M E (1959), 'Microorganisms from chicken meat related to both lactobacilli and aerobic sporeformers', *Journal of Applied Bacteriology*, **22**, 368–376.

WINDHORST H-W (2005), 'Development of organic egg production and marketing in the EU', *World's Poultry Science Journal*, **61**, 451–462.

10

Detection and enumeration of pathogens in meat, poultry and egg products

E. de Boer, Food and Consumer Product Safety Authority (VWA), The Netherlands

10.1 Introduction

The analysis of meat, poultry and egg products for the presence of pathogens is essential in ensuring the safety of these foods for the consumer and may include the following:

- Testing by manufacturers of raw materials, food ingredients and foods in national or international trade for acceptance or rejection.
- Monitoring the effectiveness of treatment processes and critical control points during production and the final quality of the end product; testing to ensure that the food will still be acceptable at the end of the expected shelf-life. Microbiological testing for Hazard Analysis Critical Control Point (HACCP) purposes should involve the enumeration of indicator organisms rather than detection of pathogens. However, pathogen testing is often included in HACCP verification systems.
- Testing by the food producer or a control authority to check whether a batch of food or food ingredients meets the relevant legal requirements. Recently, new microbiological criteria for foodstuffs were laid down in a European Union Regulation (EU, 2005). This regulation includes criteria for both food safety and process hygiene that are applicable to the main groups of food products.
- Routine monitoring or surveillance studies for specific products that may present a potential risk or about which there is little documented information. The information obtained from such studies is important to food producers and law enforcers.
- Investigation of food suspected of causing human illness or foods related to consumer complaints.

Conventional and modern microbiological methods for foodborne pathogens are designed to detect or enumerate specific target organisms. The nature and concentration of both target and non-target organisms may vary considerably between samples of different food matrices. A false-positive result is obtained when a non-target organism is mistakenly identified as the one being sought; a false-negative result arises when a target organism does not give a characteristic or 'typical' reaction in the test. The choice of method should be given careful consideration. A method that has been evaluated for a particular type of food is not necessarily applicable to other food matrices. Results of a microbiological test are always method-defined, i.e. when examining the same sample by different methods, different results may be obtained. The physiological state of a microorganism may affect its detection and microbial damage may vary in type and degree. Different methods will recover different proportions of the microbial population. Although methods have been laid down in (inter)national standards or legal requirements, the laboratory remains responsible for evaluating the performance of the method for the type of sample under investigation, and for seeking alternatives when necessary and possible (Lightfoot and Maier, 1998). If an individual laboratory changes a standard method, this modified method must be validated against the original.

There are limitations with all microbiological tests, and these must be considered before any action can be taken, following a report from the laboratory (Roberts and Greenwood, 2003). Both the method of analysis and the sampling method are major factors in the results obtained. When microbiological criteria are included in food legislation or in a specification, the analytical methods and sampling plans should be clearly identified.

In this chapter, conventional culture methods and modern rapid methods for detecting pathogens in foods will be reviewed. Principles of the techniques, standardised and alternative methods, and examples of commercial systems will be given. Methods for the detection of *Salmonella*, *Campylobacter*, *Escherichia coli* O157:H7 and *Listeria monocytogenes* will be described in some detail, because these organisms are the major pathogens associated with meat, poultry and egg products. Finally, relevant factors in the choice of method are considered.

10.2 Cultural methods

10.2.1 General considerations

Despite considerable advances in rapid diagnostic methods, conventional testing, using specific culture media, remains the foundation for the detection of foodborne pathogens in most microbiological laboratories. However, there have been various attempts to make the usual agar-based methods more convenient and user-friendly by developing formats other than those involving the traditional petri dish. Examples are the Petrifilm™ system

(3M), the use of dipslides and the hydrophobic grid membrane filter technique. In parallel, commercial identification kits have been developed to simplify and automate the identification of individual microorganisms; examples are the API® systems (bioMérieux) and BBL™-Crystal™ Identification Systems (Becton-Dickinson).

The recognition of colonies of the target organism amongst those of the accompanying flora is essential, but not always easy to achieve. In comparison with traditional culture media, newer systems use different selective and diagnostic properties. The development of chromogenic and fluorogenic culture media has facilitated the rapid detection and identification of many pathogens (Manafi, 2000). The use of multiple enrichment broths and plating media will increase the likelihood of detection. However, laboratories should weigh the cost of the increased diagnostic effort against the expected gain in sensitivity of detection and in relation to the objectives of their analyses (Davies *et al.*, 2000).

In selecting appropriate methods, it is important to use published standard methods, where they are available. With standardised methods, complete instructions are given for the entire analysis, down to the composition of media and their pH values. Standard methods are available from ISO (International Standards Organisation), CEN (Comité Européan de Normalisation; European Committee for Normalisation), IDF (International Dairy Federation), AOAC (Association of Official Analytical Chemists), Nordic Committee on Food Analysis (NMKL) and from individual standards organisations in each country. Technical committees, comprising experts from the industrial, technical and business sectors, develop ISO standards. Standardised methods have been published for the major bacterial pathogens occurring in foods, including *Salmonella*, *Campylobacter*, *E. coli* O157 and *L. monocytogenes*, which are the main pathogens in red meat, poultry and egg products. ISO standard methods for other pathogens that may occur in these foods are presented in Table 10.1.

10.2.2 *Salmonella*

In contaminated foods, *Salmonella* may be present in relatively low numbers, usually distributed unevenly, possibly sublethally injured by food processing or by intrinsic factors associated with the food itself, and mostly in the presence of high numbers of closely-related, competing organisms. These conditions complicate the isolation of *Salmonella* from food samples by means of cultural methods. The isolation procedure generally involves four basic steps (Andrews, 1996; Baylis *et al.*, 2000a; Van der Zee, 2003). Firstly, pre-enrichment in a non-selective, nutritious medium is necessary to facilitate the recovery and growth of *Salmonella* to a level of at least 10^5 cfu/ml. Selective enrichment of a small portion of the pre-enrichment culture allows further growth of *Salmonella* and, at the same time, inhibits competing organisms. Subsequent isolation on selective agar media will restrict the

Table 10.1 ISO methods for pathogens other than those of greatest concern in red meat, poultry and egg products

Organism	ISO standard	Enrichment medium	Isolation medium
Bacillus cereus	7932		Mannitol–egg yolk–polymyxin
Clostridium perfringens	7937		Tryptose–sulphite–cycloserine
Shigella	21567	Shigella bouillon	MacConkey Xylose–lysine–deoxycholate Hektoen enteric
Staphylococcus aureus	6888		Baird–Parker Rabbit plasma–fibrinogen
Yersinia enterocolitica	10273	Peptone–sorbitol–bile broth Irgasan–ticarcillin–chlorate broth	Cefsulodin–irgasan–novobiocin Salmonella–Shigella–deoxycholate–calcium chloride

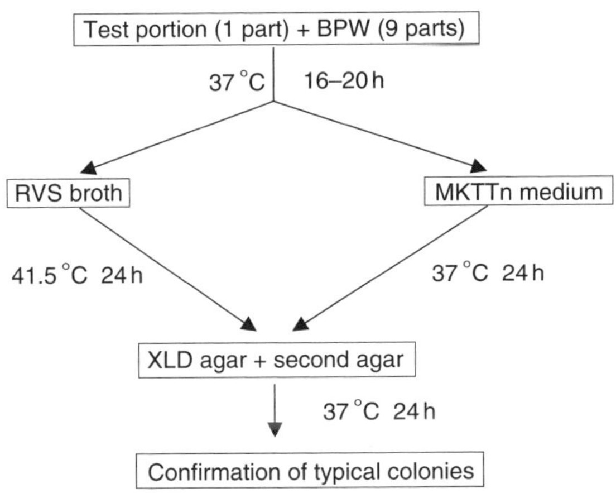

Fig. 10.1 Detection of *Salmonella* in food according to ISO 6579 (ISO, 2002). BPW, buffered peptone water; RVS, Rappaport–Vassiliadis medium with soya broth; MKTTn, Muller–Kauffmann tetrathionate–novobiocin broth; XLD, xylose–lysine–deoxycholate agar.

growth of competing, non-target organisms and will result in presumptive *Salmonella* colonies. Finally, confirmation, using biochemical and serological tests, will demonstrate the presence of *Salmonella*.

The protocol of the international standard, ISO 6579, for the detection of *Salmonella* spp. in food and animal feeding stuffs is shown in Fig. 10.1

(ISO, 2002). In the first step, buffered peptone water (BPW), a non-selective pre-enrichment medium, is mixed with the sample in a 1:10 dilution. Since many microbiological criteria for foods include the absence of *Salmonella* in 25 g of product, it is usual to mix that amount of food sample with 225 ml of BPW. For specific foodstuffs, such as cocoa and acidic foods, modifications of BPW are used, including the addition of skim-milk powder, casein or brilliant green, or a double-strength buffer. Generally, such modifications are not needed for the analysis of meat and poultry products. The time and temperature of incubation seem to be more critical than the choice of pre-enrichment medium (De Boer, 1998). Incubation of the sample suspension in BPW is generally carried out at 37 °C for 16–20 hours. A longer incubation period may result in the overgrowth of *Salmonella* by competing microflora and/or acidification of the medium below pH 4.5, which strongly reduces the likelihood of *Salmonella* isolation. However, a shorter period of incubation will not allow adequate repair of injured organisms and a sufficient increase in the number of *Salmonella* cells. After pre-enrichment, the culture is transferred to Rappaport–Vassiliadis broth with soya (RVS), in the proportion 0.1 ml to 10 ml broth, and to Muller–Kauffmann tetrathionate-novobiocin (MKTTn) broth, using 1 ml to 10 ml broth. As a result of the last revision of ISO 6579 in 2002, the selective Rappaport–Vassiliadis (RV) broth was replaced by RVS, in which tryptone is substituted by soya peptone. This modification improved the recovery rates for *Salmonella* in several studies (Van der Zee, 2003).

The second selective enrichment medium, MKTTn broth, replaced selenite-cystine (SC) broth, because of the toxicity of selenite. The use of two selective enrichment broths is considered necessary for the isolation of all serotypes of *Salmonella*, including *S.* Typhi and *S.* Paratyphi. RVS broth and MKTTn broth are incubated for 24 hours at 41.5 °C and 37 °C, respectively. After enrichment, the cultures are used to inoculate two selective media: xylose–lysine–deoxycholate (XLD) agar and another solid selective medium chosen by the laboratory. Preferably, this second agar medium should be complementary to XLD agar and especially appropriate for isolating lactose-positive *Salmonella* or *S.* Typhi and *S.* Paratyphi. The XLD agar is incubated at 37 °C and examined after 24 hours. Colonies of presumptive *Salmonella* are subcultured and their identity is confirmed by means of appropriate biochemical tests. Cultures identified as *Salmonella* on this basis are then serotyped by slide agglutination, after auto-agglutinable strains have been eliminated. A *Salmonella* reference laboratory should carry out any necessary definitive typing.

The isolation procedure for *Salmonella*, as described in the US Food and Drug Administration–Bacteriological Analytical Manual (FDA-BAM) (FDA, 2003a) differs from the ISO 6579 method in some respects. The BAM procedure gives more detail on *Salmonella* detection in specific foods, especially low-moisture foods. Also, lactose broth is commonly used as the pre-enrichment medium and both RV medium and tetrathionate broth are employed as selective enrichment media. Hektoen Enteric agar, XLD agar and bismuth-sulphite agar are described in relation to selective isolation.

A method described by the US Department of Agriculture–Food Safety Inspection Service (USDA–FSIS, 2004) for the isolation of *Salmonella* from meat, poultry and egg products uses BPW for pre-enrichment, modified RV medium for selective enrichment and xylose–lysine Tergitol 4 agar or double-modified lysine–iron agar for selective plating. In addition to standard methods, many other cultural procedures have been described for detecting *Salmonella*, in which modifications of the usual pre-enrichment, enrichment and plating media are used.

Although many attempts have been made to shorten the period of pre-enrichment, 16–20 hours is generally required for satisfactory resuscitation and multiplication of the target organisms. No single protocol is possible for the pre-enrichment of *Salmonella* in all foods and, consequently, modifications of the standard pre-enrichment medium, BPW, are often used (Andrews, 1996). Several additions to BPW, including ammonium–iron (III)-citrate, ferrioxamine E and G, and novobiocin in combination with cefsulodin, have been suggested to facilitate the isolation of *S.* Enteritidis from eggs and egg products (Van der Zee, 2003). A number of modifications of tetrathionate-based media, including Muller–Kauffmann broth, are also in use. The selectivity of these media depends on the presence of the enzyme tetrathionate reductase in salmonellas. RV broth, on the other hand, contains malachite green and magnesium chloride as selective agents, and several modifications of this medium have been described.

The use of 'motility enrichment' has been found effective for the rapid detection and isolation of *Salmonella* and some other pathogens. This method, based on the motility of most salmonellas, indicates the presence of the organism by a swarm zone, after inoculation of the pre-enrichment culture into a selective, semi-solid medium, followed by overnight incubation. In several comparative studies, motility enrichment in modified semi-solid Rappaport–Vassiliadis (MSRV) medium has been confirmed as a very effective procedure for the isolation of *Salmonella* from foods (Worcman-Barninka *et al.*, 2001). Results obtained with MSRV medium are mostly better than or equal to those obtained with RV broth. Only occasionally have lower isolation rates been reported, possibly because of the presence of non-motile salmonellas (De Boer, 1998). With MSRV medium, *Salmonella* detection is simple and cheap, and both negative and positive results are available up to 24 hours earlier than is possible with the standard ISO method. An amendment to ISO 6579, including the use of MSRV medium for the detection of *Salmonella* spp. in animal faeces and samples from the primary livestock-production stage, will be published in 2007 (ISO, 2007). Diagnostic *Salmonella* Medium (DIASALM) uses a sucrose–bromcresol purple indicator to demonstrate migrating salmonellas as a pink zone within a green medium (Van Netten *et al.*, 1991). DIASALM and

MSRV are equivalent as combined enrichment and isolation media for *Salmonella*.

The optimum plating medium for isolating all *Salmonella* serotypes has not yet been developed and the search for new isolation media continues. This also means that it is necessary to use at least two plating media in parallel. Non-fermentation of lactose, production of H_2S and motility are characteristics of *Salmonella* that are used in many of the present isolation media. However, about 1% of salmonellas do ferment lactose, and human infections by these strains have been described. To avoid missing lactose-positive strains, a suitable, lactose-independent plating medium, such as bismuth-sulphite agar, should be used. The existence of H_2S-negative and non-motile strains also makes the use of a second plating medium necessary. Table 10.2 shows the characteristics of some commonly used *Salmonella* isolation media (Waltman, 2000).

Quantitative determination of *Salmonella* may be needed in some instances, e.g. for the evaluation of *Salmonella* control programmes in the poultry industry. Direct plate counts of *Salmonella* are not usually feasible, because of the low numbers present in relation to those of the competing flora and the frequent presence of injured *Salmonella* cells. The technique of choice for quantitative estimation of salmonellas in naturally-contaminated samples is the Most Probable Number (MPN) technique. MPN methods involve serial sample dilutions, each of which is used to inoculate a number of replicate tubes of BPW. After incubation as before, the steps that follow are those described in the standard *Salmonella* isolation method (Humbert *et al.*, 1997). The MPN technique is rather impractical and laborious, but better alternative culturing methods are not yet available.

10.2.3 *Campylobacter*

Campylobacter spp. present in foods and causing human infection belong to the 'thermotolerant' or 'thermophilic' group. The most frequently encountered species are *C. jejuni* and *C. coli*. Further species have also been described, including *C. lari*, *C. upsaliensis* and others. Consideration of the following factors is essential, when choosing methods for detecting and isolating these rather fastidious organisms.

(1) The *Campylobacter* species in question are able to grow at 42 °C, but not at 25 °C. Since many campylobacters show better growth at 37 °C than at 42 °C (Corry *et al.*, 2003), 'thermotolerant' is a better term than 'thermophilic' in this case. Also, the incubation temperature may have an effect on the genotypes of *C. jejuni* obtained (Scates *et al.*, 2003).
(2) *Campylobacter* requires a microaerobic atmosphere consisting of 5% O_2, 10% CO_2 and 85% N_2 for optimum growth. This atmosphere is usually created in a closed jar system, using either gas-generating envelopes or evacuation and replacement of the atmosphere with an

Table 10.2 Characteristics of *Salmonella* isolation media

Agar medium	Selective substance(s)	Diagnostic characteristic(s)	Appearance of *Salmonella* colonies
Bismuth–sulphite	Bismuth, sulphite Brilliant green	H_2S production (+)	Black, metallic sheen
Brilliant green	Brilliant green	Lactose/sucrose (–)	Red
Deoxycholate–citrate	Deoxycholate, citrate	Lactose (–) H_2S production (+)	Colourless, black centre
Hektoen enteric	Bile salts	Lactose/sucrose/salicin (–) H_2S production (+)	Blue–green, black centre
MLCB[a]	Brilliant green Crystal violet	Lysine (+), mannitol (+) H_2S production (+)	Blue–purple, black centre
Salmonella–Shigella	Bile salts Brilliant green, citrate	Lactose (–) H_2S prodution (+)	Colourless, black centre
SM®-ID[b]	Bile salts	Glucuronate (+) β-galactosidase with chromogenic substrate (–)	Pink–red
XLD[c]	Deoxycholate	Lactose/sucrose (–) Xylose (+), lysine (+) H_2S production (+)	Red, black centre
XLT4[d]	Tergitol 4, deoxycholate	Lactose/sucrose (–) Xylose (+), lysine (+) H_2S production (+)	Red, black centre

[a] mannitol–lysine–crystal violet–brilliant green.
[b] bioMérieux.
[c] xylose–lysine–deoxycholate.
[d] xylose–lysine–tergitol.

appropriate gas mixture. As an alternative to a microaerobic atmosphere, the enrichment broth can be incubated in screw-capped bottles or flasks filled almost to the top, leaving a headspace of less than 2 cm, before tightly closing the caps or sealing the tops (ISO, 2006a).

(3) Composition of *Campylobacter* isolation media: campylobacters grow relatively slowly and are easily overgrown by competing organisms, so culture media still rely on antibiotics to suppress the growth of non-target organisms. Typically, cefoperazone, trimethoprim, rifampicin, vancomycin, polymyxin B, cycloheximide and amphotericin B are used in various combinations. Unlike many other bacteria, campylobacters do not ferment common sugars, so that pH indicators or other indicators of fermentation reactions cannot be used to demonstrate *Campylobacter* growth. A combination of ferrous sulphate, sodium metabisulphite and sodium pyruvate (FBP supplement), each at 0.25 or 0.5 g per litre, is often added to isolation media to counteract the toxic effect of oxygen. Addition of lysed horse or sheep blood helps to neutralise toxic oxygen deratives. On presently available media, *Campylobacter* colonies are frequently small and colourless. The larger, more visible colonies of other organisms commonly overgrow the campylobacters and may be more easily observed.

The protocol for the international standard, ISO 10272, for the detection of thermotolerant *Campylobacter* spp. in foods and animal feeding stuffs is shown in Fig. 10.2 (ISO, 2006a). In the first step, the test portion is added to a liquid enrichment medium, Bolton broth, in a 1 : 10 dilution. In the last revision of ISO 10272 in 2006, Bolton broth replaced both Preston broth and Park and Sanders broth, which were the enrichment media in the first version, published in 1995. Plates inoculated from enrichment cultures in Bolton broth resulted in more *Campylobacter* growth than those inoculated

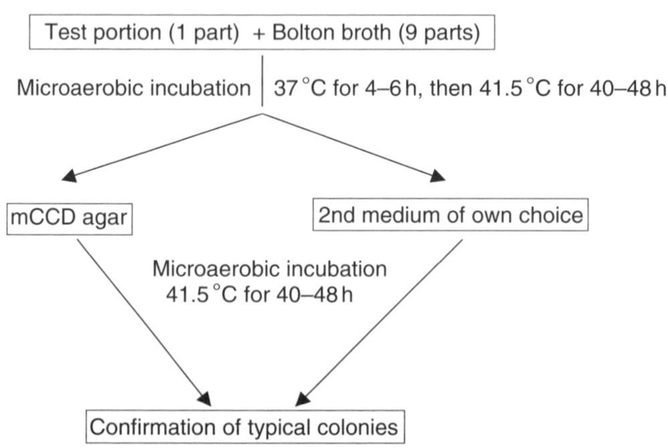

Fig. 10.2 Detection of *Campylobacter* in food according to ISO 10272–1 (ISO, 2006a). mCCD: modified charcoal–cefoperazone–deoxycholate agar.

with Preston broth (Baylis *et al.*, 2000b; Paulsen *et al.*, 2005). In other studies, Park and Sanders broth was found to be superior to Preston broth for recovering *Campylobacter* spp. from poultry products (De Boer and Humphrey, 1991; Tangvatcharin *et al.*, 2005).

Microaerobic incubation of Bolton broth is initially at 37 °C for 4–6 hours to allow repair of injured cells. Then, incubation is continued at 41.5 °C. This temperature matches that recommended for incubation in the ISO standards for *Salmonella* enrichment and the detection of *E. coli* O157.

From the cultures obtained after enrichment, two solid selective media are inoculated. In the revised standard (ISO 10272:2006), modified charcoal–cefoperazone–deoxycholate agar (mCCD agar) has replaced Karmali agar as the first-choice plating medium. Karmali agar is a variant of mCCD agar and these media do not differ significantly in their ability to recover *Campylobacter* spp. from chicken products (De Boer *et al.*, 1998). The choice of the second medium is open but, preferably, it should be one based on a principle different from that of mCCD agar. Table 10.3 shows the

Table 10.3 Composition of *Campylobacter* isolation media

Medium	Base	Blood or charcoal	Antibiotics
mCCDA[a]	Nutrient broth, FBP[c]-supplement, agar	Charcoal	Cefoperazone Amphotericin B
Karmali	Columbia agar base, haemin, sodium pyruvate	Charcoal	Cefoperazone Vancomycin Cycloheximide
Preston	Nutrient broth, FBP-supplement, agar	Lysed horse blood	Polymyxin B Rifampicin Trimethoprim Cycloheximide
Skirrow	Columbia agar base	Lysed horse blood	Vancomycin Trimethoprim Polymyxin B
CAT[b]	Nutrient broth, FBP-supplement, agar	Charcoal	Cefoperazone Teicoplanin Amphotericin B
Campy–Cefex	Brucella agar, FBP-supplement	Lysed horse blood	Cefoperazone Cycloheximide
Abeyta–Hunt–Bark	Heart Infusion agar, FBP-supplement	–	Cefoperazone Rifampicin Amphotericin B
CampyFood ID[d]	Columbia agar base, reducing mixture	Horse serum	Cefoperazone Vancomycin Amphotericin B

[a] modified charcoal–cefoperazone–deoxycholate.
[b] cefoperazone–amphotericin–teicoplanin.
[c] ferrous sulphate, sodium metabisulphite and sodium pyruvate.
[d] bioMérieux.

composition of the media most commonly used for isolating campylobacters. However, many of these media show little difference in composition and ability to isolate the organisms (Oyarzabal *et al.*, 2005), and there is still a need for media that are more selective and differential to facilitate isolation and enumeration.

The agar plates are incubated in a microaerobic atmosphere at 41.5 °C for 40–48 hours. Typical or suspect colonies are subcultured onto the non-selective Columbia blood agar. Some workers have difficulty in picking colonies from mCCD agar, because they tend to stick to the medium. Also, technical expertise is needed to distinguish between campylobacters and other organisms, which frequently grow on many existing media. Moreover, campylobacter colonies may differ considerably in morphology and size, depending on the growth conditions: dry plates sometimes favour the growth of only pinpoint colonies, while excessive humidity during incubation may lead to the formation of large, spreading colonies.

Table 10.4 shows the characteristics used in confirming and identifying thermotolerant campylobacters. It is essential that suspect colonies are examined for typical cell morphology (small, thin, curved rods) and motility ('corkscrew'-type). The oxidase test, which is positive for campylobacters, distinguishes the organisms from other, oxidase-negative, enteric bacteria. The ability to grow at 25 °C in a microaerobic atmosphere, for which thermotolerant campylobacters are negative, applies to other campylobacters and *Arcobacter* spp. Aerobic growth at 41.5 °C should also be negative and this test discriminates between thermotolerant campylobacters and oxidase-positive, aerobic, Gram-negative organisms, such as *Pseudomonas* spp. These tests are usually sufficient to confirm that an isolate belongs to the group of thermotolerant *Campylobacter* spp. Additional tests, using commercial latex agglutination methods, may provide a more definitive

Table 10.4 Characteristics of thermotolerant *Campylobacter* spp.

Characteristics	*C. jejuni*	*C. coli*	*C. lari*	*C. upsaliensis*
Characteristic 'corkscrew' motility	+	+	+	+
Microaerobic growth at 25 °C	–	–	–	–
Aerobic growth at 41.5 °C	–	–	–	–
Oxidase	+	+	+	+
Catalase	+	+	+	– or slight
Nalidixic acid	S*	S*	R/S**	S
Cephalothin	R	R	R	S
Hydrolysis of: hippurate	+	–	–	–
indoxyl acetate	+	+	–	+

+, positive; –, negative; S, sensitive; R, resistant.
* An increase in resistance to nalidixic acid has been shown for *C. jejuni* and *C. coli*. ** Both sensitive and resistant *C. lari* strains are known.

confirmation. For further identification to species level, testing is recommended for sensitivity to nalidixic acid and cephalothin, and hydrolysis of hippurate and indoxyl acetate (Table 10.4). Antibiotic-sensitivity tests are becoming less reliable, because of the increase in resistance of many *Campylobacter* strains.

The US FDA-BAM (FDA, 2001) describes a method for isolating *Campylobacter* spp. from food and water that is not fundamentally different from ISO 10272. This uses Bolton broth for enrichment and plating on either mCCD agar or Abeyta–Hunt–Bark agar. In addition to standard methods, many other procedures have been developed for isolating *Campylobacter* by means of cultural methods, in which different enrichment protocols, incubation conditions and plating media have been advocated (Corry *et al.*, 2003).

Quantitative estimates of *Campylobacter* may be useful in monitoring the effects of particular control measures and in establishing human exposure in risk-assessment studies. Quantitative estimation frequently relies on the MPN technique, although direct plating is clearly simpler. In the latter approach, sample suspensions or dilutions are inoculated directly onto selective agar plates (e.g. Line *et al.*, 2001). The plates are incubated microaerobically and the numbers of typical *Campylobacter* colonies are counted. Part 2 of ISO 10272 describes a colony count technique for *Campylobacter* in food and animal feeding stuffs (ISO, 2006b). Plates containing mCCD agar are inoculated with a specified quantity of the initial suspension of the test sample or appropriate dilutions. After incubation at 41.5 °C for 40–48 hours, suspect colonies are subjected to confirmatory tests and the number of campylobacters per ml or gram of the test sample is calculated from the number of confirmed colonies per plate.

Semi-quantitative detection can also be useful for estimating the level of *Campylobacter* contamination in a sample. Josefsen *et al.* (2002) describe a method in which ten-fold dilutions are prepared by transferring 1 ml of the primary sample suspension in BPW to 9 ml of enrichment broth. After incubation and subsequent plating, any plates showing typical colonies are recorded. Counts are determined in orders of magnitude from a modified MPN table. This method is part of the NMKL standard for detecting and enumerating *Campylobacter* in foods (NMKL, 2005).

10.2.4 *Escherichia coli* O157

Sensitive methods are needed to determine the presence of Shiga toxin-producing *E. coli* (STEC) O157 in foods or environmental samples, as this pathogen may only be present in small numbers, together with substantial levels of competing organisms. The ability to detect small numbers of STEC in foods is essential to ensure a safe food supply, because the infective dose is thought to be as low as <10 organisms (Baylis *et al.*, 2001).

Selective enrichment prior to plating on a selective isolation medium is a common part of most methods for detecting *E. coli* O157. The selective agents usually include novobiocin or acriflavin to inhibit the growth of Gram-negative organisms and a combination of vancomycin, cefsulodin and cefixime to suppress growth of *Aeromonas* and *Proteus* spp. (Blackburn and McCarthy, 2000). However, the growth of injured cells may be inhibited by some selective agents. For freeze-injured *E. coli* O157, it has been reported that food samples must be allowed to stand at room temperature for three hours prior to selective enrichment (Nakagawa *et al.*, 2000). Also, pre-enrichment in a non-selective broth, such as BPW or Universal Pre-enrichment Broth (Zhao and Doyle, 2001), may be necessary for detecting heat-, freeze-, acid- or salt-stressed cells. For selective enrichment, an incubation temperature of 41–42 °C is normally used. This range is critical: lower temperatures will result in reduced selectivity. Unlike other *E. coli* strains, the growth of O157 is often much less at 44–45 °C.

Immunomagnetic separation (IMS), following selective enrichment and subsequent spread-plating of the concentrated target cells onto a selective plating medium, appears to be the most sensitive and cost-effective method for isolating *E. coli* O157 from raw foods. Immunocapture-based separation and concentration techniques include immunological binding (capture), followed by physical separation of the target organisms from a mixed enrichment culture, and result in concentration of the cells. IMS increases sensitivity by concentrating *E. coli* O157 relative to background organisms that may mimic or overgrow the target cells on a solid selective medium. Characteristic features of most strains of this organism, compared with other *E. coli* strains, are the absence of sorbitol fermentation within 24 hours and the lack of β-glucuronidase (GUD) activity. Detection of the organism has been greatly facilitated by the use of sorbitol–MacConkey (SMAC) agar, and the selectivity of SMAC agar has been improved by adding cefixime and tellurite (CT-SMAC). Sorbitol-negative colonies, indicative of typical strains of O157, are colourless on this medium. Since some strains are sensitive to tellurite and/or are sorbitol fermenting, the use of a second isolation medium, such as one of the newer chromogenic media, is recommended. Such media exploit the characteristic absence of GUD activity in almost all strains of O157. The chromogenic compound 5-bromo-4-chloro-3-indolyl-β-D-glucuronide is usually included in these media.

The international standard, ISO 16654, specifies a horizontal method for detecting *E. coli* O157 (ISO, 2001), as shown in Fig. 10.3. Selective enrichment involves modified tryptone–soya broth containing novobiocin (mTSB + N) for a minimum of six hours at 37 °C, and subsequently for a further 12–18 hours at the same temperature. Immunomagnetic separation is used to concentrate the target cells. Selective isolation requires plating on CT-SMAC agar and any other preferred selective medium. Sorbitol-negative colonies on CT-SMAC agar and colonies typical of O157 on the

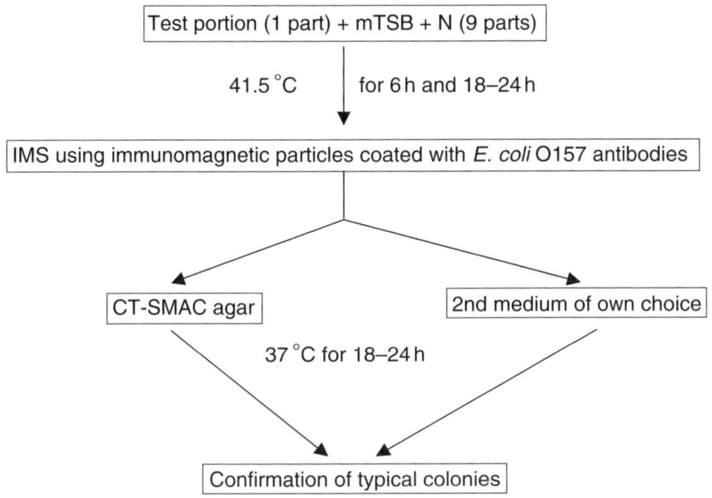

Fig. 10.3 Detection of *Escherichia coli* O157 in food according to ISO 16654 (ISO, 2001). mTSB + N, modified tryptone-soya broth with novobiocin; IMS, immunomagnetic separation; CT-SMAC, sorbitol–MacConkey agar plus cefixime and tellurite.

second medium are subjected to a test for indole formation to confirm the presence of *E. coli*, and then tested for agglutination with *E. coli* O157 antiserum. Specialised laboratories can do further tests for virulence determinants.

In addition to mTSB + N, as described in the standard method, many other enrichment broths are used in cultural methods for *E. coli* O157. These include tryptone–soya broth, *E. coli* (EC) broth and BPW supplemented with selective agents, such as novobiocin, acriflavin, bile salts, vancomycin, cefixime and cefsulodin. The efficacy of these media has been tested in several studies and shown to depend on the type of sample being examined and the enrichment conditions (De Boer and Heuvelink, 2000; Heuvelink, 2003). Modified EC broth with novobiocin is used successfully in many laboratories. Recently, a new 'acidic enrichment procedure' was described and it was shown to be effective for detecting O157 in a comparative study (Grant, 2005). This method includes enrichment under strongly acidic conditions to reduce the number of competitors, followed by raising the pH above 7.0 to permit optimum growth of the target organism in medium without inhibitors. Several chromogenic isolation media for O157 have been described (Manafi and Kremsmaier, 2001) and some of them, such as CHROMagar O157 (CHROMagar™–Dr A Rambach) and Rainbow™ Agar O157 (Biolog Inc.) have been manufactured commercially. The chromogenic media also need the addition of inhibitory compounds, such as cefixime, novobiocin and tellurite, to obtain sufficient selectivity. Variations in incubation temperature and the use of agar plates that have been stored for several weeks may cause fluctuations in typical colour development on chromogenic media.

Selective agar media for O157 have been compared in several studies. In most cases, CT-SMAC agar proved to be the most effective for isolating typical sorbitol-negative strains. However, other Enterobacteriaceae, such as some *Hafnia*, *Klebsiella* and *Enterobacter* strains, can give suspect colonies on CT-SMAC agar. Procedures for the cultural detection of *E. coli* O157 should include CT-SMAC agar and, preferably, a second isolation medium, based not on sorbitol fermentation but, for example, on the GUD reaction.

10.2.5 *Listeria monocytogenes*

Numerous procedures have been proposed for detecting *L. monocytogenes*. Since direct plating is often unsuccessful for isolating the organism from foods, the usual procedures consist of selective enrichment, followed by isolation on a selective agar. At present, there is no enrichment medium that selects *L. monocytogenes* from bacterial populations containing other *Listeria* spp. In all the various enrichment broths, *L. monocytogenes* does not grow well, when other (often faster-growing) *Listeria* spp. are present. The enrichment procedure for *L. monocytogenes* usually includes a primary and a secondary enrichment step. Primary or pre-enrichment broths contain reduced amounts of selective agents to allow resuscitation of injured cells. The commonly-used enrichment media contain acriflavin, nalidixic acid and cycloheximide as selective agents, together with a phosphate buffering system. A few media include lithium chloride as an additional selective agent and sodium pyruvate to favour the resuscitation of injured cells (Beumer and Curtis, 2003).

On many of the selective plating media, colonies of *L. monocytogenes* cannot be distinguished from those of other *Listeria* spp. This is the case with the frequently-used plating media, modified Oxford (MOX) and PALCAM. These contain aesculin and ferric citrate, and *Listeria* spp. show black zones around the colonies, due to the colour of hydrolysed aesculin. When *L. monocytogenes* is present among high levels of other *Listeria* spp., many colonies have to be tested to identify the target organism. Numerous attempts have been made to produce a solid selective medium for the specific isolation of *L. monocytogenes*, but this has proved difficult. Recently, some chromogenic media have been described that are selective for *Listeria* and also allow differentiation of *L. monocytogenes* from any other listerias present. These include Agar Listeria according to Ottaviani and Agosti (ALOA) and Rapid'L.Mono® medium (Biorad Laboratories Inc.). In ALOA medium, the chromogenic compound X-glucoside is added as a substrate for the detection of β-glucosidase activity, which is common in *Listeria*. The differentiation of *L. monocytogenes* from other *Listeria* spp.

is based on the production of a phosphatidylinositol-specific phospholipase C (PIPLC) by the former, which can hydrolyse the added substrate, resulting in an opaque, clear-cut halo surrounding the colonies. All non-pathogenic listerias produce colonies with a typical bluish appearance. ALOA was found to be superior to Oxford and PALCAM media, when samples containing both *L. monocytogenes* and *L. innocua* were examined (Vlaemynck *et al.*, 2000). Rapid'L.Mono™ medium is also based on the chromogenic detection of PIPLC, which is demonstrated by hydrolysis of x-inositol phosphate contained in the agar and blue staining of *L. monocytogenes* colonies, without a yellow halo (xylose-negative), as described by Lauer *et al.* (2005). In an evaluation of chromogenic media for detecting *Listeria* spp. in foods, ALOA, Rapid'L.Mono™ medium and Oxford agar were equally effective in recovering *L. monocytogenes*. However, recovery of other *Listeria* spp. was poorer on Rapid'L.Mono™ medium in comparison with Oxford and ALOA media (Greenwood *et al.*, 2005).

The protocol of the international standard, ISO 11290, for detecting *L. monocytogenes* in food and animal feeding stuffs is shown in Fig. 10.4 (ISO, 1996). This method includes a two-stage enrichment procedure, with streaking of cultures on ALOA and a second agar medium from the primary enrichment (half-strength Fraser broth) after one day of incubation, and from the secondary enrichment (Fraser broth) after incubation for two days. Recently, ALOA replaced Oxford and PALCAM media in this standard. One of the latter may be used as the second plating medium, for which there is a free choice. This standard method was found to be satisfactory for detecting *L. monocytogenes*, but may not be optimal for all types of food sample (Beumer and Curtis, 2003). In some cases, streaking of the

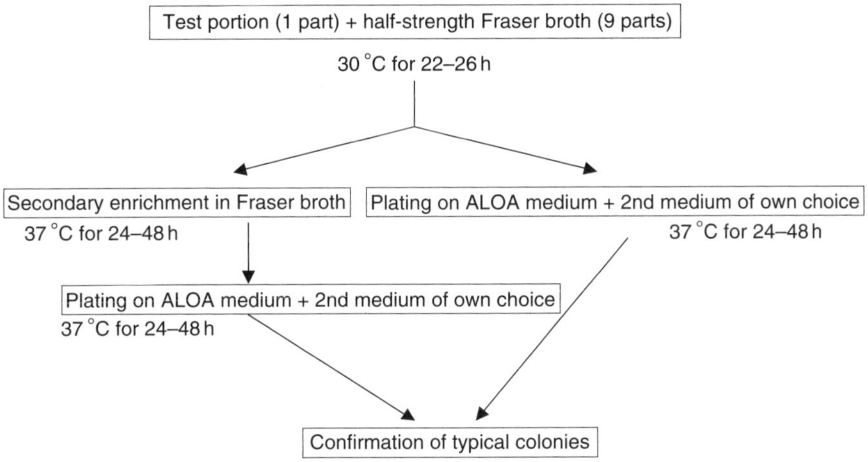

Fig. 10.4 Detection of *Listeria monocytogenes* in food according to ISO 11290–1 (ISO, 1996). ALOA, Agar Listeria according to Ottaviani and Agosti.

secondary enrichment broth after 48 hours appears to be unnecessary. ISO 11290-1 has been validated and showed an overall sensitivity of 85.6 %, with a specificity of 97.4 % (Scotter *et al.*, 2001a).

For confirmation of *Listeria* spp., typical colonies are tested for catalase (positive), Gram-staining (Gram-positive, slim, short rods) and motility (positive). If the morphological and physiological characteristics and the catalase reaction are indicative of *Listeria* spp., the strain is tested for haemolysis, hydrolysis of rhamnose and xylose, and the CAMP test, using *S. aureus* and *R. equi* as test strains. Table 10.5 shows typical reactions for the different *Listeria* spp. Several commercial kits are available for identifying *Listeria* spp. However, an additional test for haemolysis or use of the CAMP test is often necessary to confirm *L. monocytogenes*.

Enumeration of *L. monocytogenes* according to ISO 11290-2 includes the following stages:

(1) preparation of the initial sample suspension in BPW;
(2) resuscitation at 20 °C for one hour;
(3) surface plating of the sample suspension and subsequent decimal dilutions in duplicate onto ALOA medium;
(4) incubation of the plates at 37 °C and examination after 24 and 48 hours;
(5) confirmation of presumptive *L. monocytogenes*;
(6) calculation of the final number of *L. monocytogenes* (ISO, 1998; Scotter *et al.*, 2001b).

The method favoured by the USDA-FSIS (2005) for isolating *L. monocytogenes* from red meat, poultry, egg products and environmental samples involves primary enrichment in modified University of Vermont broth, secondary enrichment in Fraser broth and isolation on MOX agar.

The US FDA-BAM (FDA, 2003b) describes a method for detecting *L. monocytogenes* that involves non-selective enrichment in buffered *Listeria* enrichment broth at 30 °C for four hours, followed by addition of

Table 10.5 Differentiation of *Listeria* spp.

Species	Haemolysis	Production of acid from		CAMP test	
		Rhamnose	Xylose	*S. aureus*	*R. equi*
L. monocytogenes	+	+	–	+	–
L. innocua	–	V	–	–	–
L. ivanovii	+	–	+	–	+
L. seeligeri	+	–	+	+	–
L. welshimeri	–	V	+	–	–
L. grayi	–	V	–	–	–

+, positive reaction; –, negative reaction; V, variable reaction.

selective agents (acriflavin, nalidixic acid, cycloheximide) and continued incubation at 30 °C for a total of 48 hours. The enrichment culture is streaked at 24 and 48 hours onto aesculin-containing media, such as MOX, PALCAM or lithium chloride–phenylethanol–monolactam agar, with added aesculin and iron and, preferably, one of the newer chromogenic media.

The aesculin-containing media have been compared in several studies (Capita *et al.*, 2001b; Scotter *et al.*, 2001a; Beumer and Curtis, 2003). The results of these studies show that the media do not differ significantly in productivity, but results for each medium depend largely on the number and state of both target and non-target organisms in the sample. For optimum detection of *L. monocytogenes*, a combination of an aesculin-containing medium and a chromogenic medium is recommended (Leclercq, 2004).

10.3 Rapid methods

10.3.1 Introduction

The main limitation of conventional culture methods is that they are time consuming and take several days to yield results. There is a great need for more rapid methods of detecting foodborne pathogens. These methods would enable food producers to avoid releasing contaminated foods onto the market and allow them to sell products labelled as being free from specific pathogens.

Numerous rapid methods for detecting foodborne pathogens have already been developed and several have been commercialised. The two major categories of such methods are immunological or antibody-based assays and molecular or genetic methods, such as those involving the polymerase chain reaction (PCR). These methods will be discussed in the following paragraphs. Other types of rapid method used in food microbiology that will not be discussed separately include impedance techniques, bioluminescence, flow cytometry and direct epifluorescent microscopy (De Boer and Beumer, 1999). Of these techniques, only impedance methods are used for detecting pathogens, such as *Salmonella* (Blivet *et al.*, 1998) and *Campylobacter* (Moore and Madden, 2002), in foods. However, use of impediometry is limited, because of the need to re-calibrate the equipment for each food category and deficiencies in detecting low numbers of organisms or when the target organisms are injured (Wawerla *et al.*, 1999).

Preferably, the validation of new rapid methods should be done according to ISO 16140. This standard describes the general principles involved and gives a technical protocol for the validation of alternative methods in the field of microbiological analysis of food, animal feeding stuffs and environmental and veterinary samples (ISO, 2003). The availability of naturally-contaminated samples is often a problem in validation studies for methods relating to foodborne pathogens (Rijpens and Herman, 2002).

Approval of commercially-available test systems may be obtained from one of the certification bodies, such as MicroVal, AOAC, AFNOR or NORDVAL.

10.3.2 Immunological methods

Immunological methods rely on the specific interaction between an antibody and an antigen associated with the target organism. The choice and types of antigen and antibody are critical, as this will markedly affect the specificity of the method. The antigen may be a specific lipopolysaccharide on the outer cell wall, a protein on the flagella of certain motile bacteria or a product or toxin elaborated by the organism during growth (Baylis, 2003). The development of monoclonal antibodies has greatly improved the specificity of immunoassays. Immunoassay techniques are simple to carry out and offer the potential for high sample throughput. Most immunological tests require an inoculum of more than 10^4 cfu/ml for detecting specific organisms, making a preceding enrichment step necessary. Examples of commercially-available immunoassays for detecting foodborne pathogens are shown in Table 10.6. The choice of test system depends on the requirements of the user.

Agglutination tests

In agglutination tests, an antigen reacts with its corresponding antibody, resulting in visible clumping of bacterial cells. With latex agglutination tests, latex particles are coated with antibodies that agglutinate specific antigens and form a more easily visible precipitate. Agglutination tests are frequently used for initial confirmation of specific pathogens. Since antibodies to the target organism may cross-react with other organisms and autoagglutination may occur, these must be considered as screening tests and further confirmation will usually be necessary. Commercial agglutination tests are available for most pathogens (Microgen Bioproducts Ltd, Oxoid, Remel Inc. and others).

Enzyme-linked immunosorbent assay (ELISA)

The ELISA is a widely used immunological technique for detecting foodborne pathogens. Essentially, the technique involves the use of antibodies that are specific for a target molecule. The antibodies are immobilised on a solid support system to capture target antigens. The antigen–antibody complex is then detected by reaction with an antibody–enzyme conjugate and a chromogenic substrate. Microtitre plates are often used as solid supports, but paper membranes and polystyrene dipsticks can also be used. The sandwich ELISA is the most common form for commercially-available kits. In this technique, the target antibody is attached to the surface of a solid support, e.g. a microtitre well. An enriched food sample is added to the well and, if the target antigen is present, it will bind to the antibody.

Table 10.6 Commercially-available immunoassays for detecting pathogens in foods

Trade name (producer)	Organism*	Reference
Reveal® (Neogen Corporation)	S, E, L	Bird *et al.* (2001); Chapman and Ashton (2003)
EHEC-TEK, Salmonella-TEK (bioMérieux)	S, E	Van Poucke (1990); Blackburn and McCarthy (2000)
Assurance GDS™ (BioControl Systems Inc.)	S, C, E, L	Feldsine *et al.* (2002, 2005)
Singlepath® (Merck KGaA)	S, C, E, L	Hochel *et al.* (2004)
Transia Plate (Raisio Diagnostics Or)	S, C, E, L	Wicker *et al.* (2001)
RapidChek® (Strategic Diagnostics Inc.)	S, E, L	
Rapid Test (Oxoid)	S, L	Capita *et al.* (2001a)
VIA™ (Biotrace International)	S, C, E, L	Paula *et al.* (2002); Hughes *et al.* (2003); Briggs *et al.* (2004)
Salmonella Screen (VICAM)	S	Tan and Shelef (1999)
Bioline Salmonella Selecta (Bioline ApS)	S	Bolton *et al.* (2000)
Detex™ (Molecular Circuitry)	E	Henry *et al.* (2001)
VIDAS® (bioMérieux)	S, C, E, L	Borck *et al.* (2002); McMahon *et al.* (2004); Silbernagel *et al.* (2004)
EiaFoss™ (Foss)	S, C, E, L	Petersen and Madsen (2000); Borck and Pedersen (2005)
Pathatrix (Matrix MicroScience)	S, E, L	Arthur *et al.* (2005)
ImmunoCard STAT® (Meridian Bioscone Inc.)	E	Mackenzie *et al.* (2000); Chapman and Ashton (2003)
VIP® (BioControl Systems Inc.)	S, E, L	Chapman and Ashton (2003); Feldsine *et al.* (2005)

* S, *Salmonella*; C, *Campylobacter*; E, *E. coli* O157; L, *Listeria*.
This table does not give a complete overview of all commercial test systems and rapid changes in the availability of these systems may occur. The latest information given by the relevant companies should be followed. Websites of companies are mentioned in Section 10.6.

After washing to remove unbound material, a second antibody, containing an enzyme label, is added to the well. The antibody will bind to the target antigen, creating an antibody sandwich. After washing, a colourless substrate is added, and this is converted by the enzyme to a coloured product. Finally a 'stop' solution is added to prevent further enzyme activity and any colour change that has occurred is determined (McCarthey, 2003). ELISA tests normally take 1–2 hours before the results can be read. An initial enrichment step is usually necessary to increase the number of target organisms, resulting in a total test time of 20–24 hours. Pre-enrichment for repairing injured cells will require an additional 16–20 hours. ELISA test kits are available for *Salmonella*, *Campylobacter*, *E. coli* O157, *Listeria* and *L. monocytogenes* (Table 10.6).

Immunomagnetic separation (IMS)

In IMS methods, samples are mixed with immunomagnetic particles (or beads) coated with antibodies for the target organism. The target cells in the sample bind to the beads. Once captured and concentrated, the presence of target organisms can be determined using a variety of detection methods, such as direct plating on an agar medium, PCR, ELISA and DNA probes. The most important aspect of IMS is to capture and concentrate target bacteria, while eliminating interference from the food matrix. The use of IMS has been especially useful for detecting *E. coli* O157 in foods and is also included in the relevant ISO standard (ISO, 2001). Several commercial IMS systems are available, including Dynabeads® (Dynal Invitrogen) for *Salmonella*, *E. coli* O157 and *Listeria* (Shaw *et al.*, 1998; De Boer and Heuvelink, 2000), Captivate™ (IDG) O157 for *E. coli* O157 (Tutenel *et al.*, 2003) and A-Beads™ (Aureon Biosystems) for *Salmonella* and *E. coli* O157.

Dynabeads® anti-*Salmonella* and automated IMS, using a BeadRetriever™ (Dynal Invitrogen), have been used to detect *Salmonella* in cattle and on carcasses (Fegan *et al.*, 2005). Although IMS techniques provide simple, rapid, sensitive and low-cost methods for the isolation of target organisms from mixtures of bacteria, the food matrix often interferes with bacterial capture from a food sample by immunomagnetic particles; this may result in a low binding or capture efficiency (Varshney *et al.*, 2005).

The Pathatrix system is a novel, patented technology, which also relies on the use of antibody-coated paramagnetic particles that selectively bind and purify the target organisms from complex food matrices (Table 10.6). It analyses the entire sample by re-circulating it through a 'capture phase', in which the antibody-coated magnetic beads are immobilised. By providing heat to the system, the organism can be cultured and captured simultaneously, thus increasing the sensitivity of the method. Pathatrix, in combination with PCR, was shown to be a sensitive and rapid method for detecting *E. coli* O157 in ground beef (Arthur *et al.*, 2005).

In recent years, nanoparticles have opened up new dimensions in IMS, because of advantages over microbeads. Magnetic nanoparticles from 50 to 150 nm are 5–20-fold faster in reaction kinetics than 1–100 μm magnetic microbeads. Magnetic nanoparticles do not interfere with chemoluminescence and fluorescence in immunoassays, or with PCR reactions, and have the potential to be combined with bacterial detection techniques for the design of new immunoassay systems (Varshney *et al.*, 2005).

Other immunoassays

Several commercial immunoassays for detecting pathogens in foods are based on the immunochromatographic principle and make use of lateral-flow devices or dipsticks (Table 10.6: Reveal®, VIP®, RapidChek®, Singlepath®, ImmunoCard STAT®, *Listeria* Rapid Test, *Salmonella* Rapid Test). A lateral-flow device comprises a nitrocellulose membrane, onto which an antibody specific for the target antigen is immobilised. A small amount of enrichment culture is added to the sample port of the device and the liquid migrates by capillary diffusion through the membrane. During this flow, any target organism expressing a specific antigen binds to the relevant antibody, forming a complex. This complex further migrates towards the capture-binding protein, where it becomes immobilised and a visible line is formed in a viewing window, indicating the presence of the target antigen. A control line should be formed in a second viewing window (Baylis, 2003; Chapman and Ashton, 2003).

Recently, an immunochromatographic strip was developed for the rapid detection of *E. coli* O157 in enriched samples. The strip showed a visible signal within 10 minutes in the presence of O157, and this pathogen could be detected at a minimum level of 1.8 cfu/ml after enrichment (Jung *et al.*, 2005).

Capillary columns have also been used as a solid support system for immunoassays. A capillary immunoassay has been developed for rapid detection of *Salmonella* in foods. This consists of an antibody-modified capillary column for capturing the target bacteria and hosting the enzymatic reaction, and an optical or electrochemical detector for signal measurement (Kim *et al.*, 2005). Capillary columns are suitable for automation and the consumption of immunoreagents is minimised, because of the small size of the capillaries. The VIDAS® system and the EiaFoss™ system (Table 10.6) are automated immunoassays showing good results in detecting the major pathogens.

Although traditional culture methods for detecting non-O157 STEC are currently not available, some immunological tests have been developed for Shiga toxins, including the Premier® EHEC test (Meridian Bioscience Inc.) (Bettelheim and Beutin, 2003), Duopath® Verotoxins (Merck KGaA) (Park *et al.*, 2003), Ridascreen® Verotoxin (R-Biopharm Rhône Ltd.) (Bettelheim and Beutin, 2003), VTEC-RPLA Screen (Denka Seiken Co. Ltd.) (Bettelheim, 2001) and VTEC-RPLA (Oxoid).

10.3.3 Molecular methods

In the past 15–20 years, there has been a significant increase in the development of molecular biological methods for the detection and characterisation of pathogens in foods. The basis of these tests is a nucleic acid (DNA or RNA) target sequence, which is specific for the organism being sought. Unique nucleic acid sequences of bacterial species can be exploited to determine the presence of that species in a sample. Depending on the desired level of detection (genus, species, strain), different regions of the genome can be used as targets (Scheu *et al.*, 1998). Genes associated with virulence are often used for identifying pathogens, for example Shiga toxin-encoding genes in *E. coli*.

Nucleic acid hybridisation

This is the specific binding between a DNA or RNA molecule present in the target organism and a DNA probe that has a sequence complementary to that of the target organism. DNA probes usually contain 15–30 nucleotides. The specificity of a hybridisation assay is determined by the nucleotide sequence of the probe. In these molecular methods, the first step is usually lysis of the bacterial cells to free the nucleic acids, so that they can hybridise with the DNA probe. When the hybrid is formed, different detection techniques may be used. Radioactive and fluorescent probes allow direct detection of hybrids and are employed for a wide range of pathogens, usually to confirm the identity of cultured organisms. Indirect detection of target DNA involves enzyme reporters. Removal of unbound reporter probe can be achieved by immobilising either the target DNA or the probe, prior to hybridisation. Solid supports for immobilisation can be membranes or polymer particles. Detection of the reporter follows a washing step to remove unbound probe. A number of hybridisation assays for foodborne pathogens have been described (Hill, 1996) and some have been commercialised. The GENE-TRAK® and GeneQuence™ systems (Neogen Corporation) use pathogen-specific probes for DNA or RNA targets in the bacteria and a colorimetric system for detecting the specific probe-target hybrids; assays are available for *Salmonella*, *Campylobacter*, *E. coli* O157, *L. monocytogenes* and *S. aureus*. The AccuProbe® *Campylobacter* system (Gen-Probe) is a rapid DNA-probe test for identifying thermotolerant campylobacters. The use of DNA-probe hybridisation assays for identification purposes is limited, because the detection limit is 10^4–10^5 bacterial cells, which indicates the need for a selective enrichment step (Smith *et al.*, 2000). DNA-probe hybridisation may be combined with the PCR for sensitive and specific pathogen detection.

Polymerase chain reaction

Because of their greater sensitivity, DNA-based methods that involve an amplification step are often used. The best-known method of amplification is the PCR technique. PCR methods have the potential for rapid and sensitive detection of foodborne pathogens. In these methods, double-stranded DNA is first denatured into single strands and specific, short DNA fragments (primers) are annealed to these strands, followed by extension of primers complementary to the single-stranded DNA with a thermostable DNA polymerase (Fig. 10.5). The end product, or amplicon, is traditionally visualised as a band on an ethidium bromide-stained electrophoresis gel.

A potential problem with the technique is the presence of PCR inhibitors in some food samples, which may result in false-negative results. The degree of inhibition depends strongly on the type of food. Several methods of sample preparation have been described, including filtration, centrifugation, enzyme treatment, sample dilution, immunomagnetic separation

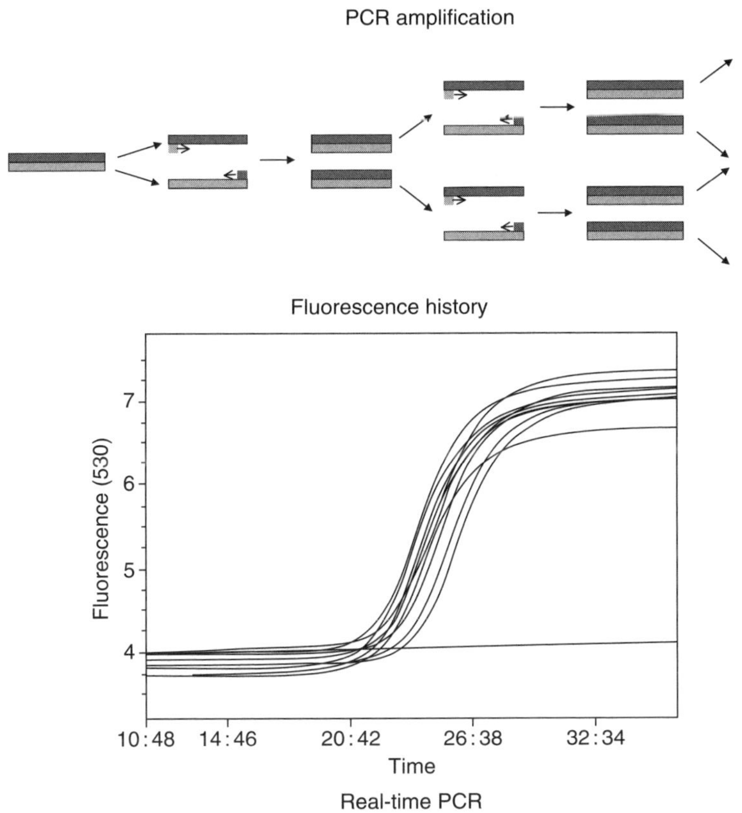

Fig. 10.5 Polymerase chain reaction (PCR) and real-time PCR.

(Smith *et al.*, 2000) and flotation (Wolffs *et al.*, 2004). Since there are many different food matrices, it is difficult to find universal DNA extraction procedures that are suitable for all types of food. Each food matrix presents its own challenges, depending on composition, and these must be addressed on a case-by-case basis (McKillip and Drake, 2004).

Another potential limitation of PCR is the detection of DNA derived from non-viable organisms (Wolffs *et al.*, 2005). Enrichment of the food sample prior to PCR amplification significantly reduces the likelihood of detecting non-viable organisms and also increases the number of live target cells, while reducing any inhibitory effects of the food matrix. However, by including an overnight enrichment step, the detection time is significantly increased. In recently-published PCR protocols, shorter enrichment periods have been suggested (Croci *et al.*, 2004). To be reliable, PCR primers and reaction conditions must be thoroughly evaluated and optimised, appropriate sample preparation procedures must be developed and the use of several controls is essential to monitor possible contamination or inhibition of the reaction (Rijpens and Herman, 2002; Hoorfar *et al.*, 2004, Table 10.7). Because of a possible non-homogeneous distribution of temperature conditions in the thermal cycler, it is recommended that two positive controls are used in each test: one placed at the centre of the block and the other towards the edge (De Medici *et al.*, 2003). Further confirmation of amplicon identity is necessary and can be obtained by a nested PCR reaction, restriction-enzyme digest analysis, hybridisation assay or sequence analysis.

PCR can target unique genetic sequences, such as bacterial virulence genes. Various PCR assays specific for different genes have been reported, such as *Salmonella* genes *invA*, 16SrRNA, *phoP*, *phoE* and *fimA*, *L. monocytogenes* genes *hly*, *prfA* and *iap* and *E. coli* O157 genes *eaeA*, *stx-1* and *stx-2* (Kawasaki *et al.*, 2005).

Table 10.7 Most important test controls in PCR assays for pathogens (Malorny *et al.*, 2003b; Hoorfar *et al.*, 2004; ISO, 2005a)

Positive process control	A negative sample, spiked with the target pathogen and processed in the same way as the test samples
Negative process control	A negative sample, which is processed in the same way as the test samples
Internal amplification control	Non-target DNA added to each reaction mixture; this is amplified simultaneously with the target DNA, resulting in an amplicon size distinguishable from that of the target DNA
Positive PCR control	Reaction mixture containing the target DNA in a defined amount or copy number
Negative PCR control	Reaction carried out with DNA-free water, without PCR inhibitors

The use of RNA-based detection methods, such as reverse-transcriptase polymerase chain reaction (RT-PCR) and nucleic acid sequence-based amplification (NASBA) (Uyttendaele *et al.*, 1997), might restrict detection to viable bacteria only. These techniques are based on the conversion of RNA to copy DNA (cDNA), using a reverse transcriptase. However, RNA is a very labile molecule that is quickly and easily degraded, particularly once the organism is dead. This property makes it much more difficult to handle RNA than DNA.

Real-time PCR

Since the introduction of real-time PCR in the mid-1990s, the technique has found many microbiological applications. Standard PCR assays are end-point analyses in which amplification products are detected at the end of the process. In real-time PCR, the PCR products are detected and monitored during amplification in the same reaction vessel, with the help of fluorescent compounds (Fig. 10.5). As the number of targets increases during amplification, fluorescence also increases. By observing the point at which fluorescence crosses a threshold level (Ct value), a cycle number can be determined for samples with different initial DNA concentrations. If the initial concentration is high, the threshold level will be crossed earlier than at a low concentration. By measuring the Ct value for samples containing known concentrations, standard curves can be constructed. These can then be used for quantification, a technique that has been applied to several different sample types and targets (McKillip and Drake, 2004).

Two general approaches are used to obtain a fluorescent signal from synthesis of the target sequence (Saunders, 2004). The first uses a fluorescent dye, such as SYBR® Green I (Molecular Probes Invitrogen), that binds to double-stranded DNA and undergoes a conformational change, which results in an increase in fluorescence. The dye is only fluorescent when bound to DNA, and this binding characteristic is used to monitor the process of amplification, as a PCR product is generated. Non-specific binding and interference from primer–dimer products may occur at the end of the cycles. Sequence confirmation for the amplified product involves a melting-curve analysis, following the PCR. At the end of the PCR run, the temperature in the thermal chamber is raised slowly and the fluorescence in each tube is measured. As soon as the double-stranded DNA starts to denature, the SYBR® Green I dye is released, resulting in a decrease in fluorescence. Because each double-stranded DNA product has its own, characteristic melting temperature, depending on its length and guanine–cytosine content, melting-curve analysis can be compared with analysing a PCR product by length in gel electrophoresis.

The second approach is the use of fluorescent resonance energy transfer (FRET). Currently, there are a number of fluorescent-probe systems available. The most commonly-used are hydrolytic or TaqMan® probes (Roche Molecular Systems Inc.), hybridisation probes and molecular beacons. All these systems have been developed for a number of targets and systems. TaqMan® probes consist of a fluorescent reporter on the 5′ end and the quenching dye on the 3′ end. When the probe anneals to the amplicon, the *Taq* DNA polymerase cleaves the probe, separating the reporter dye from the quencher. As a result of probe hydrolysis, the intensity of fluorescence increases, being directly proportional to the concentration of amplicon. This system is available commercially as the TaqMan® system (Applied Biosystems).

Hybridisation probes use two specially-designed, sequence-specific probes for each amplicon, labelled with two different fluorescent dyes. These oligonucleotides are designed to lie head-to-tail when annealed. Detection is based on the generation of a fluorescent signal, when the two probes bind next to each other to the target sequence. This system is commercially available as the LightCycler® (Roche Diagnostics).

Molecular beacon probes are single-stranded nucleic acid molecules with a stem-and-loop structure, which quenches fluorescence when the probe is alone in solution. In the presence of complementary target DNA or RNA, the beacon unfolds and hybridises and fluorescence is detected (Goel *et al.*, 2005; Hanna *et al.*, 2005). Currently, only a few applications of molecular-beacon probes have been described for detecting pathogens in foods (Chen *et al.*, 2000; McKillip and Drake, 2000). Molecular beacons are more costly than most other FRET-based methods, a feature that may delay extended use for rapid pathogen detection (McKillip and Drake, 2004).

Several real-time PCR instruments are currently available, including ABI-PRISM® 7000 (Applied Biosystems), RotorGene™ 3000 (Corbett Research), iCycler iQ® (Bio-Rad Laboratories) and LightCycler®. Most of these instruments offer 2–4 optical channels to allow for multiplexing capability, and interactive software for user-friendly data analyses, including quantitative PCR, melting curves, etc. (McKillip and Drake, 2004). Due to the complexity of the ingredients in food samples, the applicability of real-time PCR for pathogen detection needs to be verified, and sample preparation procedures must be optimised for each food commodity. In the past few years, a number of DNA primers and probes that are specific for foodborne pathogens have been developed and sample preparation procedures for certain types of food have been reported. New probe systems are constantly being developed, such as the Scorpions™ technology (DXS), which is a variant of the molecular beacon technique (Whitcombe *et al.*, 1999) and the use of D-LUX™ (Invitrogen Corporation) light upon extension primers (Knemeyer *et al.*, 2000). Examples of PCR and real-time PCR assays for detecting pathogens in foods are shown in Table 10.8. For the simultaneous detection of different pathogens, several multiplex real-time PCR assays have been developed, combining the detection of *Salmonella* and *L. monocytogenes* (Wang *et al.*, 2004), *Salmonella*, *L. monocytogenes* and *E. coli* O157:H7 (Kawasaki *et al.*, 2005) and *E. coli* O157:H7, *Salmonella* and *Shigella* (Li and Mustapha, 2004).

Table 10.8 Examples of PCR and real-time PCR assays for detecting bacterial pathogens in foods

Organism	Detection technique	Detection time (h)	Detection limit	Food matrix	Reference
Salmonella	IMS, real-time PCR	13	25 cfu/25 g	Ground beef	Mercanoglu and Griffiths (2005)
Salmonella	PCR	24	5 cells/25 g	Raw meat	Malorny *et al.* (2003a)
Salmonella	Real-time PCR	24	10^3–10^4 cfu/ml	Several foods	Malorny *et al.* (2004)
Salmonella	PCR	5	1–10 cells/25 g	Raw meat	Croci *et al.* (2004)
Salmonella	Real-time PCR	12	1 cell/ml	Meat products	Ellingson *et al.* (2004)
S. Enteritidis	Real-time PCR	24	1000 cfu/ml	Poultry	De Medici *et al.* (2003)
C. jejuni	Immunocapture PCR	8	1 cell/ml	Chicken skin	Waller and Ogata. (2000)
C. jejuni	Real-time PCR	22	1000 cfu/ml	Chicken rinse	Josefsen *et al.* (2004)
Campylobacter	Real-time PCR	24	10 cfu/ml	Poultry	Perelle *et al.* (2004)
C. jejuni	Real-time PCR	24	1 cfu/25 g	Chicken skin	Oliveira *et al.* (2005)
E. coli O157	IMS, real-time PCR	6	1600 cells/ml	Ground beef	Fu *et al.* (2005)
Y. enterocolitica	Real-time PCR, flotation	2	410 cfu/ml	Pork	Wolffs *et al.* (2004)

Microarrays
A microarray is an orderly arrangement of nucleic acid sequences on a platform, and this provides a medium for matching known and unknown sequences. An array of oligonucleotides is synthesised either *in situ* (on-chip) or by conventional means, followed by on-chip immobilisation. The array is exposed to labelled sample DNA and hybridised; then, the identity of complementary sequences is determined quantitatively. DNA microarray technology has shown its diagnostic value in areas of clinical and environmental microbiology and is likely to have potential for use in food microbiology to detect pathogens (Bodrossy and Sessitsch, 2004; McKillip and Drake, 2004; Anjum *et al.*, 2005). A DNA microarray for *Campylobacter* spp. was developed recently and has been used for direct detection of the organisms in chicken faeces (Keramas *et al.*, 2004).

Commercially-available, PCR-based detection systems
Several commercial test systems for detecting foodborne pathogens that are based on the PCR principle are currently available. These include the BAX® system (DuPont Qualicon) (Bailey and Cosby, 2003, Silbernagel *et al.*, 2003), Probelia® (Bio-Rad Laboratories) (Fach *et al.*, 1999; Wan *et al.*, 2003), LightCycler® Detection Kit (Roche Diagnostics) (Cheung *et al.*, 2004), TaqMan® Pathogen Kit (Applied Biosystems) (Cheung *et al.*, 2004) and Assurance GDS® (BioControl Systems Inc.) (Arthur *et al.*, 2005). In most of these systems, the PCR assay is simplified by combining all the reagents, including primers, enzymes and probes into one reagent mix. The PCR assays are automated and the operator does not require any skill in molecular biology. A recent study (Arthur *et al.*, 2005) showed that the LightCycler® *E. coli* (eae) Detection Kit, Assurance GDS™ for *E. coli* O157:H7 and BAX® System *E. coli* O157:H7 MP method did not differ in their ability to detect *E. coli* O157:H7 in ground beef. Culture-based systems detected more positive samples than those involving PCR, but the detection times (21–48 hours) were at least nine hours longer, compared to 7.5–12 hours. Further evaluation and comparison of systems is needed for various food matrices, to inform the choices made by routine-testing laboratories and ensure that the tests used meet their specific requirements.

Standardisation
Despite the rapidity with which PCR methods can detect foodborne pathogens, the food industry has been relatively slow to adopt the technique. One reason could be the lack of standard criteria for validating PCR sample-preparation methods, reaction components and assembly and amplification conditions for pathogens in different food matrices (McKillip and Drake, 2004). Standardised methods will encourage the use of PCR-based systems and their full acceptance alongside traditional diagnostic procedures in routine microbiology laboratories (Josefsen *et al.*, 2004).

Recently, international standards for qualitative PCR methods were developed within ISO and CEN, and these include general requirements and definitions for PCR (ISO, 2005a), requirements for sample preparation (ISO, 2005b), amplification and detection (ISO, 2005c) and performance testing of thermal cyclers (ISO, 2004). A standard on the general requirements for real-time PCR is in preparation.

As part of a European research project (FOOD-PCR), criteria for a standardised diagnostic PCR have been described (Malorny *et al.*, 2003b) and standardised PCR detection methods have been developed for thermotolerant campylobacters (Lübeck *et al.*, 2003), *Salmonella* (Malorny *et al.*, 2003a), *L. monocytogenes* (D'Agostino *et al.*, 2002) and *E. coli* O157 (Abdulmawjood *et al.*, 2003). The performance of these assays was evaluated in international collaborative trials. The diagnostic specificity and sensitivity of the methods were respectively 100 % and 96.7 % for *Campylobacter* in chicken carcass-rinse samples (Josefsen *et al.*, 2004), 97.5 % and 97.5 % for *Salmonella* in meat samples (Malorny *et al.*, 2003c), 100 % and 92.2 % for *E. coli* O157 in cattle carcass-swabs (Abdulmawjood *et al.*, 2004) and 81.8 % and 89.4 % for *L. monocytogenes* in raw milk (D'Agostino *et al.*, 2002).

10.4 Factors in the choice of technique

The main advantage of rapid tests, including immunoassays and PCR-based techniques, is their potential for quickly identifying samples that are negative for the pathogen(s) of interest. Positive samples still need to be verified by culture to obtain isolates for confirmatory tests, subtyping, antimicrobial-resistance testing, etc. Reassuring food processors that a contaminated product will not be released for commercial purposes requires the use of a reliable test system that does not give false-negative results. Equally, numbers of false-positive results must be low to prevent needless rejection of products. Particularly in the case of perishable products, the test method must yield rapid results. Factors to consider when choosing a test system include: sensitivity, specificity, reproducibility, standardisation, validation, speed, automation and computerisation, sample matrix, simplicity of sample preparation, costs of equipment, reagents, other consumables and technical support, throughput, flexibility, ease-of-use, space requirement for instruments and training of operatives (De Boer and Beumer, 1999). An evaluation of the most important factors in choosing culture-, immuno- or nucleic acid-based diagnostic techniques is shown in Table 10.9.

10.4.1 Culture techniques: advantages and disadvantages

Advantages

Culture techniques are very sensitive and inexpensive, and they can give both qualitative and quantitative information on the nature and number of the microorganisms present in a food sample.

Table 10.9 Factors affecting choice of analytical technique

	Culture	Immunoassay	Nucleic acid-based
Sensitivity	+++	++	+++
Specificity	+++	++	+++
Detection time	+	+++	+++
Analytical costs	+++	+	+
Labour needed	+	+++	++
Skills needed	++	+++	+
Automation	+	+++	+++

Scores: +++, high; ++, medium; +, low.

Disadvantages
Conventional culture methods require several days to yield results, because they rely on the ability of an organism to multiply and form visible colonies. These techniques are labour-intensive and require skilled laboratory staff.

10.4.2 Immunoassays: advantages and disadvantages

Advantages
Compared to conventional plating methods, the use of immunoassays can reduce the time needed to demonstrate a negative result to 24–30 hours. ELISA techniques usually take 1–2 hours to complete, but prior enrichment prolongs the total test time to approximately one day. The use of an ELISA reduces the labour requirement and allows a high sample throughput.

Disadvantages
The sensitivity of immunoassays is relatively low. The ELISA technique generally needs a test sample with approximately 10^4–10^5 bacteria/ml. Therefore, selective enrichment of the food sample is necessary to obtain the sensitivity needed. When injured cells of the target organism are likely to be present, inclusion of a non-selective pre-enrichment stage may significantly increase the detection rate, whilst increasing the total test time. Non-specific reactions can be caused by competing flora or by components of the food sample, and these lead to false-positive results. This is particularly likely when polyclonal antibodies are used in the ELISA. The costs for immunoassays are higher than those of corresponding culture tests.

10.4.3 Nucleic acid-based assays: advantages and disadvantages

Advantages
Nucleic acid-based tests for detecting microorganisms are precise, since they actually detect the genetic material of the organism and do not rely

on its physiological state, environmental influences or the expression of any particular surface antigens (Smith *et al.*, 2000). The ability to target different genes provides additional information for characterising the pathogen detected. This potential for further typing of the organism may give food producers and processors the opportunity to detect the source(s) of product contamination. DNA-based methods have shorter assay times than conventional techniques.

Disadvantages
Most PCR methods require a time consuming culturing step initially, due to the low number of organisms present in the sample, and additional steps to remove PCR inhibitors. The costs for equipment and reagents are relatively large and a high level of technical skill and/or support for the development of test protocols is required (Logan and Edwards, 2004). The international standardisation of nucleic acid-based methods for detecting pathogens is still in the initial phase. Unlike cultural techniques, such methods do not provide an isolate for further study or typing.

10.5 Future trends

Further improvement of detection methods will be necessary to ensure a safe food supply. Special attention should be given to the following aspects:

1. For the release of perishable products in commerce and testing of slaughter animals for the presence of specific pathogens, there is a need for more rapid methods that give results within a maximum of a few hours. These methods must show no false-negative results and very few false-positives or, better still, none.
2. Quantitative assessment of pathogens can be necessary, e.g. in monitoring the effects of control measures and in establishing human exposure in risk assessment studies. Quantifying some pathogens, such as *Salmonella* and *E. coli* O157, still depends on labour-intensive MPN methods. Better quantitative methods, including real-time PCR procedures, must be developed.
3. More information is needed on the prevalence of pathogens at different stages of the food chain, including the primary production sector. Current methods for detecting pathogens in foods may not be optimal for other matrices, such as animal faeces, feathers, environmental samples, etc. Recently, methods have been described for detecting *Campylobacter* in faeces (Lund *et al.*, 2003, 2004; Rudi *et al.*, 2004) and *E. coli* O157 in faeces and environmental samples (Ibekwe and Grieve, 2003).

4. Most methods are developed for detecting only one particular pathogen in a specified matrix. Very few are available for detecting several pathogens simultaneously.
5. In addition to testing for well-established pathogens, more attention should be given to emerging types not generally sought in routine analyses.
6. Standardisation of newly developed methods will stimulate their use. Recommendations have been made for harmonising PCR testing of animal faecal samples (Malorny and Hoorfar, 2005).
7. Numerous commercial test systems are available for detecting food-borne pathogens; however, the number of comparative studies on the performance of these systems, when used with different matrices, is currently insufficient to inform user selection properly.

For cultural methods, further improvement can be obtained by using the more recently developed chromogenic components in plating media, thus facilitating recognition of colonies of the target organism and reducing the need for confirmation. Other improvements could include the further development of commercial culture-based systems, such as Salmosyst® (Merck KGaA), S.P.R.I.N.T *Salmonella* (Oxoid), SimPlate® (BioControl Systems Inc.), FlexiPlate™ (Himedia Laboratories) and Compact Dry (Hyserve). There is a particular need for a universal pre-enrichment broth for the simultaneous enrichment of cells of injured pathogens belonging to different genera (Zhao and Doyle, 2001).

Easy-to-use, rapid agglutination tests for the identification of presumptive pathogens will remain of value in many routine laboratories (Baylis, 2003). Further automation of immunoassays will result in greater sample throughput and reduced labour requirements. Dipsticks available for rapid screening of pathogens in enrichment cultures by lateral migration of an antigen–antibody complex will continue to be developed (Fung, 2002). For rapid detection techniques, including immunoassays and molecular methods, there is a great need to replace traditional enrichment procedures by faster and more efficient techniques that separate and concentrate cells of the relevant pathogen to levels that allow direct detection in food samples. Several methods, including centrifugation and immunomagnetic separation, are used successfully and must be further developed and evaluated (Benoit and Donahue, 2003). Biosensors, too, have potential for detecting pathogens in food samples, but their sensitivity needs to be increased (Radke and Alocilja, 2005).

The current technology of real-time PCR can be improved, resulting in more sophisticated high-throughput platforms and portable devices. Improvement can also be achieved in probe design, leading to increased sensitivity and specificity. Combinations of techniques based on different principles may result in greater sensitivity. PCR-based systems employing an immunomagnetic step prior to detection produced fewer false-positive results (Arthur *et al.*, 2005).

Further development is needed for microarrays or 'DNA chips' and, in general, for molecular tests, so that they can become more user-friendly and suitable for application in routine laboratories.

10.6 Sources of further information

Useful reference books

CORRY J E L, CURTIS G D W and BAIRD R M (eds) (2003), *Handbook of Culture Media for Food Microbiology*, 2nd edition, Progress in Industrial Microbiology, vol. 37, Amsterdam, Netherlands, Elsevier.

EDWARDS K, LOGAN J and SAUNDERS N (eds) (2004), *Real-time PCR – an Essential Guide*, Wymondham, Norfolk, UK, Horizon Bioscience.

FRATAMICO P M, BHUNIA A K and SMITS J L (eds) (2005), *Foodborne Pathogens: Microbiology and Molecular Biology*, Wymondham, Norfolk, UK, Horizon Bioscience.

McMEEKIN T A (ed) (2003), *Detecting Pathogens in Food*, Cambridge, UK, Woodhead Publishing.

ROBERTS D and GREENWOOD M (eds) (2003), *Practical Food Microbiology*, 3rd edition, Oxford, UK, Blackwell Publishing.

Useful websites

Companies

3m: www.3m.com
Applied Biosystems: www.appliedbiosystems.com
Aureon Biosystems GmbH: www.aureonbio.com
Becton Dickinson: www.bd.com
BioControl Systems Inc.: www.rapidmethods.com
Bioline ApS: www.bioline.dk
Biolog Inc: www.biolog.com
bioMérieux sa: www.biomerieux.com
Bio-Rad Laboratories: www.bio-rad.com
Biotrace International: www.biotrace.co.uk
CHROMagar: www.chromagar.com
Corbett Research: www.corbettresearch.com
Denka Seiken Co. Ltd.: www.denka-seiken.co.jp
Dupont Qualicon: www.qualicon.com
DXS: www.dxsgenotyping.com
Dynal Invitrogen: www.invitrogen.com
Foss: www.foss.dk
Gen-Probe: www.gen-probe.com
Himedia Laboratories: www.himedialabs.com
Hyserve GmbH: www.hyserve.com
IDG: www.lab-m.com
Matrix MicroScience: www.matrixmsci.com
Merck KGaA: www.merck.de
Meridian Bioscience Inc: www.meridianbioscience.com
Microgen Bioproducts Ltd: www.microgenbioproducts.com
Molecular Probes Invitrogen: http://probes.invitrogen.com/
Neogen Corporation: www.neogen.com
Oxoid: www.oxoid.com

Raisio Diagnostics OY: www.raisiodiagnostics.com
R-Biopharm Rhône Ltd.: www.r-biopharmrhone.com
Remel Inc.: www.remel.com
Roche: www.roche.com
Strategic Diagnostics Inc.: www.sdix.com
VICAM: www.vicam.com

Organisations
AOAC: www.aoac.org
European Commission, DG Health and Consumer Protection: www.europa.eu.int/comm/dgs/health_consumer/
FDA Bacteriological Analytical Manual: www.cfsan.fda.gov/~ebam/
ISO: www.iso.org
USDA-FSIS: www.fsis.usda.

10.7 References

ABDULMAWJOOD A, BÜLTE M, COOK N, ROTH S, SCHÖNENBRÜCHER H and HOORFAR J (2003), 'Toward an international standard for PCR-based detection of *Escherichia coli* O157, Part 1. Assay development and multi-center validation', *Journal of Microbiological Methods*, **55**, 775–786.

ABDULMAWJOOD A, BÜLTE M, ROTH S, SCHÖNENBRÜCHER H, COOK N, HEUVELINK A E and HOORFAR J (2004), 'Development, validation, and standardization of polymerase chain reaction-based detection of *E. coli* O157', *Journal of AOAC International*, **87**, 596–603.

ANDREWS W H (1996), 'Evolution of methods for the detection of *Salmonella* in foods', *Journal of AOAC International*, **79**, 4–12.

ANJUM M F, MAROONEY C, FOOKES M, BAKER S, DOUGAN G, IVENS A and WOODWARD M J (2005), 'Identification of core and variable components of the *Salmonella* enterica subspecies I genome by microarray', *Infection and Immunity*, **73**, 7894–7905.

ARTHUR T M, BOSILEVAC J M, NOU X and KOOHMARAIE M (2005), 'Evaluation of culture- and PCR-based detection methods for *Escherichia coli* O157:H7 in inoculated ground beef', *Journal of Food Protection*, **68**, 1566–1574.

BAILEY J S and COSBY D E (2003), 'Detection of *Salmonella* from chicken rinses and chicken hot dogs with the automated BAX PCR system', *Journal of Food Protection*, **66**, 2138–2140.

BAYLIS C L (2003), 'Immunological techniques: immunochromatography, enzyme-linked immunofluorescent assays and the agglutination techniques', in *Detecting Pathogens in Food*, ed. McMeekin T A, Cambridge, UK, Woodhead Publishing, 217–240.

BAYLIS C L, MACPHEE S and BETTS R P (2000a), 'Comparison of methods for the recovery and detection of low levels of injured *Salmonella* in ice cream and milk powder', *Letters in Applied Microbiology*, **30**, 320–324.

BAYLIS C L, MACPHEE S, MARTIN K W, HUMPHREY T J and BETTS R P (2000b), 'Comparison of three enrichment media for the isolation of *Campylobacter* spp. from foods', *Journal of Applied Microbiology*, **89**, 884–891.

BAYLIS C L, HEUVELINK A E, HOFSTRA H and DE BOER E (2001), 'Practical considerations and difficulties associated with the detection of verocytotoxin-producing *Escherichia coli* (VTEC) in foods', in *Verocytotoxigenic E. coli*, eds Duffy G, Garvey P and McDowell D A, Trumbull, Connecticut, USA, Food & Nutrition Press, 57–89.

BENOIT P W and DONAHUE D W (2003), 'Methods for rapid separation and concentration of bacteria in food that bypass time-consuming cultural enrichment', *Journal of Food Protection*, **66**, 1935–1948.

BETTELHEIM K A (2001), 'Development of a rapid method for the detection of verocytotoxin-producing *Escherichia coli* (VTEC)', *Letters in Applied Microbiology*, **31**, 31–35.

BETTELHEIM K A and BEUTIN L (2003), 'Rapid laboratory identification and characterization of verocytotoxigenic (Shiga toxin producing) *Escherichia coli* (VTEC/STEC)', *Journal of Applied Microbiology*, **95**, 205–217.

BEUMER R R and CURTIS G D W (2003), 'Culture media and methods for the isolation of *Listeria monocytogenes*', in *Handbook of Culture Media for Food Microbiology*, 2nd edition, eds Corry J E L, Curtis G D W and Baird R M, Progress in Industrial Microbiology, vol. 37, Amsterdam, The Netherlands, Elsevier, 79–90.

BIRD C B, HOERNER R J and RESTAINO L (2001), 'Comparison of the Reveal 20 Hour Method and the BAM culture method for the detection of *Escherichia coli* O157:H7 in selected foods and environmental swabs: collaborative study', *Journal of AOAC International*, **84**, 737–751.

BLACKBURN C DE W and McCARTHY J D (2000), 'Modifications to methods for the enumeration and detection of injured *Escherichia coli* O157:H7 in foods', *International Journal of Food Microbiology*, **55**, 285–290.

BLIVET D, SALVAT G, HUMBERT F and COLIN P (1998), 'Development of a new culture medium for the rapid detection of *Salmonella* by indirect conductance measurements', *Journal of Applied Microbiology*, **84**, 399–403.

BODROSSY L and SESSITSCH A (2004), 'Oligonucleotide microarrays in microbial diagnostics', *Current Opinions in Microbiology*, **7**, 245–254.

BOLTON F J, FRITZ E, POYTON S and JENSEN T (2000), 'Rapid enzyme-linked immunoassay for detection of *Salmonella* in food and feed products: performance testing programme', *Journal of AOAC International*, **83**, 299–303.

BORCK B and PEDERSEN K (2005), 'Pulsed-field gel electrophoresis types of *Campylobacter* spp. in Danish turkeys before and after slaughter', *International Journal of Food Microbiology*, **101**, 63–72.

BORCK B, STRYHN H, ERSBOLL A K and PEDERSEN K (2002), 'Thermophilic *Campylobacter* spp. in turkey samples: evaluation of two automated enzyme immunoassays and conventional microbiological techniques', *Journal of Applied Microbiology*, **92**, 574–582.

BRIGGS J, DAILIANIS A, HUGHES D and GARTHWAITE I (2004), 'Validation study to demonstrate the equivalence of a minor modification (TECRA ULTIMA protocol) to AOAC Method 998.09 (TECRA *Salmonella* Visual Immunoassay) with the cultural reference method)', *Journal of AOAC International*, **87**, 374–379.

CAPITA R, ALONSO-CALLEJA C, MORENO B and GARCÍA-FERNÁNDEZ M C (2001a), 'Occurrence of *Listeria* species in retail poultry meat and comparison of a cultural/immunoassay for their detection', *International Journal of Food Microbiology*, **65**, 75–82.

CAPITA R, ALONSO-CALLEJA C, PRIETO M, DEL CAMINO GARCÍA-FERNÁNDEZ M and MORENO B (2001b), 'Comparison of PALCAM and modified Oxford plating media for isolation of *Listeria* species in poultry meat following UVM II or Fraser secondary enrichment broths', *Food Microbiology*, **18**, 555–563.

CHAPMAN P A and ASHTON R (2003), 'An evaluation of rapid methods for detecting *Escherichia coli* O157 on beef carcasses', *International Journal of Food Microbiology*, **87**, 279–285.

CHEN W, MARTINEZ G and MULCHANDANI A (2000), 'Molecular beacons: a real-time polymerase chain reaction assay for detecting *Salmonella*', *Analytical Biochemistry*, **280**, 166–172.

CHEUNG P-Y, CHAN C W, WONG W, CHEUNG T L and KAM K M (2004), 'Evaluation of two real-time polymerase chain reaction pathogen detections kits for *Salmonella* spp. in food', *Letters in Applied Microbiology*, **39**, 509–515.

CORRY J E L, ATABAY H I, FORSYTHE S J and MANSFIELD L P (2003), 'Culture media for the isolation of campylobacters, helicobacters and arcobacters', in *Handbook of Culture Media for Food Microbiology*, 2nd Edition, eds Corry J E L, Curtis G D W and Baird R M, Progress in Industrial Microbiology, vol. 37, Amsterdam, Netherlands, Elsevier, 271–316.

CROCI L, DELIBATO E, VOLPE G, DE MEDICI D and PALLESCHI G (2004), 'Comparison of PCR, Electrochemical Enzyme-Linked Immunosorbent Assays, and the standard culture method for detecting *Salmonella* in meat products', *Applied and Environmental Microbiology*, **70**, 1393–1396.

D'AGOSTINO M, WAGNER M, VAZQUEZ-BOLAND J A, KUCHTA T, KARPISKOVA R, HOORFAR J, NOVELLA S, SCORTTI M, ELLISON J, MURRAY A, FERNANDES I, KUHN M, PAZLAROVA J, HEUVELINK A and COOK N (2002), 'A validated PCR-based method to detect *Listeria monocytogenes* using raw milk as a food model – Towards an international standard', *Journal of Food Protection*, **67**, 1646–1655.

DAVIES P R, TURKSON P K, FUNK J A, NICHOLS M A, LADELY S R and FEDORKA-CRAY P J (2000), 'Comparison of methods for isolating *Salmonella* bacteria from faeces of naturally infected pigs', *Journal of Applied Microbiology*, **89**, 169–177.

DE BOER E (1998), 'Update on media for isolation of Enterobacteriaceae from foods', *International Journal of Food Microbiology*, **45**, 43–53.

DE BOER E and BEUMER R R (1999), 'Methodology for detection and typing of foodborne microorganisms', *International Journal of Food Microbiology*, **50**, 119–130.

DE BOER E and HEUVELINK A E (2000), 'Methods for the detection and isolation of Shiga toxin producing *Escherichia coli*', *Journal of Applied Microbiology; Symposium Supplement*, **88**, 133S–143S.

DE BOER E and HUMPHREY T J (1991), 'Comparison of methods for the isolation of thermophilic campylobacters from chicken products', *Microbial Ecology in Health and Disease*, **4** (Special Issue) S43.

DE BOER E, VAN BEEK P and PELGROM K (1998), 'Comparison of culture media for the isolation of campylobacters from chicken meat', in *Campylobacter, Helicobacter and Related Organisms*, eds Lasticova A J, Newell D G and Lasticova E E, Proceedings 9th International Workshop, Cape Town, 1997, University of Cape Town, South Africa, 370–372.

DE MEDICI D, CROCI L, DELIBATO E, DI PSQUALE S, FILETICI E and TOTI L (2003), 'Evaluation of DNA extraction methods for use in combination with SYBR Green I real-time PCR to detect *Salmonella* enterica serotype Enteritidis in poultry', *Applied and Environmental Microbiology*, **69**, 3456–3461.

ELLINGSON J L E, ANDERSON J L, CARLSON S A and SHARMA V K (2004), 'Twelve hour real-time PCR technique for the sensitive and specific detection of *Salmonella* in raw and ready-to-eat meat products', *Molecular and Cellular Probes*, **18**, 51–57.

EU (2005), 'Commission Regulation (EC) No 2073/2005 of 15 November on microbiological criteria for foodstuffs', *Official Journal of the European Union*, **L338**, 1–26.

FACH P, DILASSER F, GROUT J and TACHE J (1999), 'Evaluation of a polymerase chain reaction-based test for detecting *Salmonella* spp. in food samples: Probelia *Salmonella* spp.', *Journal of Food Protection*, **62**, 1387–1393.

FEGAN N, VANDERLINDE P, HIGGS G and DESMARCHELIER P (2005), 'A study of the prevalence and enumeration of *Salmonella* enterica in cattle and on carcasses during processing', *Journal of Food Protection*, **68**, 1147–1153.

FELDSINE P T, KERR D E, LEUNG S C, LIENAU A H, MILLER S M and MUI L A (2002), 'Assurance enzyme immunoassay eight hour method for detection of enterohemorrhagic *Escherichia coli* O157:H7 in raw and cooked beef (modification of AOAC Official Method 996.10): collaborative study', *Journal of AOAC International*, **85**, 1037–1044.

FELDSINE P T, GREEN S T, LIENAU A H and KERR D E (2005), 'Comparative validation study to demonstrate the equivalence of a minor modification to AOAC methods 996.09, Vip for EHEC and 996.10, assurance Eia EHEC with the reference culture method for the detection of *Escherichia coli* O157:H7 in beef', *Journal of AOAC International*, **88**, 1193–1196.

FDA (2001), Food and Drug Administration, *Bacteriological Analytical Manual*, '*Campylobacter*', Chapter 7, www.cfsan.fda.gov/~ebam/.

FDA (2003a), Food and Drug Administration, *Bacteriological Analytical Manual*, '*Salmonella*', Chapter 5, www.cfrsan.fda.gov/~ebam/.

FDA (2003b), Food and Drug Administration, *Bacteriological Analytical Manual*, 'Detection and enumeration of *Listeria monocytogenes* in foods', Chapter 10, www.cfrsan.fda.gov/~ebam/.

FU Z, ROGELJ S and KIEFT T L (2005), 'Rapid detection of *Escherichia coli* O157 by immunomagnetic separation and real-time PCR', *International Journal of Food Microbiology*, **99**, 47–57.

FUNG D Y C (2002), 'Predictions for rapid methods and automation in food microbiology', *Journal of AOAC International*, **85**, 1000–1002.

GOEL G, KUMAR A, PUNIYA A K, CHEN W and SINGH K (2005), 'Molecular beacon: a multitask probe', *Journal of Applied Microbiology*, **99**, 435–442.

GRANT M A (2005), 'Comparison of new enrichment procedure for shiga toxin-producing *Escherichia coli* with five standard methods', *Journal of Food Protection*, **68**, 1593–1599.

GREENWOOD M, WILLIS C, DOSWELL P, ALLEN G and PATHAK K (2005), 'Evaluation of chromogenic media for the detection of *Listeria* species in food', *Journal of Applied Microbiology*, **99**, 1340–1345.

HANNA S E, CONNOR C J and WANG H H (2005), 'Real-time Polymerase Chain Reaction for the food microbiologist: technologies, applications, and limitations', *Journal of Food Science*, **70**, R49–53.

HENRY Y M, NATRAJAN N and LAUER W F (2001), 'Detex for detection of *Escherichia coli* O157 in raw ground beef and raw ground poultry', *Journal of AOAC International*, **84**, 752–760.

HEUVELINK A E (2003), 'Review of media for the isolation of diarrhoeagenic *Escherichia coli*', in *Handbook of Culture Media for Food Microbiology*, 2nd edition, eds Corry J E L, Curtis G D W and Baird R M, Progress in Industrial Microbiology, vol. 37, Amsterdam, The Netherlands, Elsevier, 229–247.

HILL W E (1996), 'The polymerase chain reaction: applications for detection of food-borne pathogens', *Critical Reviews in Food Science and Nutrition*, **36**, 123–173.

HOCHEL I, VIOCHNA D, SKVOR J and MUSIL M (2004), 'Development of an indirect competitive ELISA for detection of *Campylobacter jejuni* subsp. *jejuni* O:23 in foods', *Folia Microbiology (Praha)*, **49**, 579–586.

HOORFAR J, MALORNY B, ABDULMAWJOOD A, COOK N, WAGNER M and FACH P (2004), 'Practical considerations in design of internal amplification controls for diagnostic PCR assays', *Journal of Clinical Microbiology*, **42**, 1863–1868.

HUGHES H D, DAILIANIS A, DUNCAN L, BRIGGS J, McKINTYRE D A and SILBERNAGEL K (2003), 'Modification of enrichment protocols for TECRA *Listeria* Visual Immunoassay method 995.22: collaborative study', *Journal of AOAC International*, **86**, 340–354.

HUMBERT F S, SALVAT G, LALANDE F and COLIN P (1997), 'Miniaturized Most Probable Number and enrichment serology technique for the enumeration of *Salmonella* spp. on poultry carcasses', *Journal of Food Protection*, **60**, 1306–1311.

IBEKWE A M and GRIEVE C M (2003), 'Detection and quantification of *Escherichia coli* O157:H7 in environmental samples by real-time PCR', *Journal of Applied Microbiology*, **94**, 421–431.

ISO (1996), ISO 11290–1: 1996, 'Microbiology of food and animal feeding stuffs – Horizontal method for the detection and enumeration of *Listeria monocytogenes* – Part 1: Detection method'. Added in 2004: Amendment 1: Modification of the isolation media and the haemolysis test, and inclusion of precision data, Geneva, Switzerland, International Organization for Standardization.

ISO (1998), ISO 11290–2: 1998, 'Microbiology of food and animal feeding stuffs – Horizontal method for the detection and enumeration of *Listeria monocytogenes* – Part 2: Enumeration method'. Added in 2004: Amendment 1: 'Modification of the enumeration medium', Geneva, Switzerland, International Organization for Standardization.

ISO (2001), ISO 16654: 2001, 'Microbiology of food and animal feeding stuffs – Horizontal method for the detection of *Escherichia coli* O157', Geneva, Switzerland, International Organization for Standardization.

ISO (2002), ISO 6579: 2002, 'Microbiology of food and animal feeding stuffs – Horizontal method for the detection of *Salmonella* spp.', Geneva, Switzerland, International Organization for Standardization.

ISO (2003), ISO 16140: 2003, 'Microbiology of food and animal feeding stuffs – Protocol for the validation of alternative methods', Geneva, Switzerland, International Organization for Standardization.

ISO (2004), ISO/TS 20836: 2004, 'Microbiology of food and animal feeding stuffs – Polymerase chain reaction (PCR) for the detection of food-borne pathogens – Performance testing of thermal cyclers', Geneva, Switzerland, International Organization for Standardization.

ISO (2005a), ISO 22174: 2005, 'Microbiology of food and animal feeding stuffs – Polymerase chain reaction (PCR) for the detection of food-borne pathogens – General requirements and definitions', Geneva, Switzerland, International Organization for Standardization.

ISO (2005b), ISO 20837: 2005, 'Microbiology of food and animal feeding stuffs – Polymerase chain reaction (PCR) for the detection of food-borne pathogens – Requirements for sample preparation for qualitative detection', Geneva, Switzerland, International Organization for Standardization.

ISO (2005c), ISO 20838: 2005, 'Microbiology of food and animal feeding stuffs – Polymerase chain reaction (PCR) for the detection of food-borne pathogens – Requirements for amplification and detection for qualitative methods', Geneva, Switzerland, International Organization for Standardization.

ISO (2006a), ISO 10272: 2006, 'Microbiology of food and animal feeding stuffs – Horizontal method for the detection and enumeration of *Campylobacter* spp. – Part 1: Detection method', Geneva, Switzerland, International Organization for Standardization.

ISO (2006b), ISO 10272: 2006, 'Microbiology of food and animal feeding stuffs – Horizontal method for the detection and enumeration of *Campylobacter* spp. – Part 2: Colony count technique', Geneva, Switzerland, International Organization for Standardization.

ISO (2007), ISO 6579: 2007, 'Microbiology of food and animal feeding stuffs – Horizontal method for the detection of *Salmonella* spp., Annex D: Detection of *Salmonella* spp. in animal faeces and in samples from the primary production stage', Geneva, Switzerland, International Organization for Standardization.

JOSEFSEN M H, LÜBECK P S, AALBÆK B and HOORFAR J (2002), 'Preston and Park-Sanders protocols adapted for semi-quantitative isolation of thermotolerant *Campylobacter* from chicken rinse', *International Journal of Food Microbiology*, **80**, 177–183.

JOSEFSEN M H, COOK N, D'AGOSTINO M, HANSEN F, WAGNER M, DEMNEROVA K, HEUVELINK A E, TASSIOS P T, LINDMARK H, KMET V, BARBANERA M, FACH P, LONCAREVIC S and HOORFAR J (2004), 'Validation of a PCR-based method for detection of food-borne thermotolerant campylobacters in a multicenter collaborative trial', *Applied and Environmental Microbiology*, **70**, 4379–4383.

JUNG B Y, JUNG S K and KWEON C H (2005), 'Development of a rapid immunochromatographic strip for detection of *Escherichia coli* O157', *Journal of Food Protection*, **68**, 2140–2143.

KAWASAKI S, HORIKOSHI N, OKADA Y, TAKESHITA K, SAMESHIMA T and KAWAMOTO S (2005), 'Multiplex PCR for simultaneous detection of *Salmonella* spp., *Listeria monocytogenes*, and *Escherichia coli* O157:H7 in meat samples', *Journal of Food Protection*, **68**, 551–556.

KERAMAS G, BANG D D, LUND M, MADSEN M, BUNKENBORG H, TELLEMAN P and CHRISTENSEN B V (2004), 'Use of culture, PCR analysis, and DNA microarrays for detection of *Campylobacter jejuni* and *Campylobacter coli* from chicken feces', *Journal of Clinical Microbiology*, **42**, 3985–3991.

KIM B, XIAO-LI S and LI Y (2005), 'Evaluation of a capillary immunoassay system for detection of *Salmonella* Typhimurium in poultry products, *Journal of Food Protection*, **68**, 1799–1803.

KNEMEYER J-P, MARME N and SAUER M (2000), 'Probes for detection of specific DNA sequences at the single-molecule level', *Analytical Chemistry*, **72**, 3717–3724.

LAUER W F, FACON J-P and PATEL A (2005), 'Evaluation of a chromogenic medium for identification and differentiation of *Listeria monocytogenes* in selected food', *Journal of AOAC International*, **88**, 511–517.

LECLERCQ A (2004), 'Atypical colonial morphology and low recoveries of *Listeria monocytogenes* strains on Oxford, PALCAM, Rapid'Lmono and ALOA solid media', *Journal of Microbiological Methods*, **57**, 251–258.

LI Y and MUSTAPHA A (2004), 'Simultaneous detection of *Escherichia coli* O157: H7, *Salmonella*, and *Shigella* in apple cider and produce by a multiplex PCR', *Journal of Food Protection*, **67**, 27–33.

LIGHTFOOT N F and MAIER E A (1998), *Microbiological Analysis of Food and Water: Guidelines for Quality Assurance*, Amsterdam, The Netherlands, Elsevier.

LINE J E, STERN N J, LATTUADA C P and BENSON S T (2001), 'Comparison of methods for recovery and enumeration of *Campylobacter* from freshly processed broilers', *Journal of Food Protection*, **64**, 982–986.

LOGAN J M J and EDWARDS K J (2004), 'An overview of real-time PCR platforms', in *Real-time PCR – an Essential Guide*, eds Edwards K, Logan J and Saunders N, Wymondham, Norfolk, UK, Horizon Bioscience, 13–29.

LÜBECK P S, WOLFFS P, ON S L W, AHRENS P, RÅDSTRÖM P and HOORFAR J (2003), 'Toward an international standard for PCR-based detection of food-borne thermotolerant campylobacters: assay development and analytical validation', *Applied and Environmental Microbiology*, **69**, 5664–5669.

LUND M, WEDDERKOPP A, WAINØ M, NORDENTOFT S, BANG D D, PEDERSEN K and MADSEN M (2003), 'Evaluation of PCR for detection of *Campylobacter* in a national broiler surveillance salmonellas in Denmark', *Journal of Applied Microbiology*, **94**, 929–935.

LUND M, NORDENTOFT S, PEDERSEN K and MADSEN M (2004), 'Detection of *Campylobacter* spp. in chicken fecal samples by real-time PCR', *Journal of Clinical Microbiology*, **42**, 5125–5132.

MACKENZIE A, ORRBINE E, HYDE L, BENOIT M, CHAN F, PARK C, ALVERSON J, LEMBKE A, HOBAN D and KENNEDY W (2000), 'Performance of the ImmunoCard STAT! *E. coli* O157:H7 test for detection of *Escherichia coli* O157:H7 in stools', *Journal of Clinical Microbiology*, **38**, 1866–1868.

MALORNY B and HOORFAR J (2005), 'Recommendations for the harmonization of PCR testing of fecal samples: Lessons from the detection of salmonellas in pigs', *Journal of Clinical Microbiology*, **43**, 3033–3037.

MALORNY B, HOORFAR J, BUNGE C and HELMUTH R (2003a), 'Multicenter validation of the analytical accuracy of *Salmonella* PCR: towards an international standard', *Applied and Environmental Microbiology*, **69**, 290–296.

MALORNY B, TASSIOS P T, RÅDSTRÖM P, COOK N, WAGNER M and HOORFAR P (2003b), 'Standardization of diagnostic PCR for the detection of foodborne pathogens', *International Journal of Food Microbiology*, **83**, 39–48.

MALORNY B, HOORFAR J, HUGAS M, HEUVELINK A, FACH P, ELLEBROEK L, BUNGE C, DORN C and HELMUTH R (2003c), 'Interlaboratory diagnostic accuracy of a *Salmonella* specific PCR-based method', *International Journal of Food Microbiology*, **89**, 241–249.

MALORNY B, PACCASSONI E, FACH P, BUNGE C, MARTIN A and HELMUTH R (2004), 'Diagnostic real-time PCR for detection of *Salmonella* in food', *Applied and Environmental Microbiology*, **70**, 7046–7052.

MANAFI M (2000), 'New developments in chromogenic and fluorogenic culture media', *International Journal of Food Microbiology*, **60**, 205–218.

MANAFI M and KREMSMAIER B (2001), 'Comparative evaluation of different chromogenic/fluorogenic media for detecting *Escherichia coli* O157:H7 in food', *International Journal of Food Microbiology*, **71**, 257–262.

McCARTHEY J (2003), 'Immunological techniques: ELISA', in *Detecting Pathogens in Food*, Ed. McMeekin, T A, Cambridge, UK, Woodhead Publishing, 241–258.

McKILLIP J L and DRAKE M (2000), 'Molecular beacon polymerase chain reaction detection of *Escherichia coli* O157:H7 in milk', *Journal of Food Protection*, **63**, 855–859.

McKILLIP J L and DRAKE M (2004), 'Real-time nucleic acid-based detection methods for pathogenic bacteria in food', *Journal of Food Protection*, **67**, 823–832.

McMAHON W A, SCHULTZ A M and JOHNSON R L (2004), 'Evaluation of VIDAS *Salmonella* (SLM) immunoassay method with Rappaport-Vassiliadis (RV) medium for detection of *Salmonella* in foods; collaborative study', *Journal of AOAC International*, **87**, 867–883.

MERCANOĞLU B and GRIFFITHS M W (2005), 'Combination of immunomagnetic separation with real-time PCR for rapid detection of *Salmonella* in milk, ground beef and alfalfa sprouts', *Journal of Food Protection*, **68**, 557–561.

MOORE J E and MADDEN R H (2002), 'Impediometric detection of *Campylobacter coli*', *Journal of Food Protection*, **65**, 1660–1662.

NAKAGAWA H, HARA-KUDO Y, KOJIMA T, IKEDO M, KODAKA H, KONUMA H and KUMAGAI S (2000), 'Detection of freeze-injured *Escherichia coli* O157:H7 cells from foods by resuscitation prior to selective enrichment', *International Journal of Food Microbiology*, **60**, 107–110.

NMKL (2005), 'Thermotolerant *Campylobacter*. Detection and enumeration in food', no. 119 rev., Esbo, Finland, Nordic Committee on Food Analysis.

OLIVEIRA T C R M, BARBUT S and GRIFFITHS M (2005), 'A robotic DNA purification protocol and real-time PCR for the detection of *Campylobacter* in foods', *Journal of Food Protection*, **68**, 2131–2135.

OYARZABAL O A, MACKLIN K S, BARBAREE J M and MILLER R S (2005), 'Evaluation of agar plates for direct enumeration of *Campylobacter* spp. from poultry carcass rinses', *Applied and Environmental Microbiology*, **71**, 3351–3354.

PARK C H, KIM H J, HIXON D L and BUBERT A (2003), 'Evaluation of the Duopath Verotoxin Test for detection of Shiga toxins in human stools', *Journal of Clinical Microbiology*, **41**, 2650–2653.

PAULA A M, GELLI D S, LANDGRAF M, DESTRO M T and FRANCO B D (2002), 'Detection of *Salmonella* in foods using Tecra *Salmonella* VIA and Tecra *Salmonella* UNIQUE rapid immunoassays and a cultural procedure', *Journal of Food Protection*, **65**, 552–555.

PAULSEN P, KANZLER P, HILBERT F, MAYRHOFER S, BAUMGARTNER S and SMULDERS F J M (2005), 'Comparison of three methods for detecting *Campylobacter* spp. in chilled or frozen meat', *International Journal of Food Microbiology*, **103**, 229–233.

PERELLE S, JOSEFSEN M, HOORFAR J, DILASSER F, GROUT J and FACH P (2004), 'A Lightcycler real-time PCR hybridisation probe assay for detecting food-borne thermophilic *Campylobacter*', *Molecular and Cellular Probes*, **18**, 321–327.

PETERSEN L and MADSEN M (2000), '*Listeria* spp. in broiler flocks: recovery rates and species distribution investigated by conventional culture and the EiaFoss method', *International Journal of Food Microbiology*, **58**, 113–116.

RADKE S M and ALOCILJA E C (2005), 'A high density microelectrode array biosensor for detection of *E. coli* O157:H7', *Biosensors and Bioelectronics*, **20**, 1662–1667.

RIJPENS N P and HERMAN L M F (2002), 'Molecular methods for identification and detection of bacterial food pathogens', *Journal of AOAC International*, **85**, 984–995.

ROBERTS D and GREENWOOD M (2003), *Practical Food Microbiology*, 3rd edition, Oxford, UK, Blackwell Publishing.

RUDI K, HØIDAL H K, KATLA T, JOHANSEN B K, NORDAL J and JAKOBSEN K S (2004), 'Direct real-time PCR quantification of *Campylobacter jejuni* in chicken fecal and cecal samples by integrated cell concentration and DNA purification', *Applied and Environmental Microbiology*, **70**, 790–797.

SAUNDERS N A (2004), 'An introduction to real-time PCR', in *Real-time PCR – an Essential Guide*, eds Edwards K, Logan J and Saunders N, Wymondham, Norfolk, UK, Horizon Bioscience, 1–12.

SCATES P, MORAN M and MADDEN R H (2003), 'Effect of incubation temperature on isolation of *Campylobacter jejuni* genotypes from foodstuffs enriched in Preston broth', *Applied and Environmental Microbiology*, **69**, 4658–4661.

SCHEU P M, BERGDORF K and STAHL U (1998), 'Detection of pathogenic and spoilage micro-organisms in food with the polymerase chain reaction', *Food Microbiology*, **15**, 13–31.

SCOTTER S L, LANGTON S, LOMBARD B, SCHULTEN S, NAGELKERKE N, IN 'T VELD P H, ROLLIER P and LAHELLEC C (2001a), 'Validation of

ISO method 11290 Part 1 – Detection of *Listeria monocytogenes* in foods', *International Journal of Food Microbiology*, **64**, 295–306.

SCOTTER S L, LANGTON S, LOMBARD B, LAHELLEC C, SCHULTEN S, NAGELKERKE N, IN 'T VELD P H and ROLLIER, P (2001b), 'Validation of ISO method 11290 Part 2 – Enumeration of *Listeria monocytogenes* in foods', *International Journal of Food Microbiology*, **70**, 121–129.

SHAW S J, BLAIS B W and NUNDY D C (1998), 'Performance of the Dynabeads anti-Salmonella system in the detection of *Salmonella* species in foods, animal feeds, and environmental samples', *Journal of Food Protection*, **61**, 1507–1510.

SILBERNAGEL K, JECHOREK R, CARVER C, BARBOUR W M and MROZINSKI P (2003), 'Evaluation of the BAX system for detection of *Salmonella* in selected foods: collaborative study', *Journal of AOAC International*, **86**, 1149–1159.

SILBERNAGEL K M, CARVER C N, JECHOREK R P and JOHNSON R L (2004), 'Evaluation of VIDAS *Listeria monocytogenes* II (LMO2) immunoassay method for the detection of *Listeria monocytogenes* in foods: collaborative study', *Journal of AOAC International*, **87**, 1123–1132.

SMITH T J, O'CONNOR L, GLENNON M and MAHER M (2000), 'Molecular diagnostics in food safety: rapid detection of food-borne pathogens', *Irish Journal of Agricultural and Food Research*, **39**, 309–319.

TAN W and SHELEF L A (1999), 'Automated detection of *Salmonella* spp. in foods', *Journal of Microbiological Methods*, **37**, 87–91.

TANGVATCHARIN P, CHANTHACHUM S, KOPAIBOON P, INTTASUNGKHA N and GRIFFITHS M W (2005), 'Comparison of methods for the isolation of thermotolerant *Campylobacter* from poultry', *Journal of Food Protection*, **68**, 616–620.

TUTENEL A V, PIERARD D, VANDEKERCHOVE D, VAN HOOF J and DE ZUTTER L (2003), 'Sensitivity of methods for the isolation of *Escherichia coli* O157 from naturally infected bovine faeces', *Veterinary Microbiology*, **94**, 341–346.

USDA-FSIS (2004) 'Laboratory guide book, MLG 4.03: Isolation and identification of *Salmonella* from meat, poultry and egg products', www.fsis.usda.gov/Science/Microbiological_Lab_Guidebook/index.asp.

USDA-FSIS (2005), 'Laboratory guide book, MLG 8.03: Isolation and identification of *Listeria monocytogenes* from red meat, poultry, egg and environmental samples', www.fsis.usda.gov/Science/Microbiological_Lab_Guidebook/index.asp.

UYTTENDAELE M, BASTIAANSE A and DEBEVERE J (1997), 'Evaluation of the NASBA nucleic acid amplification system for assessment of the viability of *Campylobacter jejuni*', *International Journal of Food Microbiology*, **37**, 13–20.

VAN DER ZEE H (2003), 'Media for the isolation of *Salmonella*', in *Handbook of Culture Media for Food Microbiology*, 2nd edition, eds Corry J E L, Curtis G D W and Baird R M, Progress in Industrial Microbiology, vol. 37, Amsterdam, Netherlands, Elsevier, 195–208.

VAN NETTEN P, VAN DER ZEE H and VAN MOOSDIJK A (1991), 'The use of a diagnostic semisolid medium for the isolation of *Salmonella* enteritidis from poultry', in *Quality of Poultry Products. III Safety and Marketing Aspects*, ed Mulder R W A W, Spelderholt Jubilee Symposia, Doorwerth, Netherlands, 59–66.

VAN POUCKE L S (1990), 'SALMONELLA-TEK, a rapid screening method for *Salmonella* species in food', *Applied and Environmental Microbiology*, **56**, 924–927.

VARSHNEY M, YANG L, XIAO-LI S and LI Y (2005), 'Magnetic nanoparticle-antibody conjugates for the separation of *Escherichia coli* O157:H7 in ground beef', *Journal of Food Protection*, **68**, 1804–1811.

VLAEMYNCK G, LAFARGE V and SCOTTER S (2000), 'Improvement of the detection of *Listeria monocytogenes* by the application of ALOA, a diagnostic, chromogenic isolation medium', *Journal of Applied Microbiology*, **88**, 430–441.

WALLER D F and OGATA S A (2000), 'Quantitative immunocapture PCR assay for detection of *Campylobacter jejuni* in foods', *Applied and Environmental Microbiology*, **66**, 4115–4118.

WALTMAN W D (2000), 'Methods for cultural isolation of *Salmonella*', in *Salmonella in Domestic Animals*, eds Wray C and Wray A, Oxford, UK, CABI Publishing, 355–372.

WAN J, KING K, FORSYTH S and COVENTRY M J (2003), 'Detection of *Listeria monocytogenes* in salmon using the Probelia polymerase chain reaction system', *Journal of Food Protection*, **66**, 436–440.

WANG X, JOTHIKUMAR N and GRIFFTIHS M W (2004), 'Enrichment and DNA extraction protocols for the simultaneous detection of *Salmonella* and *Listeria monocytogenes* in raw sausage meat with multiplex real-time PCR', *Journal of Food Protection*, **67**, 189–192.

WAWERLA M, STOLLE A, SCHALCH B and EISGRUBER H (1999), 'Impedance microbiology: applications in food hygiene', *Journal of Food Protection*, **62**, 1488–1496.

WHITCOMBE D, THEAKER J, GUY S P, BROWN T and LITTLE S (1999), 'Detection of PCR products using self-probing amplicons and fluorescence', *Natural Biotechnology*, **17**, 804–807.

WICKER C, GIORDANO M, ROUGIER S, SORIN M-L and ARBAULT P (2001), '*Campylobacter* detection in food using an ELISA-based method', *Journal of Medical Microbiology*, **291**, 31.

WOLFFS P, KNUTSSON R, NORLING, B and RÅDSTRÖM P (2004), 'Rapid quantification of *Yersinia enterocolitica* in pork samples by a novel sample preparation method, flotation, prior to real-time PCR', *Journal of Clinical Microbiology*, **42**, 1042–1047.

WOLFFS P, NORLING B and RÅDSTRÖM P (2005), 'Risk assessment of false-positive quantitative real-time PCR results in food, due to detection of DNA originating from dead cells', *Journal of Microbiological Methods*, **60**, 315–323.

WORCMAN-BARNINKA D, DESTRO M T, FERNANDES S A and LANDGRAF M (2001), 'Evaluation of motility enrichment on modified semi-solid Rappaport-Vassiliadis medium (MSRV) for the detection of *Salmonella* in foods', *International Journal of Food Microbiology*, **64**, 387–393.

ZHAO T and DOYLE M P (2001), 'Evaluation of Universal Pre-enrichment Broth for growth of heat-injured pathogens', *Journal of Food Protection*, **64**, 1751–1755.

11

Techniques for identifying foodborne microorganisms

S. M. Russell and P. C. Vasavada, University of Georgia, USA

11.1 Introduction

There is a growing need worldwide for rapid, accurate and reproducible techniques for identifying microorganisms that are associated with foods, such as red meat, poultry and eggs. The Joint Food and Agriculture Organisation/World Health Organisation Expert Committee on Food Safety stated that foodborne illness was the most widespread human health problem in the contemporary world and a major cause of lost productivity (FAO/WHO, 1984). In industrialised countries, it is estimated that up to 10 % of the population may suffer foodborne illness each year (Käferstein *et al.*, 1997). The situation is much worse in developing countries, where infant diarrhoea is often a cause of severe illness and death (http://www.cdc.gov/mmwr/preview/mmwrhtml/00018677.htm). Moreover, public health authorities in industrialised societies are now being challenged by new or newly emerging agents of foodborne illness. Some of these agents cause serious life-threatening diseases, especially now that a larger proportion of the population can be made up of people that are particularly vulnerable, e.g. elderly or immunocompromised individuals, including those receiving chemotherapy. Between 1988 and 1993, the number of cases of salmonellosis reported in Australia increased by 100 % and by 500 % in Japan (Käferstein *et al.*, 1997). Therefore the emphasis in this chapter is on techniques now available for the rapid characterisation and identification of foodborne pathogens. However, the same general approach is also relevant to the organisms responsible for product spoilage that are discussed in Chapters 5 and 9 although, in routine analyses, these are rarely identified definitively.

11.2 Purification of isolates

11.2.1 Composition of the growth medium

Before an isolate can be characterised and identified, it must be purified by plating on a non-selective agar medium, and subsequently grown in a liquid medium to provide an inoculum for the tests being undertaken. Microbial growth media should always contain a proper balance of available nutrients. Depending on the type of organisms being studied, the following ingredients may be included: (i) a suitable nitrogen source (peptone from animal or plant material is commonly used); (ii) yeast, meat or malt extract to provide growth factors that may be lost during manufacture of peptone and specific, essential amino acids; (iii) a carbohydrate (glucose usually) to provide energy; and (iv) a buffer to ensure that any acids produced during growth are neutralised and do not prevent further growth (Atlas, 1993; Holbrook, 2000).

11.2.2 Cultivation on an agar medium

To obtain a pure culture and make an initial assessment of an unknown isolate, the streak-plate method is normally used. The objective is to spread ever-decreasing numbers of cells over the surface of the agar in order to obtain discrete, well-separated colonies. This also enables the characteristics of the colonies to be noted as an aid to the identification process (Holbrook, 2000). Depending on the composition of the agar medium, the following may be observed for single colonies: colony shape and size, colour (e.g. pH change), pigment production, haemolysis, proteolysis, lipolysis, etc. A very popular method for streaking plates is to obtain a loopful of culture and use a tortuous streaking motion to spread it onto a small area of the plate towards the edge. Then, another sterile loop is wiped once through the area streaked originally and a tortuous streak is made in another, adjacent, small area of the plate towards the edge. This is done repeatedly until five separate streaks have been made, making sure not to cross over into any area with a heavier inoculum. The plate is incubated at the appropriate temperature for the organism in question (often that used originally for isolation). This procedure should result in excellent separation of colonies. Once purified, the organism may be transferred to an agar slope, incubated and stored for characterisation and identification at a later date.

11.3 Storage

11.3.1 Short-term storage

Most microbes of concern to the food industry may be maintained for short periods of time on slopes of nutrient-rich, double-strength plate count agar or tryptic–soy agar (TSA), held under refrigeration or at room temperature.

However, if cultures are to be maintained for longer than a few days, they must be kept in a different form. For enteric organisms like *Salmonella*, the US Centers for Disease Control (CDC) recommend blood agar base, TSA or brain–heart infusion agar (see: http://www.cdc.gov/ncidod/dbmd/diseaseinfo/cholera/ch10.pdf). Media containing carbohydrates, such as Kligler iron agar or triple-sugar–iron agar, should not be used, because the organisms will convert the carbohydrates to acid, which will then reduce the viability of the culture.

Agar slopes are prepared by autoclaving the medium in tubes and, while still in liquid form, leaning the tubes at an angle of 45° until solid. A culture is picked from the purification plate using a sterile loop and (i) stabbed deep into an agar slope once or twice, so that the loop reaches the butt of the tube, and then (ii) streaked onto the surface of the agar. The slope is incubated overnight at an appropriate temperature. The tube should be sealed with a screw-cap or bung and over-wrapped with Parafilm. Cultures should be stored at room temperature (22–25 °C) in the dark. Sterile mineral oil may be used to prevent the culture from drying out. When prepared in this way, the culture should remain viable for many months or even years.

11.3.2 Long-term storage

Lyophilisation (freeze-drying) has long been used to preserve bacterial cultures for prolonged storage. However, most food microbiology laboratories do not have easy access to a freeze-dryer. Therefore, a newer, easier method has been developed that utilises porous beads. To preserve a culture by this method, a well-isolated colony is removed from the purification plate with a sterile loop and mixed with the beads in a cryo-preservative solution. The beads are then frozen (preferably rapidly, using a –70 °C freezer) and stored in the frozen state. Asha *et al.* (2005) reported that *Clostridium difficile* could be recovered after storage on Protect Cryobeads™ (Hardy Diagnostics) for 1–2 years. The authors found that a high proportion of their isolates failed to survive under traditional storage conditions.

11.4 Preliminary examination

Once an organism has been isolated on an agar plate and purified, prior to testing, certain preliminary tests are needed to categorise the organism. This is essential, because the procedures used subsequently for characterising and identifying the isolate may differ significantly according to the Gram-stain reaction, presence or absence of oxidase and catalase activity, and mode of glucose dissimilation. Likewise, there are certain visual characteristics that should be taken into account, so that the relevant tests can be used to identify the organism. The following include examples of immediate observations that can be made by simply evaluating the morphology or reactions of colonies growing on agar plates, as described by Koneman *et al.* (1992).

11.4.1 Colony characteristics on an agar plate

The main variations in colony morphology are shown in Fig. 11.1. Other characteristics that may vary from one organism to another are:

- diameter (mm), where appropriate;
- colour: e.g. white, yellow, black, buff, orange, salmon-pink;
- surface: glistening, dull, ground-glass appearance;
- density: opaque, translucent, transparent;
- consistency: butyrous, viscid, membranous, brittle, hair-like.

11.4.2 Reactions on agar media

Even on a non-differential agar medium, such as plate count agar, some organisms produce characteristic pigments. Examples are the fluorescent

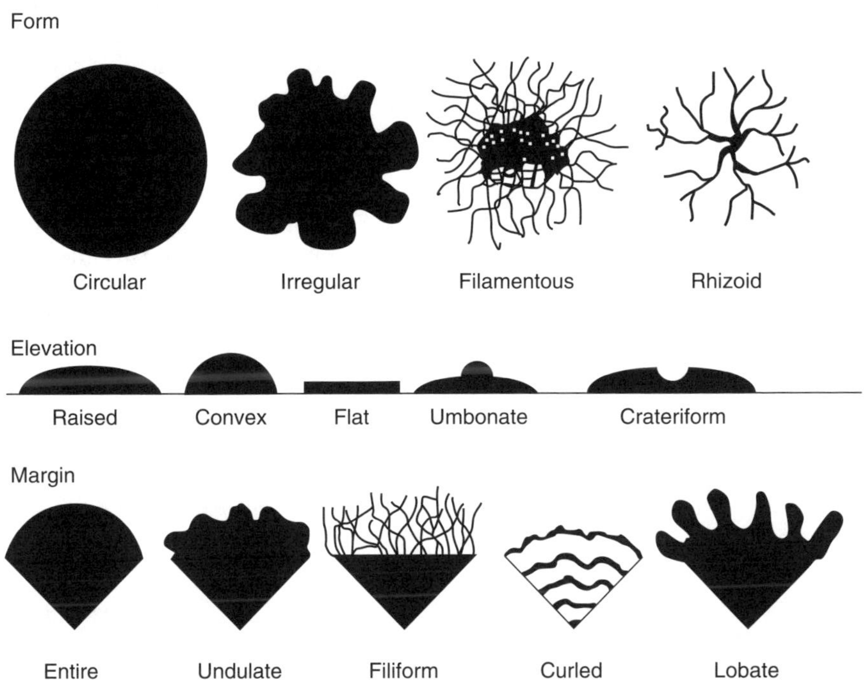

Fig. 11.1 Variations in colony morphology. (Source: http://www.austin.cc.tx.us/microbugz/03morphology.html)

pigment formed by some *Pseudomonas* spp., the blue–green pigment, pyocyanin, of *Ps aeroginosa* that diffuses into the agar and the non-diffusable, brown pigment produced by some strains of *Bacillus subtilis*, which is confined to the colonies. Sometimes, a substrate may be included to detect a specific enzyme reaction. On egg-yolk agar, for example, several types of reaction may be observed. These include a zone of precipitation around each colony due to lecithinase or an iridescent sheen over the colonies (pearly layer) and narrow zone of precipitation in the medium from lipase activity. Alternatively, a zone of clearing around the colony indicates proteolysis.

An organism growing on a specialised, differential medium may exhibit certain colour changes due to one or more metabolic by-products formed during growth. Dyes, pH indicators and enzyme-reaction indicators are often added to media to produce visual effects that assist in identifying specific organisms (Koneman *et al.*, 1992). The reader is also referred to the sections on chromogenic media below.

11.4.3 Odour production

Many microbial strains produce characteristic odours in culture media that may assist in identification, as shown in Table 11.1 (Koneman *et al.*, 1992). Although most organisms cannot be distinguished solely on the basis of odour production, different odours can reflect differences between species that are due to the production of specific metabolites.

11.4.4 Gram-stain reaction

Gram-staining is a time-honoured technique for separating bacteria into two major groups on the basis of their cell-wall composition. Broadly, this is indicted by the ability of some organisms, but not others, to retain a specific dye, following a decolourisation process. Those that retain the dye are termed Gram-positive, the others Gram-negative. There are many modifications of the technique, but all are time consuming. As a more rapid alternative, a colony can be mixed with a dilute solution of potassium

Table 11.1 Some organisms that produce characteristic odours in culture

Microorganism	Odour
Pseudomonas aeruginosa	Grape juice
Proteus spp.	Burned chocolate
Eikenella corrodens	Bleach
Alcaligenes faecalis	Freshly-cut apples
Streptomyces spp.	Musty basement
Clostridium spp.	Faecal, putrid

hydroxide and observed for a string of soapy material, which is used to indicate a Gram-negative organism. No such material is formed when a Gram-positive strain is tested in this way (Koneman *et al.*, 1992).

11.4.5 Catalase test

Many bacteria produce the enzyme catalase, which causes the release of oxygen from hydrogen peroxide. To determine catalase activity, one drop of 3 % hydrogen peroxide is placed directly onto the colony or onto a colony smeared on a microscope slide. If the organism is catalase-positive, effervescence will occur, as oxygen bubbles are liberated.

11.4.6 Cytochrome oxidase test

Cytochromes are key components of electron transport systems. To determine the presence of cytochrome oxidase, the culture is spread onto a paper strip impregnated with 1 % aqueous tetramethyl-p-phenylenediamine dihydrochloride or oxalate. Rapid production of a blue colour indicates a positive result. The test is particularly relevant to the differentiation of Gram-negative bacteria.

11.4.7 Mode of glucose utilisation

The test distinguishes between two basic types of carbohydrate metabolism. The organism is stab-inoculated into two tubes of a glucose-containing agar medium with a pH indicator. One tube is incubated in air, the other anaerobically or with an agar seal. An organism with an oxidative type of metabolism will utilise the glucose only in the presence of air. By contrast, a fermentative organism will produce a positive result in both tubes. Some organisms do not utilise carbohydrates as energy sources and thus both tubes will be negative.

11.5 Identification systems

11.5.1 Miniaturised test kits

Traditional methods for identifying pathogenic bacteria and the bacteria, yeasts and moulds responsible for food spoilage are extremely labour-intensive and time consuming. For example, detection and identification of a pathogen, such as *Salmonella*, using conventional methods, may require up to seven days (Andrews *et al.*, 1984). Plate counts of psychrotrophs require an incubation period of, e.g. 10 days at 7 °C (Gilliland *et al.*, 1984), even without attempting to identify any of the resultant organisms. Since a highly perishable product, such as raw poultry, should be shipped to market within 24 hours of processing (Anonymous, 1988), the food is likely to be consumed well before the test results become available. Therefore, numerous miniaturised and rapid test kits, as well as automated systems, have been developed to address this need.

API® kits

Among the various biochemical kits that have been developed for microbial identification, the API® kits (bioMérieux) are aimed specifically at the following: anaerobes (API 20A), *Campylobacter* spp. (API CAMPY), corynebacteria (Rapid CORYNE), Enterobacteriaceae (API 20E), Gram-negative non-Enterobacteriaceae (API NFT), *Lactobacillus* spp. (API 50CH), *Listeria* spp. (API LISTERIA), *Staphylococcus* and *Micrococcus* spp. (API Staph-IDENT), *Streptococcus* spp. (Rapid STREP) and yeasts (API 20C). To use one of these kits, the test organism is suspended in saline and a strip containing various biochemical substrates for assimilation tests is inoculated with the suspension. The strip is incubated at a specific temperature for a specified period of time, After incubation, the reactions are determined manually or the strip may be read automatically. Some reactions require the addition of particular reagents to yield a result. The whole series of reactions is used to obtain a number. By comparing the number to a database, the organism can be identified.

Gooch and Hill (1982) reported that the API 20E system correctly identified the relvant organisms to genus and species level in 90.2 % of cases, when compared with conventional methodology. Cox *et al.* (1984) found that the advantages associated with the API 20E system, as opposed to other rapid kits, are as follows: the tests are easy to inoculate and read, there is an extensive database and results are obtained relatively quickly. The disadvantages of the system are that test strips are difficult to stack in the incubator and to handle, and reactions are not always easy to interpret (Cox *et al.*, 1984).

Micro-ID® System

The Micro-ID® System (Remel Inc.) is a rapid method for identifying Enterobacteriaceae. Each test strip contains 20 filter-paper discs impregnated with reagents. An inoculum of the test bacterium is distributed among a series of wells, each of which contains one of the discs. The reagents present in the disc include a substrate that reacts with a bacterial enzyme and a system for detecting the metabolic product(s) of this reaction to yield an easily identifiable colour change within four hours of incubation. The series of reactions is used to obtain a number that is compared with a database to identify the organism.

Gooch and Hill (1982) observed that the Micro-ID® System successfully identified test organisms to genus and species level in 93.5 % of cases, in comparison with conventional tests. When compared with other rapid methods, Cox *et al.* (1984) reported that the Micro-ID® System was easier to inoculate, had a shorter incubation time, required little time for the addition of reagents and was relatively easy to read. A disadvantage was that the reactions were not always easy to interpret.

IDS RapID™ Systems

To identify microbes with the IDS RapID™ Systems (Remel Inc.), a colony of each test organism is mixed with an inoculation fluid to yield a suspension of appropriate density. This suspension is then used to inoculate a test strip containing 10–18 substrates, depending on the type of bacterium or yeast to be identified. Upon addition of the inoculum, the dehydrated reagents present in the wells are rapidly re-hydrated. As the test organism utilises some of the substrate in the different reaction mixtures, colour changes occur. The resultant pattern of positive and negative reactions is compared with those in a database to identify the organism. Strips are available for Enterobacteriaceae, non-fermentative organisms, *Haemophilus* spp., *Leuconostoc* spp., *Listeria* spp., *Neisseria* spp., *Pediococcus* spp. and *Streptococcus* spp., as well as for anaerobes, urinary tract bacteria and yeasts.

Celig and Schreckenberger (1991) investigated the efficacy of the RapID ANA 11 in identifying anaerobes. At genus level, strains were identified correctly for 96 % of Gram-negative bacilli, 94 % of *Clostridium* spp., 83 % of non-sporing, Gram-positive bacilli and 97 % of cocci. In relation to species identification, the figures were 86 %, 76 %, 81 % and 97 % respectively.

Biolog Microbial Identification System

The Biolog Microbial Identification System (Biolog) utilises 96-well microtitre plates that are pre-loaded with reagents for metabolic tests (Bochner, 1996). Each well in a series contains a different carbon source and a redox indicator system. The dry chemicals are rehydrated by inoculating with a bacterial cell suspension of specific turbidity. During incubation, the bacterium utilises the carbon sources in some of the wells. If a particular compound is utilised, a redox reaction occurs and the colour of the medium changes from clear to purple in 4–24 hours. The colour changes may be determined by manual inspection or by means of a micro-plate reader in conjunction with a computer. Once the plates have been read, the MicroLog™ software converts this information into a number that is compared to a database for identification purposes. The system may be used to identify many species of Gram-positive and Gram-negative bacteria. Also, a micro-plate has been developed specifically for identifying *Escherichia* and *Salmonella* spp.

In addition, Biolog has developed MT MicroPlates™, which are 96-well microtitre plates containing the same nutrient base and colour-forming reagents as before, but without any carbon sources. These plates may be tailored for specific purposes by adding different carbon sources, which allows certain metabolic capabilities to be studied. Klingler *et al.* (1992)

evaluated the Biolog System for its ability to differentiate various bacterial isolates. Of 39 strains tested from the American Type Culture Collection, 98 % were identified correctly to genus level and 76 % to species level within 4–24 hours. Overall, the system is easy to use, relatively inexpensive and able to identify many different strains; however, it is important that the microtitre plate is inoculated properly and this may require practice in order to obtain consistent results. The person conducting the test must be careful to avoid removing any nutrient material with the swab used to collect the colony, because such material would be transferred to the wells. Thus, nutrient carry-over could support growth of the test organism and cause false-positive colour changes, so that the organism is either identified incorrectly or not at all.

BBL™ Crystal™ system

This system, supplied by Becton–Dickinson Microbiology Systems, is a relatively new means of identifying bacteria. Two such test kits are available: The Rapid Stool/Enteric ID Kit (RS/E Kit) and the Enteric/Nonfermenter ID Kit (E/NF Kit). Both make use of modified conventional and chromogenic substrates (30 in total) that are contained within a novel type of plate (Holmes *et al.*, 1994). Each kit comprises a plastic base containing the reaction wells, to which, following inoculation, is clipped a lid with dehydrated substrates on the tips of plastic prongs. After being incubated for three hours in the case of the RS/E Kit and 18–20 hours for the E/NF Kit, the results are interpreted visually and recorded manually. They are then converted to a 10-digit profile number that is compared with a database to obtain an identification.

When compared with the API® and VITEK® systems (see below), Crystal™ performed well. Robinson *et al.* (1995) reported that, of 512 Gram-negative bacilli (381 Enterobacteriaceae and 131 non-enteric bacilli), 95.5 % were identified correctly to both genus and species levels by Crystal™. In fact, 93.9 % of strains were identified within 24 hours, without any need for supplementary testing. Moreover, identification errors associated with the system were infrequent and appeared to be distributed randomly among the genera included in the trial.

Both RS/E and E/NF kits were evaluated for their ability to identify Gram-negative organisms by Holmes *et al.* (1994). Using 203 and 266 strains respectively, the kits identified 91 % and 93 % of Enterobacteriaceae correctly. The systems were found to be safe and easy to use.

Overall, when compared with conventional methods, these miniaturised systems for identifying microbes are relatively accurate, require less incubator space, need no time or labour for media preparation, and results can be obtained much more rapidly. Also, the consistency of the materials used in the kits is very good and errors associated with improper preparation or sterilisation of conventional media are avoided. However, the kits are subject to inoculation error, because a specified concentration of microbial cells must be used in the inoculum or the results are likely to be unreliable. It should be noted that cultures which cannot be emulsified in saline are difficult – and in some cases impossible – to identify with the systems in question.

11.6 Commonly-used selective and differential media

The media discussed below are included here because they provide a presumptive identification of the organisms being isolated and, at best, require little or no use of confirmatory tests.

11.6.1 Techniques for faecal indicator organisms

Petrifilm™

A technique used for estimating bacteria on the surfaces of equipment, etc. involves the use of a dry-medium film supplied by 3M. The basic system consists of standard methods (SM) nutrients and a cold water-soluble gelling agent contained in two adjacent films (Swanson *et al.*, 1992). The bottom film is coated with the SM nutrients and the top film is coated with the gelling agent plus triphenyltetrazolium chloride, which facilitates counting by staining colonies red.

To use Petrifilm™ for direct-contact enumeration, 1 ml of sterile 0.1 % peptone water is placed on the covering paper and the film bearing the medium is allowed to contact the liquid, allowing the medium to re-hydrate within 30 minutes. After this time, the film is lifted off and the hydrated medium can be used to take a sample by pressing it onto the test surface. The medium is then incubated and colony counts expressed as cfu per $10\,cm^2$. Populations of coliforms and *E. coli* can be estimated with Petrifilm Coliform Count Plates and *E. coli* Count Plates, respectively. In comparison with conventional plate-count methods, this technique is inexpensive, requires less labour and results can be obtained in 24 hours instead of 48 hours. Disadvantages are that the peptone water must be prepared separately, the medium on the film tends to smear and results can be obtained much more rapidly with other techniques.

Chromogenic media

Chromogenic media have gained in popularity as a means of distinguishing between different organisms and allowing the presumptive identification of certain types that are associated with foods. Differentiation is based on colour changes that occur, when microbial metabolic by-products react with chromogenic agents in the media. Examples are CPS ID2 (bioMérieux) and CHROMagar™ (Dr A Rambach – BBL™ CHROMagar™ Orientation). Both media are sold in a ready-to-use form for the isolation, enumeration and identification of *E. coli* and *Enterococcus* spp. in a single step. The media contain two different chromogenic substrates: CPS ID2 permits the detection of β-glucuronidase and indole produced by *E. coli* (blue colour), tryptophan deaminase produced by *Proteus* (brown colour) and β-glucosidase produced by enterococci. CMO can detect enzymes associated with lactose metabolism, from which a pink to red colour is formed (Hengstler *et al.*, 1997). Colonies growing on these media are stained pink, red, blue, blue–green or purple, or their natural colour is unchanged, if they do not produce any of the relevant enzymes. The two formulations were designed originally as non-selective media for isolating, enumerating and differentiating urinary tract pathogens. However, they are now used for *E. coli* and *Enterococcus* spp. in food samples, without the need for confirmatory testing.

Most investigations involving these media have been on urine samples. For example, Hengstler *et al.* (1997) isolated 266 strains of *E. coli* from this type of sample on CPS ID2 and CMO, whereas only 260 were isolated on blood agar and 248 on MacConkey agar. One strain (0.4 %) did not develop the expected colour on CMO and 23 strains (8.7 %) failed to develop it on CPS ID2. In addition, 266 strains of enterococci were isolated and these produced small, blue-green colonies on both media. The *Klebsiella–Enterobacter–Serratia* (KES) and the *Proteus–Morganella–Providencia* groups could be identified in both cases. Thus, 58 of the 64 KES strains produced the expected colour on CPS ID2 and 63 of 66 strains did so on CMO. It was concluded that one of the greatest advantages in using these media is the easy recognition of particular organisms among mixed bacterial populations (Hengstler *et al.*, 1997).

Merlino *et al.* (1996) evaluated CMO for the differentiation and presumptive identification of Gram-negative bacilli and *Enterococcus* spp., using a multi-point inoculation technique. The strains tested included 1404 Gram-negative bacilli and 74 strains of enterococci. It was found that 99.3 % of 588 *E. coli* strains produced a pink-to-red colony. However, four of the *E. coli* strains were found to be negative for *o*-nitrophenyl-β-D-galactopyranoside and therefore there was no colour change. Strains of *Proteus mirabilis* and *P. vulgaris* were easily differentiated on CMO. Strains of *P. mirabilis* ($n = 184$) produced clear colonies with diffusible brown pigment around the periphery, whereas *P. vulgaris* appeared bluish–green with a slight brown background. All 26 strains of *Aeromonas hydrophila* produced clear to pink colonies, when incubated at 35–37 °C. At room temperature, on the other hand, the colour changed to blue after 2–3 hours. Colonies of *Serratia marcescens* ($n = 29$) appeared aqua blue, which became darker at room temperature. All 74 enterococci produced a blue colour, but there was no differentiation of *Klebsiella*, *Enterobacter* and *Citrobacter* spp., although these organisms could be distinguished readily from other Enterobacteriaceae. Thus, it was concluded that CMO medium allowed easy recognition of some organisms among a mixed microflora (Merlino *et al.*, 1996).

In a study of urine samples (D'Souza *et al.*, 2004), samples were plated in parallel on sheep blood agar, MacConkey agar and CMO, and the results compared, following incubation. In most cases, there was complete agreement between all three media. The same was observed for 400 single-pathogen cultures and nine mixed cultures with respect to the number of organisms recovered in each case and their identity. The use of CMO reduced the inoculation time by >50 % and the work-up time by >20 %.

11.6.2 Chromogenic media for pathogens

Chromogenic media have been developed for several of the major food-borne pathogens and have facilitated the isolation and differentiation of these organisms. Some have been certified by the Association of Official Analytical Chemists International (AOACI) for the analysis of certain foods, when used as part of an official method. An example is CHROMagar™ 0157 (CMA 0157), which distinguishes between *E. coli* 0157 and other strains of *E. coli* by means of a highly specific, chromogenic substrate for β-glucuronidase. On this medium, 0157 strains produce mauve-coloured colonies, while other Gram-negative organisms are inhibited or produce colourless, blue, green or blue-green colonies. The effectiveness of the medium was demonstrated by Vetterli (2004), and it reduced the time needed to detect positive samples, as well as thc amount of labour and materials required to confirm the absence of the target organism.

It is difficult to distinguish between *Listeria monocytogenes* and other *Listeria* spp. on primary isolation, but RAPID'L mono™ agar (Biorad) achieves this, because (i) *L. monocytogenes* produces the enzyme phosphatidylinositol phospholipase C (PIPLC) and (ii) the organism cannot metabolise the xylose present. The only *Listeria* spp. with PIPLC activity are *L. monocytogenes* and *L. ivanovii.* Incorporation of xylose in the medium differentiates between the two, and *L. ivanovii*, a xylose utiliser, produces colonies with distinct yellow haloes, while *L. monocytogenes* has blue colonies without such haloes. Other, non-pathogenic listerias produce white colonies. Use of RAPID'L mono™ agar permits the presumptive identification of *L. monocytogenes* within 24 hours, but detection of other listerias is poor (Greenwood *et al.*, 2005) and not all strains appeared able to grow on the medium or express PIPLC activity during a seven-day incubation period (Gracieux *et al.,* 2003).

Another chromogenic medium for listerias is CHROMagar™ Listeria (CMAL) on which the organisms produce turquoise colonies in as little as 24 hours at 36 ± 1 °C, when inoculated from a primary enrichment broth. This is due to the chromogenic substrate 5 bromo-4-chloro-3 indoxyl-β-D-glucopyranoside for detecting β-D-glucosidase activity that occurs in all *Listeria* spp. (Reissbrodt, 2004). The medium utilises the cleavage of L-α-phosphatidylinositol by PIPLC to distinguish the pathogenic species, *L. monocytogenes* and *L. ivanovii*, by a zone of white precipitation around each colony. The supplier (BBL) claims that CMAL has a 99–100 % sensitivity and 100 % specificity.

Strains of *Staphylococcus aureus* can be isolated on the recently-developed Petrifilm™ Staph Express Count Plate (3M). The test system contains a water-soluble gelling agent and chromogenic, modified Baird-Parker medium that is selective and differential for the target organism. Suspect colonies are red-violet in colour and, since *Staph. aureus* can degrade DNA, which is included in the medium, together with an appropriate dye, colonies of this organism are surrounded by pink zones. When mechanically separated poultry meat was tested by the Petrifilm™ system and conventional Baird–Parker medium, results were comparable, as with other foods of animal origin (Ingham *et al.*, 2003).

BBL™ CHROMagar™ Staph aureus (CSA) is another chromogenic medium, on which *Staph. aureus* forms mauve-coloured colonies; however, the coagulase test is still required for confirmation. With clinical samples, Flayhart *et al.* (2004) demonstrated an overall sensitivity of 99.5 % and a specificity of 98 %.

In the case of *Salmonella*, it is often difficult to distinguish suspect colonies from those of competing organisms, but on BBL™ CHROMagar™ Salmonella, the rose-violet colonies are readily observed among other bacteria, including coliforms, that are either blue-green or colourless (Eigner *et al.*, 2001). In comparison with conventional isolation media, the time needed to confirm suspect colonies was reduced by about 24 hours.

Chromogenic media are also available for yeasts and filamentous fungi, and differentiation of *Candida* spp. For example, BBL™ CHROMagar™ Candida includes a chromogenic substrate that enables *C. albicans*, *C. tropicalis* and *C. krusei* to produce colonies of different colours. These range from light to medium green, dark to metallic blue and light mauve to mauve respectively. The colonies are also flat with a whitish border. Other yeasts may appear light to dark mauve.

11.7 Automated microbial identification

Microbial identification systems based on substrate utilisation and analysis of cellular fatty acids (Table 11.2) and metabolic products, using gas chromatography, have been available for many years (Ewing, 1973; Welch, 1991; Harris and Humber, 1993; Clontz, 1996; Odumeru *et al.*, 1999).

11.7.1 Substrate utilisation

VITEK®

The VITEK® AutoMicrobic System (AMS, bioMérieux) is one of the most versatile methods available for the rapid identification of microorganisms. To identify an isolate, a colony is mixed with a 0.45 % saline solution in a

Table 11.2 Automated systems for microbial identification (Olson, 1996)

System	Basis	% GN* correct
bioMérieut VITEK®	Substrate utilisation	75–97
Biolog	Substrate utilisation	56–60
MIDI-Sherlock®	Gas chromatography of Cellular fatty acids	81–93

* GN: Gram-negative organisms.

test tube to yield a specific turbidity, depending on the type of organism being identified. The suspension is used to inoculate a transparent card. One end of a small plastic tube is inserted into the test card and the other end is placed in the suspension in the test tube. The card and tube containing the suspension are placed in a filler stand. The filler stand is then transferred to a filler, which uses a vacuum to pull the suspension up into each of 30 test wells that contain specific compounds. The card is sealed, inserted into a holder and placed in the AMS.

As the test organism utilises certain substrates in the card, the optical density of each reaction mixture changes. This is measured hourly and the series of readings is compared to a standard database in order to identify the organism. Isolates can usually be identified in 4–18 hours and as many as 120 samples may be analysed simultaneously.

Eight different test cards are available, depending on the microbe to be identified and include the following: Gram Negative (GNI), Gram Positive (GPI), Nonfermenter (NFC), Yeast (YBC), Bacillus (BAC), Anaerobe (ANI), Bioburden Enumeration (BIO) and the Assay Card (ASC). The BIO card is used to enumerate microbial populations in liquid samples, while the ASC is used to measure the strength or efficacy of antibiotics, vitamins, biocides or preservatives by monitoring microbial growth in the presence of these substances.

Bailey *et al.* (1985) studied the AMS to determine its ability to identify stock cultures and freshly collected strains of Enterobacteriaceae from ground beef, processed chickens, frozen pot pies and commercial poultry feeds. The system correctly identified 135/136 (99.3 %) of the stock cultures and 160/163 (98.2 %) of the fresh isolates to species level. Robinson *et al.* (1995) compared the Crystal™, API® 20E and VITEK®, and reported that, of 381 Enterobacteriaceae tested, the AMS identified 96.1 % correctly.

Twelve laboratories evaluated the GNI card in relation to Enterobacteriaceae and found that the AMS correctly identified 96.7 % of *Salmonella* strains, 97.0 % of *E. coli* and 93.0 % of the other genera included (Knight *et al.*, 1990). The AMS and the GNI card have been approved by AOACI as a screening method for the presumptive identification of *Salmonella*, *E. coli* and Enterobacteriaceae isolated from foods.

The VITEK AMS is able to identify many organisms accurately. The database is extensive, numerous samples can be tested at any one time and the cost per test is low; however, the initial cost for the instrumentation is relatively high.

VITEK® 2

The VITEK® 2 is similar to the VITEK® in that, following primary isolation, a microbial suspension is prepared in a tube of saline and adjusted to the required density, using a DensiChek densitometer. The inoculum tube is then placed in a rack termed a cassette. The sample identification number is entered into the Smart Carrier *via* a barcode or keypad and linked electronically to the barcode provided on each test card. All information entered at the bench is then transferred to the instrument in a memory chip attached to the cassette. This provides a tracking system from the bench to the final report.

Schreckenberger *et al.* (2005) compared the VITEK® Legacy, VITEK® 2 Colorimetric and Phoenix™ (Becton-Dickinson Diagnostics) systems for identifying fermentative and non-fermentative bacteria. A total of 417 Gram-negative isolates was evaluated. The authors concluded that the VITEK® 2 was 91.7 % accurate to species level, without additional testing, and that the VITEK® 2 and Phoenix™ were the most accurate systems of those evaluated. When additional tests were used, all three systems were found to be 95 % accurate. The lowest mis-identification rates were observed with the VITEK® Legacy and VITEK® 2 systems (Schreckenberger *et al.*, 2005).

Graf *et al.* (2000) showed that the VITEK® 2 system was also a suitable means of identifying yeasts. Using the VITEK® ID-YST card, the VITEK® 2 system was able to identify yeasts and related organisms within 15 hours, due to a sensitive, fluorescence-based technology. The ID-YST card covers 47 biochemical reactions, while the database is very comprehensive and includes 51 different taxa. Graf *et al.* (2000) investigated the reliability of the VITEK® ID-YST card for identifying organisms normally seen in clinical settings. A total of 241 strains, representing 21 species isolated from clinical specimens, was studied. The tests were performed on 24–55-hour cultures on Sabouraud gentamicin–chloramphenical agar. Each strain was tested in parallel with the ID 32C strip for comparison. This was combined with a microscopical examination of cell morphology and an agglutination test for *C. krusei*. In total, 222 strains (92.1 %) were identified unequivocally, including 11 isolates (4.6 %) identified with the aid of simple additional tests. Ten further strains (4.1 %), for which results were uncertain, could not be identified unequivocally, even with supplementary tests. Also, four strains (1.7 %) were mis-identified and five (2.1 %) could not be identified at all (Graf *et al.*, 2000). Nevertheless, the authors concluded that the VITEK® 2 system was a rapid and accurate method for identifying medically important yeasts and yeast-like organisms.

11.7.2 Headspace gas analysis

Gas chromatography (GC) was shown to be an effective means of characterising microorganisms chemically (Larsson and Mardh, 1977). To identify bacteria using GC, the gaseous atmosphere above an actively-growing microbial population is sampled and analysed to determine the volatile by-products evolved, as substrates in the growth medium are metabolised. Because different species produce characteristic metabolic by-products, the chromatographic patterns obtained can be used to differentiate between them. Once these volatile substances have been determined, a database is used to identify the organism. The technique has been used traditionally for rapid identification of bacteria associated with human infections; however, studies have been conducted more recently, in which the same technique has been used as a means of characterising spoilage odours produced by bacteria growing on fresh poultry (Viehweg *et al.*, 1989). This method is expensive and is only recommended for certain, specific applications.

11.7.3 Cellular fatty acid analysis

Microbes may also be identified by extracting cellular fatty acids, analysing them by GC and then comparing the resultant patterns with a database. For many years, the analysis of short-chain or volatile fatty acids has been used to identify anaerobic bacteria (Sasser, 1990a). Researchers have used fatty acids that are 9–20 carbons long to determine genus and species, especially for non-fermentative, Gram-negative bacteria. Since the development of fused-silica capillary columns, it has become feasible to use GC of whole-cell fatty-acid methyl esters to identify many species of bacteria (Sasser, 1990a).

More than 300 fatty acids and related compounds are found in bacteria. Both the presence and absence of these compounds, and their quantification, can be used effectively to separate bacterial species. Gas–liquid chromatography of fatty-acid methyl esters (FAME) was shown to be an effective tool for identifying bacteria that are important in clinical and industrial settings (Miller, 1987; Moore *et al.*, 1987; Stockman *et al.*, 1987; Osterhout *et al.*, 1989) and the results closely parallel those of ribosomal RNA and DNA homology studies (Sasser and Smith, 1987).

MIDI has developed databases of FAME profiles for identifying aerobic and anaerobic bacteria and yeasts. The MIDI Sherlock® Microbial Identification System includes a Hewlett-Packard capillary gas chromatograph, autosampler, detector, computer, printer, Microbial Identification Software (MIS) and Library Generation Software (LGS) package. The LGS contains two cluster analysis packages that have 'tracking' capabilities. The Dendrogram and 2-D Plot programmes use data obtained from microbial fatty-acid analyses and yield easy-to-understand plots of the isolate in question and its relationship to other organisms (Sasser, 1990b).

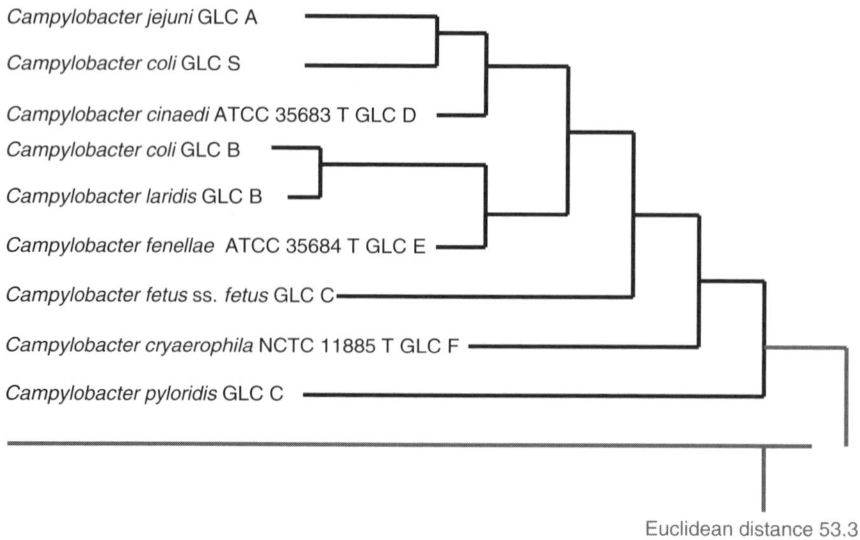

Fig. 11.2 Dendrogram of spp. *Campylobacter (*Lambert *et al.*, 1987).

A dendrogram is a tree diagram generated by the cluster analysis technique to produce pair matching and may be based on cellular fatty-acid analysis. Thus, it depicts the relatedness of strains. As an example Lambert *et al.* (1987) used this approach to classify 368 strains of *Campylobacter* spp. (Fig. 11.2). Based on fatty acid differences, *Camp. jejuni* and *Camp. coli* were more closely related than *Camp. laridis (lari)* or *Camp. fetus ss. fetus*. The dendrogram also shows that *Camp. pyloridis (pylori)* was linked at a Euclidian distance of >30 (53.3) and hence was wrongly classified in the genus *Camp*. It is now classified as *Helicobacter* (Smith and Siegel, 1996). This type of system is excellent for differentiating bacteria. However, the databases need further development and the initial cost of the instrumentation is relatively high.

Hinton *et al.* (2004) analysed the FAME profiles of *Campylobacter* isolates, using the MIDI Sherlock® MIS to demonstrate the presence of *Campylobacter* spp. on poultry carcasses and in scald tank water samples. Some strains of *Camp. jejuni* re-appeared in the same processing facility at different times of the year.

11.8 Systems for typing isolates following identification

Systems for typing isolates involve one or more techniques for distinguishing between different strains of a single bacterial species and is necessary for epidemiological investigations to track the organism in question to its point of origin. The following are techniques that have been used in the past to separate bacterial species into subtypes and, in some cases, are still commonly employed.

11.8.1 Biotyping

Biotyping involves identifying subspecies diversity on the basis of colony morphology, metabolic activities and toxin production. Using this technique, strains are referred to as 'biotypes'. The most commonly used tests to distinguish between strains depend on the ability of the organism to ferment a variety of different sugars. Usually, however, sugar fermentation alone is not sufficient and tests for a range of other metabolic capabilities are needed to assist in the differentiation. Biotyping kits have been developed for the purpose and are now readily available. The API 20E and Biolog (Toth *et al.*, 1999) are reliable examples of this type of technique.

11.8.2 Phage typing

Bacteriophages are viruses that attack and destroy bacteria. Phage typing detects differences between strains of bacteria by determining their susceptibility to attack by specific phages. Strain differences will depend on the presence or absence of certain receptors on the surface of the bacterial strain under investigation. The phages use these receptors to bind to the bacterial cell wall.

A well-established example of phage typing is the technique used for *Staph. aureus*, described by Wentworth (1963). To conduct this type of test, a pure culture of *Staph. aureus* is streaked over the surface of an agar medium. The plate is then inoculated in specific areas with 23 different phages known to attack this species and to which the organism shows a strain-dependent susceptibility. The plate is then incubated overnight. The staphylococci multiply to a visible biomass and, in so doing, create a lawn of bacteria over the whole plate. In areas where the staphylococci are attacked by phage, the cells are lysed, which causes zones of clearing to appear in the lawn. The pattern of phage susceptibility indicates the subspecies, because the phages are always inoculated onto the plate in the same sequence.

11.8.3 Serological typing

Bacterial antigens occur on the surface of the cell or the flagella. By using specific antibodies present in specially-prepared antisera, differences between individual strains may be detected. This is generally done by means of agglutination tests, including sero-agglutination in a tube and slide agglutination. In a study described by Baudart *et al.* (2000), for example, serotyping was performed by sero-agglutination. Polyvalent *Salmonella* O (somatic or cell-wall antigens) and H (flagellar antigens) antisera were used to obtain a presumptive identification and then the definitive antigenic formula was determined with the use of monovalent antisera. A discussion of somatic and flagellar antigens is provided in Evins *et al.* (1976). Because O or H antigens are not always expressed, no agglutination is observed in some cases (Baudart *et al.*, 2000).

11.8.4 Bacteriocin typing

Bacteriocin typing depends on the fact that some strains of bacteria produce compounds that are only active usually against other strains of the same species. Such compounds are termed 'bacteriocins'. Tests are conducted to determine the ability of the test strain to produce one or more bacteriocins with the ability to lyse a specific set of indicator strains of related bacteria. In addition, the sensitivity of the test strain to standard bacteriocins may be determined. The technique is not used very often, but an excellent example is described by Heddell and Mitchell (1978).

11.8.5 Protein typing

Protein typing is accomplished be determining the proteins produced by different subspecies of bacteria as metabolic by-products. The proteins are extracted from a culture of the test organism, separated by polyacrylamide gel electrophoresis (PAGE) and compared with the proteins produced by other strains of bacteria. The method is inexpensive to use; however, it is cumbersome in comparison with other methods. Senior and Voros (1990) investigated the use of protein typing for *Morganella morganii* and reported that, after analysis of cell lysates by PAGE, strains could be differentiated in 21 types on the bais of outer membrane proteins of 35–40 kDa.

11.9 Genetic methods for typing bacterial strains

Molecular subtyping of bacterial isolates has been used in epidemiological investigations since the 1980s (Holmberg *et al.*, 1984; Holmberg and Wachsmuth, 1989). The techniques involve the identification of differences between or within species by characterising cellular proteins or nucleic acids. In particular, techniques have been developed to analyse chromosomal DNA, and molecular subtyping of species is now considered an essential part of tracking foodborne infectious diseases (Swaminathan *et al.*, 2001). Widespread use of molecular typing has resulted in the development of numerous techniques and protocols for subtyping bacteria (Swaminathan and Matar, 1993). Some of these will be described below.

11.9.1 Pulsed field gel electrophoresis (PFGE)

Application of the technique to strains of *S.* Typhimurium involves separating the DNA within the bacterium of interest into segments, using restriction enzymes. The DNA molecules are then dispensed into an agarose matrix and an electric field is applied across the matrix. The DNA will elongate and migrate towards the anode. The agarose matrix contains a highly irregular network of molecules with pores of various dimensions, large open areas and regions of different densities (Achtman and Morelli, 2001). During electrophoresis, as the electric current is pulsed through the agarose, DNA migration will depend on molecular size. In conventional electrophoresis, involving a constant electric field, DNA molecules will migrate to a distance that is inversely proportional to the logarithm of their length. Therefore, for smaller molecules, relatively small differences in length result in large differences in mobility. Because of the logarithmic relationship, the sensitivity (separation of molecules within the matrix) decreases as the size of the DNA increases. In addition, with a continuous electrical charge, large DNA molecules migrate abnormally fast (Achtman and Morelli, 2001). Therefore, PFGE was developed to solve these problems and allow the electrophoretic separation of larger molecules in agarose gels, as described originally by Schwartz and Cantor (1984). The original method involved pulsed, alternating, orthogonal electric fields being applied to a gel. In this way, large DNA molecules were trapped every time the direction of the electric field changed and they would only begin to migrate after re-orientating along the new field axis. The larger the DNA molecule, the longer the time required for re-orientation (Achtman and Morelli, 2001). Therefore, DNA molecules with a re-orientation time less than the duration of the electric pulse could be fractionated according to their size. The limit of resolution for PFGE depends on the following factors (Achtman and Morelli, 2001):

1. the uniformity of the two electric fields;
2. the duration of the electric pulses;
3. the ratio of the pulse times for each of the alternating electric fields;
4. the angles of the two electric fields to the gel;
5. the ratio of the strengths of the two electric fields.

Currently, the methods used for PFGE allow for resolution of DNA fragments up to 5000 kb in length. By permitting the separation of large molecules in agarose, Achtman and Morelli (2001) pointed out that PFGE has extended the size range of molecules amenable to molecular analysis and has profoundly altered the study of genes and genomes. For some bacterial species, intact chromosomes can be separated from each other by PFGE, which allows for gene mapping using Southern hybridisation. Application of PFGE can reveal chromosome-length polymorphisms, thus facilitating evolutionary and population studies in a number of bacterial species.

The CDC demonstrated clearly the value of this method, while investigating an outbreak of *E. coli* 0157:H7 in contaminated hamburgers that were served by a fast-food restaurant chain in the western USA in 1993. Scientists with the CDC used PFGE to characterise clinical and food isolates of the organism and were able to demonstrate its value in tracking the sources of the infections (Barrett *et al.*, 1994).

In 1995, the CDC, with the assistance of the Association of Public Health Laboratories, selected various state public health laboratories and designated them area laboratories for a national molecular subtyping network for foodborne bacterial disease surveillance (Swaminathan *et al.*, 2001). This network later became known as PulseNet. Using PFGE, these laboratories were able to standardise and use PFGE subtyping and pattern analysis techniques to track and determine rapidly the origins of outbreaks of foodborne illness caused by bacterial agents.

11.9.2 Polymerase chain reaction (PCR)

PCR is a technique used to amplify a specific target segment of DNA as a means of detecting the presence of a particular organism among a complex mixture of other organisms. The reaction produces millions of copies of a specific DNA segment within a few hours. The reaction mixture contains a heat-stable DNA poylmerase, free nucleotides and a pair of 'primers' (short DNA sequences complemntary to the specific sequence of the target DNA) in a volume of 25–100 μl (Lantz *et al.*, 1994).

The PCR technique is based on the repetition of three steps, all conducted in succession, each at a different controlled temperature. Each step requires about 60 seconds to perform. In the first step, the two strands of the DNA target sequences are separated using heat denaturation (Lantz *et al.*, 1994). As the temperature is lowered to the annealing temperature (45–65 °C), each primer will anneal (hybridise) to only one of the separated strands. The primer sequence is determined by the nucleotide sequences flanking that of the target DNA being amplified. The third step in the procedure is the synthesis of the complementary strand at the ends of each primer. The heat-stable DNA polymerase begins to synthesise new target DNA (PCR products) by adding free nucleotides to the primers. Newly synthesised PCR products can then serve as templates in the subsequent rounds of amplification. After 30–40 cycles of heat denaturation, annealing and primer extension, target DNA sequences will have been amplified by a factor of 10^9. By this exponential amplification, yielding (PCR product)^n, where n is the number of cycles, it becomes possible to detect a specific DNA region by gel electrophoresis or by a computerised DNA detection system. Lantz *et al.* (1994) observed that, in a few cases, food samples containing whole bacteria could be used directly as PCR samples, because the repeated cycles of heat denaturation will release bacterial DNA. However, a procedure for sample preparation is required before carrying out PCR on the majority of food samples, because they contain substances that can inhibit the PCR and/or the incidence of pathogenic bacteria is low.

Loeffelholz *et al.* (1992) reported that PCR had a sensitivity of 97 % and a specificity of 99.7 %, while traditional culturing had a sensitivity of only 85.7 %, but a specificity of 100 %, when detecting *Chlamydia trachomatis*. In fact, PCR is an extremely sensitive and specific method for detecting pathogenic bacteria in clinical samples. In relation to food samples, however, this technology is still somewhat difficult to use. It requires extensive staff training, is expensive and cannot differentiate between live bacterial cells and those that are dead. Therefore, PCR would be unsuitable for analysing cooked foods that originally contained viable pathogenic bacteria.

11.9.3 DNA/RNA hybridisation

Gene-Trak® Systems (Neogen Corporation) has developed DNA hybridisation assays that can be used to screen samples for the presence of *Salmonella*, *Campylobacter*, *Listeria*, *E. coli*, *Staph. aureus* and *Yersinia enterocolitica*. Although each individual procedure differs slightly, a general description of the methodology is presented below.

To conduct DNA hybridisation assays, the sample must be enriched first in a non-selective medium to increase the number of target organisms present to a point at which they can be detected (Gene Trak, 1991). To conduct a *Salmonella* assay, for example, all bacteria in the sample to be analysed are lysed with a Lysis Solution and the *Salmonella* rRNA target strands are released into this solution. Two DNA probes, each with a distinct function, are then added to the solution. Both probes are homologous to unique rRNA sequences of *Salmonella* and they hybridise to adjacent regions on the same target rRNA molecule. The capture probe contains a poly dA tail, which allows the hybrid molecules to be captured onto a solid support that binds to the tail. The detector probe is labelled at both ends using fluorescein. This probe binds to the end of the hybrid molecule that is not bound to the solid support. Thus, the final hybrid molecule contains a strand of target *Salmonella* rRNA that has two probes attached. The hybridisation reaction is carried out at 65 °C. The next step in the assay is to capture the hybrid onto a solid support. The support used is a plastic 'dipstick' that is coated with poly dT. The poly dA tail on the hybrid molecule attaches to the poly dT on the dipstick and the molecule is captured. The remaining step in the assay is to detect the strand of *Salmonella* rRNA that has been captured on the solid support. The detector probe on the other end of the captured hybrid is first treated with a polyclonal, anti-fluorescein antibody (anti-Fl), conjugated to the enzyme horseradish peroxidase (HRP). This conjugate then binds to the fluorescein molecules on the detector probe. The complex is allowed to react with a substrate of HRP, hydrogen peroxide, in the presence of a chromogen. A blue colour develops in proportion to the amount of enzyme conjugate bound to the complex and thus is also in proportion to the amount of *Salmonella* rRNA captured. The reaction is stopped with sulphuric acid, which changes the colour that has developed from blue to yellow. The colour intensity is measured by determining the absorbance at 450 nm with a photometer. An absorbance in excess of a pre-determined cut-off value indicates a positive result for the presence of the test organism (Gene Trak, 1991).

Rose *et al.* (1991) compared the use of DNA hybridisation for detecting *Salmonella* in meat and poultry products to conventional detection methods. The authors reported that the DNA hybridisation procedure was more sensitive than the cultural methods. There were no false-positive or false-negative results from using colourimetric DNA hybridisation. Another DNA hybridisation assay used for detecting *E. coli* was evaluated by Hsu *et al.* (1991). Using this assay, the authors were able to detect all 233 strains of *E. coli* tested. Of the 207 strains of other species tested, only *E. fergusonii* and *Shigella* caused false-positive results. The total false-negative rate was 1.2 %, compared with 23.4 % for the conventional culture method used. It was concluded that the DNA hybridisation method was significantly more accurate than conventional means of detecting *E. coli* in foods.

11.9.4 16S RNA gene sequencing

Since the discovery of PCR and DNA sequencing, comparisons of the gene sequences of different bacterial species have shown that the 16S rRNA gene sequence is highly conserved within a species and among species of the same genus, and hence can be used as the new standard for speciation (Olsen and Woese, 1993). Using this approach, phylogenetic trees derived from base differences between species are constructed and bacteria are classified and re-classified into new genera. Numerous researchers have reported on the use of the technique for identifying bacterial strain with ambiguous biochemical profiles (Woo *et al.*, 2000, 2001a, b, c; 2002a, b, c; 2003a; Lau *et al.*, 2003), species that are rarely encountered clinically and a non-cultivable bacterium (Woo *et al.*, 2003b).

11.9.5 Multilocus enzyme electrophoresis

As indicated by Selander *et al.* (1986), numerous methods have been used to type or characterise bacterial strains in relation to systematics and epidemiology. These include serotyping, monoclonal antibody typing, biotyping, bacteriophage typing, fimbriation typing, resistotyping, cell electrophoresis, whole-protein extract electrophoresis, outer membrane protein electrophoresis and various types of carbohydrate, lipid and other chemical profiling or fingerprinting. The methods are used to detect phenotypic variation in bacterial species; however, they are unable to provide information on the frequency of alleles and multilocus genotypes that is required for analysis of the genetic structure of bacterial populations (Selander *et al.*,

1986). Multilocus enzyme electrophoresis has been used in large-scale studies to determine the genetic diversity of bacterial populations.

Selander *et al.* (1986) described a modified method based on one described earlier that was developed originally for studying mammals. Using this method, isolates were characterised by the relative electrophoretic mobilities of a large number of water-soluble, cellular enzymes. In practice, the net electrostatic charge will determine the rate of the migration of a protein during electrophoresis, which is further dependent on its amino acid sequence (Selander *et al.*, 1986). The method allows mobility variants of an enzyme, called electromorphs or allozymes, to be equated directly with alleles at the corresponding structural gene locus. The authors reported that electromorph or allozyme profiles over different loci can be equated with multi-locus genotypes, while electromorph frequencies can be equated with allele frequencies. The method is used to differentiate bacterial species and a complete description of such methods may be found in the comprehensive paper by Selander *et al.* (1986).

11.9.6 Riboprinting

A new, fully automated technique called the RiboPrinter® Microbial Characterization System (DuPont Qualicon) has been developed for characterising bacterial isolates on the basis of their DNA 'fingerprint'. The genetic fingerprint (RiboPrint®) pattern is generated from the rRNA operons and other surrounding regions of the bacterial genome. In bacteria, the rRNA genes are highly conserved; however, they are also completely unique and RiboPrint® patterns can be used to distinguish between bacterial isolates. To analyse a sample using this system, DNA is extracted from bacterial cells and is fragmented with a restriction enzyme. The DNA fragments are separated by their molecular weight, using electrophoresis, and transferred to a membrane. Then, they are hybridised with a DNA probe and a chemi-luminescent label is introduced. Light is thus emitted from the hybridised fragments and an image of the pattern generated is captured by means of a camera. A computer analyses the RiboPrint® image and compares it with a database to identify the bacterium in question.

Ribotyping is a powerful tool for tracking the source of an infection. For example, a bacterium isolated from the stool of an infected individual can be ribotyped. The ribotype is then compared to those of the same species isolated from various foods that the individual is though to have consumed in the days preceding the infection. Likewise, isolates from the animals or plants used to produce the food or relevant environmental samples may be evaluated in the same manner. In this way, the bacterium causing the infection can be tracked effectively to determine its point of origin. The same approach has been applied to various food handling situations to solve specific problems associated with production and processing (Gendel and Ulaszek, 2000; Kuntz *et al.*, 2003). Clearly, the benefit of this type of analytical system is that bacterial isolates can be characterised below species level. More than 75 different RiboPrint® patterns have been observed for *L. monocytogenes* alone. While being useful for tracing sources of human infection, the system may also be used to exonerate a company that is falsely implicated in an outbreak of foodborne illness.

Ribotyping and PFGE can be used together to provide rapid identification of microbial isolates. Fontana *et al.* (2003) used an automated ribotyping and PFGE system to identify 116 strains of *S. enterica* serotype Newport, including 64 multidrug-resistant (MDR) isolates. The technique successfully distinguished between the two, with a sensitivity of 100 % and 98 %, respectively, and a specificity of 76 % and 89 %, respectively. Clustering of PFGE patterns (but not ribotyping) linked human and bovine cases. Automated ribotyping rapidly identified the MDR strains, and PFGE detected associations that aided epidemiological investigations (http://www.cdc.gov/ncidod/EID/vol9no4/02-0423.htm).

The RiboPrint® Microbial Characterization System is fully automated; however, the initial cost of the system and the cost of analysing each sample are very high.

11.10 Sources of further information

3M	www.3m.com
Becton-Dickinson Microbiology Systems	www.bd.com
Biolog Inc.	www.biolog.com
bioMérieux sa	www.biomerieux.com
Bio-Rad Laboratories	www.bio-rad.com
CHROMagar	www.chromagar.com
DuPont Qualicon	www.qualicon.com
Hardy Diagnostics	www.hardydiagnostics.com
MIDI Inc.	www.midi-inc.com
Neogen Corporation	www.neogen.com
Remel Inc.	www.remelinc.com

11.11 References

ACHTMAN M and MORELLI G (2001), 'Pulsed field gel electrophoresis', in *Meningococcal Diseases: Methods and Protocols*, eds Pollard A J and Maiden M C J, Methods in Molecular Medicine, vol 67, Totowa, New Jersey, USA, Humana Press, 147–148.

ANDREWS W H, POELMA P L and WILSON C R (1984), 'Isolation and identification of *Salmonella* species', in *Bacteriological Analytical Manual*, 6th edition, Arlington, Virginia, USA, Association of Official Analytical Chemists, 7.01–7.17.

ANONYMOUS (1988), 'Poultry processor predicts product perishability', *Prepared Foods*, **157**, 118.

ASHA N J, FAWLEY W N, FREEMAN J and WILCOX M H (2005), 'Increased rate of DNA recovery from United Kingdom epidemic *Clostridium difficile* PCR Ribotype I strains stored cryogenically', *Journal of Clinical Microbiology*, **43**, 5794–5795.

ATLAS R M (1993), *Handbook of Microbiological Media*, Boca Raton, Florida, USA, CRC Press.

BAILEY J S, COX N A, THOMSON J E and FUNG D Y C (1985), 'Identification of Enterobacteriaceae in foods with the AutoMicrobic System', *Journal of Food Protection*, **48**, 147–149.

BARRETT T J, LIOR H, GREEN J H, KHAKHRIA R, WELLS J G and BELL B P (1994), 'Laboratory investigation of a multi-state food-borne outbreak of *Escherichia coli* O157:H7 by using pulsed-field electrophoresis and phage typing', *Journal of Clinical Microbiology*, **32**, 3013–3017.

BAUDART J, LEMARCHAND K, BRISABOIS A and LEBARON P (2000), 'Diversity of *Salmonella* strains isolated from the aquatic environment as determined by serotyping and amplification of the ribosomal DNA spacer regions', *Applied and Environmental Microbiology*, **66**, 1544–1552.

BOCHNER B R (1996), 'The Biolog MicroStation System and general procedures for identifying environmental bacteria and yeast', in *Automated Microbial Identification and Quantitation: Technologies for the 2000s*, ed. Olson W P, Buffalo Grove, IL, USA, Interpharm Press, 13–51.

CELIG D M and SCHRECKENBERGER P C (1991), 'Clinical evaluation of the RapID-ANA II panel for identification of anaerobic bacteria', *Journal of Clinical Microbiology*, **29**, 457–462.

CLONTZ L (1996), 'Substrate utilization and the automated identification of microbes', in *Automated Microbial Identification and Quantitation*, ed. Olson W P, Buffalo Grove, IL, USA, InterPharm Press, 53–80.

COX N A, FUNG D Y C, GOLDSCHMIDT M C, BAILEY J S and THOMSON J E (1984), 'Selecting a miniaturized system for identification of Enterobacteriaceae', *Journal of Food Protection*, **47**, 74–77.

D'SOUZA H D, CAMPBELL M C and BARON E J (2004), 'Practical bench comparison of BBL CHROMagar Orientation and standard two-plate media for urine cultures', *Journal of Clinical Microbiology*, **42**, 60–64.

EIGNER U, REISSBRODT R, HAMMANN R and FAHR A M (2001), 'Evaluation of a new chromogenic medium for the isolation and presumptive identification of *Salmonella* species from stool specimens', *European Journal of Clinical Microbiology and Infectious Diseases*, **20**, 558–565.

EVINS G M, LINNE L L and COLVIN H M (1976), 'Evaluation of commercial *Salmonella* O polyvalent and Vi antisera', *Journal of Clinical Microbiology*, **4**, 349–353.

EWING W H (1973), 'Differentiation of Enterobacteriaceae by biochemical reactions', *Rev. DHEW Publication No. CDC 75–8270*, Atlanta, GA, USA, DHEW.

FAO/WHO (1984), 'The role of food safety in health and development'. Report of Joint Food and Agriculture Organization/World Health Organization Expert Committee on Food Safety, *World Health Organization Technical Reporting Service*, **705**, 1–79.

FLAYHART D, LEMA C, BOREK A and CARROLL K C (2004), 'Comparison of the BBL CHROMagar *Staph aureus* agar medium to conventional media for detection of *Staphylococcus aureus* in respiratory samples', *Journal of Clinical Microbiology*, **42**, 3566–3569.

FONTANA J, STOUT A, BOLSTORFF B (2003), 'Automated ribotyping and pulsed-field gel electrophoresis for rapid identification of multidrug resistant Salmonella serotype Newport', *Emerging Infectious Diseases*, **9**, 494–504.

GENDEL S M and ULASZEK J (2000), 'Ribotype analysis of strain distribution in *Listeria monocytogenes*', *Journal of Food Protection*, **63**, 179–185.

GENE TRAK (1991), 'The Gene Trak colorimetric assay format', Framingham, MA, USA, Gene Trak Systems.

GILLILAND S E, MICHENER H D and KRAFT A A (1984), 'Psychrotrophic microorganisms', in *Compendium of Methods for the Microbiological Examination of Foods*, Washington, DC, USA, American Public Health Association, 135–141.

GOOCH W M and HILL G A (1982), 'Comparison of Micro-ID and API 20E in rapid identification of Enterobacteriaceae', *Journal of Clinical Microbiology*, **15**, 885–890.

GRACIEUX P, ROCHE S M, PARDON P and VELGE P (2003), 'Hypovirulent Listeria monocytogenes strains are less frequently recovered than virulent strains on PALCAM and Rapid'L.mono media', *International Journal of Food Microbiology*, **83**, 133–145.

GRAF B, ADAM T, ZILL E and GOBEL U B (2000), 'Evaluation of the VITEK 2 system for rapid identification of yeasts and yeast-like organisms', *Journal of Clinical Microbiology*, **38**, 1782–1785.

GREENWOOD M, WILLIS C, DOSWELL P, ALLEN G and PATHAK K (2005), 'Evaluation of chromogenic media for the detection of Listeria species in food', *Journal of Applied Microbiology*, **99**, 1340–1345.

HARRIS L and HUMBER J (1993), 'Automicrobic system for biochemical identification of *Listeria* species isolated from foods: collaborative study', *Journal of AOAC International*, **76**, 822–830.

HEDDELL G W and MITCHELL A A (1978), 'Evaluation and application of an improved bacteriocin typing method for *Klebsiella aerogenes*', *Journal of Clinical Pathology*, **31**, 16–21.

HENGSTLER K A, HAMMANN R and FAHR A M (1997), 'Evaluation of BBL CHROMagar orientation medium for detection and presumptive identification of urinary tract pathogens', *Journal of Clinical Microbiology*, **35**, 2773–2777.

HINTON A, CASON J A, HUME M E, INGRAM K D and RAD Z A (2004), 'Spread of *Campylobacter* spp. during poultry processing in different seasons', *Poultry Science*, **3**, 432–437.

HOLBROOK R (2000), 'Detection of microorganisms in foods – Principles of culture methods', in *The Microbiological Safety and Quality of Food. Volume II*, eds Lund B M, Baird-Parker A C and Gould G W, Gaithersburg, MD, USA, Aspen Publishers, 1761–1790.

HOLMBERG S D and WACHSMUTH I K (1989), 'Plasmid and chromosomal DNA analyses in the epidemiology of bacterial diseases', in *Nucleic Acid and Monoclonal Antibody Probes: Applications in Diagnostic Microbiology*, eds Swaminathan B and Prakash G, New York, USA, Marcel Dekker, 105–129.

HOLMBERG S D, WACHSMUTH I K, HICKMAN-BRENNER F W and COHEN M L (1984), 'Comparison of plasmid profile analysis, phage typing, and antimicrobial susceptibility testing in characterizing *Salmonella typhimurium* isolates from outbreaks', *Journal of Clinical Microbiology*, **19**, 100–104.

HOLMES B, COSTAS M, THAKER T and STEVENS M (1994), 'Evaluation of two BBL Crystal Systems for identification of some clinically important Gram-negative bacteria', *Journal of Clinical Microbiology*, **32**, 2221–2224.

HSU H Y, CHAN S W, SOBELL D I, HALBERT D N and GROODY E P (1991), 'A colorimetric DNA hybridization method for the detection of *Escherichia coli* in foods', *Journal of Food Protection*, **54**, 249–255.

INGHAM S C, BECKER K L and FANSLAU M A (2003), 'Comparison of the Baird–Parker Agar and 3M Petrifilm Staph Express Count Plate methods for enumeration of *Staphylococcus aureus* in naturally and artificially contaminated foods, a research note', *Journal of Food Protection*, **66**, 2151–2155.

KÄFERSTEIN F K, MOTARJEMI Y and BETTCHER D W (1997), 'Food-borne disease control: A transnational challenge', *Emerging Infectious Diseases*, **3**, 503–510.

KLINGLER J M, STOWE R P, OBENHUBER D C, GROVES T O, MISHRA S K and PIERSON D L (1992), 'Evaluation of the Biolog Automated Microbial Identification System', *Applied and Environmental Microbiology*, **58**, 2089–2092.

KNIGHT M T, WOOD D W, BLACK J F, GOSNEY G, RIGNEY R O and AGIN J R (1990), 'Gram-Negative Identification Card for identification of *Salmonella*, *Escherichia coli*, and other Enterobacteriaceae isolated from foods: collaborative study', *Journal of the Association of Official Analytical Chemists*, **73**, 729–733.

KONEMAN E W, ALLEN S D, JANDA W M, SCHRECKENBERGER P C and WINN W C (eds) (1992), *Color Atlas and Textbook of Diagnostic Microbiology*, 4th edition, Philadelphia, PA, USA, Lippincott, 36–40.

KUNTZ R L, GATES K W, SEGARS W I, MCDONALD J L, HARTEL P G and GODFREY D G (2003), 'Targeted sampling protocol as prelude to bacterial source tracking with *Enterococcus faecalis*', *Journal of Environmental Quality*, **32**, 2311–2318.

LAMBERT M A, PATTON C M, BARRETT T J and MOSS C W (1987), 'Differentiation of *Campylobacter* and *Campylobacter*-like organisms by cellular fatty acid composition', *Journal of Clinical Microbiology*, **29**, 1508–1512.

LANTZ P G, HAHN-HÄGERDAL B and RÅDSTRÖM P (1994), 'Sample preparation methods in PCR-based detection of food pathogens. Trend', *Food Science and Technology*, **5**, 384–389.

LARSSON L and MARDH P A (1977), 'Application of gas chromatography to diagnosis of microorganisms and infectious diseases', *Acta Pathologica Microbiologica Scandinavica*, Supplement, **259**, 5–15.

LAU S K P, WOO P C Y and TSE H (2003), 'Invasive *Streptococcus iniae* infections outside North America', *Journal of Clinical Microbiology*, **41**, 1004–1009.

LOEFFELHOLZ M J, LEWINSKI C A, SILVER S R, PUROHIT A P, HERMAN S A, BUONAGURIO D A and DRAGON E A (1992), 'Detection of *Chlamydia trachomatis* in endocervical specimens by polymerase chain reaction', *Journal of Clinical Microbiology*, **30**, 2847–2851.

MERLINO J, SIARAKAS S, ROBERTSON G J, FUNNELL G R, GOTTLIEB T and BRADBURY R (1996), 'Evaluation of CHROMagar Orientation for differentiation and presumptive identification of Gram-negative bacilli and Enterococcus species', *Journal of Clinical Microbiology*, **34**, 1788–1793.

MILLER L (1987), 'Covariance analysis to minimize the effect of culture conditions on fatty acid composition', *Abstracts of the 87th Annual Meeting for the American Society for Microbiology*, Atlanta, GA, USA, 2–6 March, 243.

MOORE L V H, JOHNSON J L and MOORE W E C (1987), '*Selenemonas noxia* sp. nov., and *Selenomonas artemidis* sp. nov., from the human gingival crevice', *International Journal of Systematic Bacteriology*, **36**, 271–280.

ODUMERU J A, STEELE M, FRUHNER L, LARKIN C, JIANG J, MANN E and McNAB W B (1999), 'Evaluation of Accuracy and Repeatability of Identification of Food-Borne Pathogens by Automated Bacterial Identification Systems', *Journal of Clinical Microbiology*, **37**, 944–949.

OLSEN G J and WOESE C R (1993), 'Ribosomal RNA: a key to phylogeny', *Journal of the Federation of American Societies for Experimental Biology*, **7**, 113–123.

OLSON W P (ed) (1996), *Automated Microbial Identification and Quantitation: Technologies for the 2000s*, Buffalo Grove, IL, USA, Interpharm Press.
OSTERHOUT G, WOOD L and MERZ W G (1989), 'Cellular fatty acid (CFA) analysis for the identification of medically important yeast', *Abstracts of the 89th Annual Meeting for the American Society for Microbiology*, New Orleans, CA, USA, 14–18 May, 466.
REISSBRODT R (2004), 'New chromogenic plating media for detection and enumeration of pathogenic Listeria spp. – an overview', *International Journal of Food Microbiology*, **95**, 1–9.
ROBINSON A, MCCARTER Y S and TETREAULT J (1995), 'Comparison of Crystal Enteric/Nonfermenter System, API 20E System, and Vitek AutoMicrobic System for identification of Gram-negative bacilli', *Journal of Clinical Microbiology*, **33**, 364–370.
ROSE B E, LLABRÉS C M and BENNETT B (1991), 'Evaluation of a colorimetric DNA hybridization test for detection of salmonellae in meat and poultry products', *Journal of Food Protection*, **54**, 127–130.
SASSER M (1990a), 'Identification of bacteria by gas chromatography of cellular fatty acids', *Technical Note #101*, Newark, DE, USA, MIDI.
SASSER M (1990b), ' "Tracking" a strain using the microbial identification system', *Technical Note #102*, Newark, DE, USA, MIDI.
SASSER M and SMITH D H (1987), 'Parallels between ribosomal RNA and DNA homologies and fatty acid composition in *Pseudomonas*', *Abstracts of the 87th Annual Meeting for the American Society for Microbiology*, Atlanta, GA, USA, 2–6 March, 241.
SCHRECKENBERGER P C, RISTOW K L and KRILCICH A M (2005), 'Comparison of the Vitek Legacy, Vitek 2 Colorimetric and Phoenix Systems for Identification of Fermenting and Non-Fermenting bacteria of clinical origin', *Presented at the 105th General Meeting of the American Society for Microbiology*, Washington, DC, USA, American Society for Microbiology.
SCHWARTZ D C and CANTOR C R (1984), 'Separation of yeast chromosome-sized DNAs by pulsed field gradient gel electrophoresis', *Cell*, **37**, 67–75.
SELANDER R K, CAUGANT D A, OCHMAN H, MUSSER J M, GILMOUR M N and WHITTAM T S (1986), 'Methods of multilocus enzyme electrophoresis for bacterial population genetics and systematics', *Applied and Environmental Microbiology*, **51**, 873–884.
SENIOR B W and VOROS S (1990), 'Protein profile typing – a new method of typing *Morganella morganii* strains', *The Journal of Medical Microbiology*, **33**, 259–264.
SMITH A R and SIEGEL J P (1996), 'Cellular fatty acid analysis for the classification and identification of bacteria', in *Automated Microbial Identification and Quantitation*, ed. Olson W P, Buffalo Grove, IL, USA, Interpharm Press, 179–222.
STOCKMAN L, ROBERTS G D and SMITH D H (1987), 'Identification of mycobacteria by cell wall composition with the Hewlett-Packard Microbial Identification System', *Abstracts of the 87th Annual Meeting for the American Society for Microbiology*, 135.
SWAMINATHAN B and MATAR G M (1993), 'Molecular typing methods', in *Diagnostic Molecular Microbiology*, eds Persing D H, Smith T F, Tenover F C and White T J, Washington, DC, USA, American Society for Microbiology, 26–50.
SWAMINATHAN B, BARRETT T J, HUNTER S B, TAUXE R V and THE CDC PULSENET TASK FORCE (2001), 'PulseNet: The Molecular Subtyping Network for Foodborne Bacterial Disease Surveillance, United States', *Emerging Infectious Diseases*, **7**, 382–389.

SWANSON K M J, BUSTA F F, PETERSON E H and JOHNSON M G (1992), 'Colony count methods', in *Compendium of Methods for the Microbiological Examination of Foods*, Washington, DC, USA, American Public Health Association, 75–95.

TOTH I K, BERTHEAU Y, HYMAN L J, LAPLAZE L, LÓPEZ M M, MCNICOL J, NIEPOLD F, PERSSON P, SALMOND G P C, SLETTEN A, VAN DER WOLF J M and PÉROMBELON M C M (1999), 'Evaluation of phenotypic and molecular typing techniques for determining diversity in *Erwinia carotovora* subspp. *atroseptica*', *Journal of Applied Microbiology*, **87**, 770–781.

VETTERLI K (2004), 'Comparison of BBL™ CHROMagar™O157 to sorbitol MacConkey for recovery of *E. coli* O157 in stool cultures', *Presented at the 104th General Meeting of the American Society for Microbiology, New Orleans, Louisiana, USA*, Washington, DC, USA, American Society for Microbiology.

VIEHWEG S H, SCHMITT R E and SCHMIDT-LORENZ W (1989), 'Microbial spoilage of refrigerated fresh broilers. Part VII. Production of off odours from poultry skin by bacterial isolates', *Lebensmittel-Wissenschaft und-Technologie*, **22**, 356–367.

WELCH D (1991), 'Applications of cellular fatty acid analysis', *Clinical Microbiological Reviews*, **4**, 422–438.

WENTWORTH B B (1963), 'Bacteriophage typing of the staphylococci', *Bacteriology Reviews*, **27**, 253–272.

WOO P C Y, LEUNG P K L and LEUNG K W (2000), 'Identification by 16S ribosomal RNA gene sequencing of an Enterobacteriaceae species from a bone marrow transplant recipient', *Molecular Pathology*, **53**, 211–215.

WOO P C Y, LEUNG A S P and LEUNG K W (2001a), 'Identification of slide-coagulase positive, tube-coagulase negative *Staphylococcus aureus* by 16S ribosomal RNA gene sequencing', *Molecular Pathology*, **54**, 244–247.

WOO P C Y, CHEUNG E Y L and LEUNG K W (2001b), 'Identification by 16S ribosomal RNA gene sequencing of an Enterobacteriaceae species with ambiguous biochemical profile from a renal transplant recipient', *Diagnosis of Microbiological Infectious Diseases*, **39**, 85–93.

WOO P C Y, FUNG A M Y and WONG S S Y (2001c), 'Isolation and characterization of a *Salmonella enterica* serotype Typhi variant and its clinical and public health implications', *Journal of Clinical Microbiology*, **39**, 1190–1194.

WOO P C Y, FUNG A M Y and LAU S K P (2002a), 'Diagnosis of pelvic actinomycosis by 16S ribosomal RNA gene sequencing and its clinical significance', *Diagnosis of Microbiological Infectious Diseases*, **43**, 113–118.

WOO P C Y, FUNG A M Y and LAU S K P (2002b), 'Identification by 16S ribosomal RNA gene sequencing of *Lactobacillus salivarius* bacteremic cholecystitis', *Journal of Clinical Microbiology*, **40**, 265–267.

WOO P C Y, LEUNG K W and TSOI H W (2002c), 'Thermo-tolerant *Campylobacter fetus* bacteraemia identified by 16S ribosomal RNA gene sequencing: an emerging pathogen in immunocompromised patients', *Journal of Medical Microbiology*, **51**, 740–746.

WOO P C Y, TENG J L L and LAU S K P (2003a), 'Analysis of a viridans group strain reveals a case of bacteremia due to Lancefield group G alpha-hemolytic *Streptococcus dysgalactiae* subsp. *equisimilis* in a patient with pyomyositis and reactive arthritis', *Journal of Clinical Microbiology*, **41**, 613–618.

WOO P C Y, LAU S K P, FUNG A M Y, CHIU S K, YUNG R W H and YUEN K Y (2003b), 'Gemella bacteraemia characterized by 16S ribosomal RNA gene sequencing', *Journal of Clinical Pathology*, **56**, 690–693.

12

Validation of analytical methods used in food microbiology

M. Uyttendaele and J. Debevere, Gent University, Belgium

12.1 Introduction

Microbiological tests are important in: governmental food inspection to enforce legal requirements; international trade to determine compliance with microbiological standards; commercial relationships between trading partners to ensure that agreed microbiological specifications are met; the food industry to maintain quality control and process requirements; academic laboratories for research purposes; and reference laboratories to confirm the results of other laboratories and to provide surveillance data.

The results of these tests must be reliable and therefore it is necessary to determine the performance characteristics of each analytical method. It is also important that all parties involved agree with and accept the methods employed. Commercial activities are facilitated by mutual recognition of the test methods used in relation to international trade.

Standardised methods have been developed by international, national and trade organisations, such as ISO (International Standards Organisation), AOACI (Association of Official Analytical Chemists International), CEN (Comité Européen de Normalisation), NMKL (Nordisk Metodikkomitté för Livsmeddel), AFNOR (Association Française de Normalisation), NNI (Nederlands Normalisatie Instituut), DIN (Deutsches Institut für Normung) and IDF (International Dairy Federation): see Bertram-Drogatz *et al.* (2000).

These standardised methods for detection and/or enumeration of microbial contaminants in foods usually involve traditional isolation techniques. The primary intention is to provide the user with a reliable and internationally-accepted method that allows equivalent results to be obtained in different laboratory settings, without dependence on the materials of any one manufacturer. Although, in essence, such methods serve only as analytical guidelines, historically they are recommended or accepted in many countries by governmental and trading agencies, and they are recognised as official methods for the detection or enumeration of microorganisms in foods. Thus, they are considered to be reference methods (Hitchins, 1996; Lahellec, 1998).

During the 1990s, several alternative methods have been introduced for detecting and/or enumerating microorganisms in foods as a result of recent developments in immunology, biotechnology and instrumentation. These alternative methods are often more rapid and user-friendly and open to automation. Therefore, they are of great interest to the food industry and to control laboratories and are often preferred for routine use to the classical reference methods. Before an alternative method can be used and become accepted by the interested parties, fitness for purpose must be demonstrated by an independent organisation. In practice, a validation study is needed to show that the technical performance of a new method is acceptable.

The increasing need for validated methods also arises from the concept of the 'official' laboratory, which must be accredited according to the requirements of EN/ISO/IEC 17025 (ISO, 2005) and thus needs to use either standardised methods, validated alternative methods or, if using a modified standard method or their own alternative, to carry out a proper validation. It is also evident that standardised and validated methods can only give reliable results if used in a food microbiology laboratory with an overall Quality Assurance programme in place.

The first part of this chapter defines a number of technical performance characteristics that are commonly determined in a validation study. The second part gives an overview of the validation protocols that are currently in use or suggested by AOAC International and the European validation procedure set out in EN/ISO 16140 (ISO, 2003), which will be the basis for MicroVal certification of test kits that is recognised throughout the European Union (EU). This part of the chapter discusses a number of issues that need to be addressed in developing an effective validation scheme. In a third part, some examples of validation studies on new techniques are discussed to illustrate the above-mentioned issues. Finally, the fourth part of the chapter will focus on the need for laboratory managers to be committed to Quality Assurance in order to provide controlled conditions for microbiological analyses, involving properly validated methods. The chapter concludes with an overview of future trends and developments.

12.2 Definition of performance characteristics

Validation is a demonstration that the technical performance of a particular method is comparable to that of the relevant standard method and the method has proven ability to detect or enumerate the organism or group of organisms specified. It should include one or a combination of the following:

- determination of the performance of the method using reference materials (see below) or, if these are not available, well-characterised pure cultures of relevant test strains;
- a comparative study of the appropriate methods;
- an inter-laboratory study of the method in question;
- testing to determine the effects of certain variables, e.g. period of incubation, food matrix on the performance of the method.

Validation studies typically determine some or all of the following parameters.

12.2.1 'Trueness' (bias) of quantitative methods

Trueness is the closeness of agreement between the true value or, if not known, the accepted reference value and the mean result, which is obtained when the test procedure is applied a large number of times (= systematic error) (Ellison *et al.*, 2000; ISO, 2003).

The trueness can be obtained by an analysis of certified reference materials (CRMs), obtained from a certifying body, e.g. the Community Bureau of Reference (BCR) of the European Commission (EC). The production of reference materials is based on spray-drying bacteria suspended in milk. The resultant material is then mixed with sterile milk powder and treated to ensure homogeneity and stability, before being encapsulated in gelatine. Following analysis by various, experienced laboratories, a number of reference materials were given BCR certification. Then, tables were produced showing the 95 % confidence limits for the number of capsules likely to be examined in practice (In 't Veld, 1998; In 't Veld *et al.*, 1999). However, only a limited number of CRMs are available for microbiological purposes. Non-certified reference materials with known values are available from different organisations for more, but not all, test organisms. Organisations that are active in distributing reference materials are: Cécalait (France); the working group CHEK of the Food and Consumer Product Safety Authority (VWA, Netherlands); LED Techno (Zolder, Belgium), distributing Senate™ (Bury, UK); BioTRADING, distributing Quality Testing Schemes (Mijdrecht, Netherlands) and Oxoid (Basingstoke, UK), distributing Remel Inco's Quanti-Cult®vials and loops, etc.

In circumstances where no stable reference material is available, spiked materials can be used as an alternative for recovery studies, on condition that the inoculum level used for spiking is determined independently by a standardised method and replicate testing. As a substitute, 'known values' can be obtained after many replicated measurements on naturally-contaminated samples, using the reference method, which is independent of the method to be validated. Under these conditions, however, the true value is not established exactly and trueness can only be determined approximately by examining the same sub-samples with both the reference method and the method being evaluated. Participation in proficiency testing schemes can also help in assessing laboratory performance for the parameter of trueness.

12.2.2 'Precision' (quantitative methods)

Precision is the closeness of agreement between independent test results obtained by applying the test procedure several times to the same sample under stipulated conditions (= random error) (Notermans *et al.*, 1997; Ellison *et al.*, 2000). It is usually expressed in terms of imprecision and calculated as a standard deviation from the test results. Low precision is reflected by a relatively large standard deviation.

Checks on precision should be made for routinely-used methods to ensure that the result does not vary with time due to changes in reagents, equipment, staff, etc. Distinction is made between:

- 'repeatability', which indicates the variability observed within a laboratory, over a short time, using a single operator, the same apparatus and identical test material;
- 'reproducibility', which indicates the variability observed when different laboratories analyse the same sample by the same method – intra-laboratory reproducibility also relates to the variation in results observed when one or more factors, such as incubation time, reagents, equipment and operator, are varied within the same laboratory.

12.2.3 'Accuracy' (qualitative and quantitative methods)

Accuracy is the closeness of agreement between a test result and the true value or, if not known, the accepted reference value. It is a qualitative concept and involves a combination of random components and common systematic error. For microbiological analyses, the term 'relative accuracy' is sometimes used and is defined as the degree of correspondence between the response obtained by the reference method and that yielded by the alternative method from identical samples. The term 'relative' implies that the reference method does not automatically provide the accepted reference value (ISO, 2003).

12.2.4 'Detection limit' (qualitative and quantitative methods)

'Detection limit' is the smallest number of culturable organisms that can be detected reliably in a sample. For qualitative methods, it can be defined as the smallest number of culturable organisms detectable on 50 % of occasions by both reference and alternative methods. For quantitative methods, the detection limit is higher than the critical level, the latter being the lowest level of culturable organisms that can be enumerated reliably. For instance, this can be the average of results from a blank sample (no target organism) plus three times the standard deviation of the blank and can be determined by analysing a rather high number of blank samples (ISO, 2003). Such samples account for any effect of the food matrix and/or competing flora on the test result.

12.2.5 'Linearity' (quantitative methods)

When used with a given matrix, linearity is the ability of the method to yield a result that is in proportion to the amount of analyte present in the sample. That is, an increase in analyte corresponds to a linear or proportional increase in the result (ISO, 2003).

12.2.6 'Sensitivity' and 'specificity' (qualitative methods)

Sensitivity and specificity are associated with the degree to which a method responds uniquely to the specified target organism or group of organisms, and they relate to the number of false-positive and false-negative results obtained with the validated method. Various definitions for sensitivity and specificity have been proposed, among which are the following (Notermans *et al.*, 1997).

The sensitivity of a method is indicated by the proportion of target organisms that can be detected among a known population; it can be calculated from the following equation:

$$\text{Sensitivity (\%)} = \frac{\text{number of true positives } (P)}{P + \text{number of false negatives}} \times 100$$

A failure to detect the target organism(s), when present, is a false-negative result and will lower the sensitivity of the test. In food microbiology, only a very low frequency of false-negative results can be tolerated, because of the food-safety implications.

The specificity of a method is the ability to discriminate between the target organism(s) and other organisms; it can be calculated from the formula:

$$\text{Specificity (\%)} = \frac{\text{number of true negatives } (N)}{N + \text{number of false positives}} \times 100$$

A positive result in the absence of the target organism(s) is a false-positive result and will lower the specificity of the method. For rapid screening methods, a relatively high frequency of false-positives may be acceptable, because all apparent positives are subjected to confirmatory tests.

In addition to sensitivity and specificity, as described above, the inclusivity and exclusivity of a qualitative method may also be determined. Inclusivity is the ability of the validated method to detect a wide range of strains belonging to the target organism(s), while exclusivity is the extent to which the validated method fails to detect a relevant range of non-target strains (ISO, 2003).

12.2.7 'Robustness'/'ruggedness' (qualitative and quantitative methods)

The terms 'robustness' and 'ruggedness' are applied to the sensitivity of the method to small changes in environmental conditions or methodological factors during execution, e.g. time and temperature of incubation, sources of ingredients/materials, purity and shelf-life of reagents (Andrews, 1996).

12.2.8 'Practicality'

Other considerations that can be important in adopting a new method, and are generally referred to as practicality, may include (Andrews, 1996):

- any safety hazards associated with the test procedure;
- whether the procedure is quick and easy to perform, shows possibilities for automation and allows a high throughput of samples;
- a need for the analyst to receive extensive training;
- availability of the test system and the reputation of the manufacturer (Quality Control during production, servicing arrangements, etc.).

12.2.9 Performance characteristics and standardised methods

In the validation process, determination of the performance characteristics of a method will facilitate its acceptance by international, national and regional regulators and trading partners. Standardised methods that are published by the corresponding standardisation organisations or trade bodies are considered to have been validated. In such cases, the laboratory concerned must demonstrate that validation criteria indicated in the standard can be achieved in practice. However, such methods have not necessarily been validated. In contrast to IDF and AOACI methods, which have gone through a validation process before acceptance, only recently have a number of ISO methods been subjected to a validation process that is required by the EC (Standards, Measurement and Testing, Fourth Framework Programme, Project SMT4-CT96-2098). The performance characteristics of six ISO methods were determined, namely those for *Bacillus cereus* (enumeration), *Listeria monocytogenes* (detection and enumeration), *Staphylococcus aureus* (enumeration), *Clostridium perfringens* (enumeration) and *Salmonella* (detection), and details have been published recently as amendments to the corresponding ISO methods (Schulten *et al.*, 2000; Scotter *et al.*, 2001a,b).

Precision, including limits for repeatability (*r*) and reproducibility (*R*), was determined on the basis of inter-laboratory tests and involved three types of food, minced meat, fresh cheese and dried potato, contaminated at various levels, and reference materials. The values obtained from the inter-laboratory tests may not be applicable to bacterial concentrations and food matrices other than those studied. A general indication of the repeatability limit (*r*) for *B. cereus*, when testing food samples, is $r = 2.0$ (expressed as a ratio of the test results). This means that, if a first test result of 10000 *B. cereus* per gram of food product was obtained and then the test was repeated, the ratio between the first and second test results should not be greater than 2.0. Thus, the second result should be between 5000 ($10000 \div 2$) and 20000 (10000×2) per gram. Correspondingly, a general indication of the reproducibility limit (*R*) for *B. cereus* is $R = 2.6$, again expressed as a ratio of the test results. This means that, if a test result of 10000 *B. cereus* per gram of food was observed at the first laboratory, the ratio of the respective results from the first and second laboratories should not be greater than 2.6. Therefore, the result from the second laboratory should be between 3800 ($10000 \div 2.6$) and 26000 (10000×2.6) per gram.

For reference materials, the limit values for repeatability and reproducibility are lower, because of the absence of the matrix effect: $r = 1.3$ and $R = 1.7$ (ISO, 2004). It is expected that the intra-laboratory reproducibility limit will also be lower than that reported between laboratories, since fewer variables are involved. While extensive validation studies are sometimes lacking, historically, ISO methods have been considered as internationally–accepted standard methods, because they are the outcome of open discussions between experts of the different participating nations (recommended by national committees) in dedicated working groups, including ISO/TC34/SC9 for microbiology (Lahellec, 1998).

Although these standardised methods are revised regularly to include improvements that have been made in traditional isolation methods, this process is laborious and time consuming, and such methods do not always include the latest developments. If any modification is made to a standardised method by the user, e.g. a change in the period of incubation, a reduction in the number of culture media or confirmatory tests and limited chill storage of culture media before reading plates or proceeding with the test protocol, a limited validation exercise should be carried out. This is needed to show that the modification does not affect the outcome of the test method and still guarantees a reliable result, at least for the food matrix in question. For example, it has been demonstrated recently that the semi-solid medium MSRV (modified semi-solid Rappaport–Vassiliadis) serves as a good alternative to the selective enrichment broths used in detecting motile *Salmonella* strains in animal faeces and in samples from the primary livestock-production stage. This will be included as an amendment to the EN/ISO 6579 horizontal method for the detection of *Salmonella* (ISO, 2002).

During the 1990s, numerous new methods were developed that either generate results more rapidly than the traditional culture methods and/or are easier to use, while being suitable for automation. These systems can be prepared 'in-house' or are available as commercial test kits. An extensive validation exercise should be carried out if the method is to be used on a routine basis in an official control laboratory applying for accreditation. Where an alternative method is to be used routinely without the requirement to meet external Quality Assurance criteria, e.g. in self-regulating or applied research laboratories, a less stringent validation of the alternative method may be appropriate. Thus, it is a matter for the user, in agreement with the client or auditor, to decide how extensive or stringent the validation protocol should be and how far it needs to go in relation to the number of samples, food matrices and repetitions involved. At present, within ISO TC34 SC9 /CEN TC275 WG 6, it has been acknowledged that EN/ISO 16140 (ISO, 2003) is restricted in scope to full validation of alternative methods by official certification bodies. The need for an appropriate, easy-to-handle and efficient standard was identified for the in-house evaluation/validation of a method optimised in a research laboratory prior to routine, restricted application. Also needed was laboratory verification of a further-developed, horizontal standard method applied to a defined food type. At present, a working group in ISO TC34 SC9, constituted in December 2005, focuses on the development of standards with defined terminology and minimum requirements for these types of validation studies.

12.3 Validation protocols

An appropriate procedure should be developed for each proposed method. The validation protocol will differ for a qualitative and a quantitative method, and the stringency of the criteria set for the technical characteristics of the method will depend upon its subsequent use, e.g. as a rapid screening method within the framework of a HACCP programme or an analytical procedure to detect the cause of a foodborne disease outbreak. The criteria will also depend on the scope of the method, e.g. the type of microorganism being sought and the kind of food involved.

In the past, different countries have developed their own validation schemes. Also, several standardisation organisations, such as AOACI, IDF, AFNOR, NMKL, have expanded their activities and developed validation protocols for alternative methods. This has frustrated the manufacturers of new systems, because they have to undertake different validations in different countries, in order to gain widespread acceptance of their tests. Clearly, there is a need for the harmonisation of these validation schemes.

In 2002, the European standard 'Protocol for the validation of alternative methods' was accepted by the CEN. This standard was the outcome of the MicroVal project, which started in 1993 with the aim of setting up a European validation procedure (Rentenaar, 1996; MicroVal secretariat, 1998). The first goal, to establish an internationally accepted protocol for the validation of alternative microbiological methods through standardisation, has been achieved. Through the CEN/ISO 'Vienna agreement', this European standard will also be adopted as an ISO standard (ISO, 2003), and agreement has been reached with AOACI for mutual recognition of the different validation schemes. The standard EN/ISO 16140 (ISO, 2003) was prepared by Technical Committee CEN/TC 275 'Food analysis – Horizontal methods' in collaboration with Technical Committee ISO/TC 34 'Agricultural food products' and was due for revision in 2005. Also, a European organisation has been set up for the independent certification of alternative methods based on the European standard (second goal of the MicroVal project) and pilot validation studies are on-going (Rentenaar 1996; MicroVal Secretariat, 1998). The proposed standard for validation of alternative methods (ISO, 2003) describes technical protocols for the validation of both qualitative and quantitative methods, each of which includes a methods comparison and an inter-laboratory study. There are specific recommendations for the design of the test protocol and calculation and interpretation of the data obtained, following appropriate statistical analysis. Standardisation of the validation scheme, as provided by MicroVal, represents important progress in applying consistent validation requirements. Nevertheless, the acceptance criteria are not defined clearly in the protocol. Results obtained with the alternative method are required to be 'comparable' with those of the reference method. The actual criteria will depend upon the type of method and the circumstances under consideration.

AOACI is an organisation that has a long tradition in the validation of laboratory methods. Collaborative studies are the essence of the AOACI validation process. In this type of study, competent, experienced analysts, working independently in different laboratories, use a specified method to analyse homogenous samples for a particular microorganism. Although no standardised protocol is available for the collaborative study, various recommendations are given regarding the minimum number of food types to be tested and the number of samples to be analysed for each food type. An associate referee, under the guidance of a general referee, and assisted by a statistical consultant, is responsible for the actual development of the study protocol, and this must be approved by the Methods Committee on Microbiology and Extraneous Materials. The associate referee also conducts ruggedness testing and a pre-collaborative study to determine the applicability of the method for detecting the target organism(s) in a wide range of food matrices and different conditions. The associate referee is required to be an expert on a particular target organism, type of method or food matrix, or a combination of these. The general referee must be a recognised authority in the field of interest (Andrews, 1996; De Smedt,

1998). A number of issues need to be addressed in the preparation of an effective validation scheme.

12.3.1 Choice of reference method

The reference method should be one that is accepted internationally. Usually, an internationally standardised method is chosen (ISO, AOACI or IDF method) or, if none is available, it is possible to use certain nationally-recognised methods or a method that has been published in a scientific journal and used successfully for several years by different laboratories, but without official recognition (Andrews, 1996; ISO, 2003). The choice of reference method is important, because the method is supposed to give the 'true' result. Indeed, for qualitative methods, the alternative under validation is considered to produce a false-positive result, if a positive is obtained while the reference method shows a negative result. This implies that the reference method should reveal all contaminated samples as true positives. For example, in validation studies on rapid systems for detecting *Salmonella*, the ISO method can be chosen as a reference method. However, it has been shown for certain food matrices that a modified ISO method, using a semi-solid medium, such as DIASALM (LabM) or MSRV (Oxoid), instead of selective enrichment in a broth, leads to a higher number of confirmed *Salmonella*-positive samples than the original ISO method (De Zutter *et al*., 1991; van der Zee *et al*., 2002). Therefore, this modified ISO method is equally suitable as a reference method for comparative studies (Poppe and Duncan, 1996).

However, for organisms that may be difficult to culture, e.g. *Campylobacter* in environmental samples or frozen/acidified/fermented foods and/or with organisms for which an appropriate cultural method is lacking, e.g. non-O157 enterohaemorrhagic *Escherichia coli* serotypes, it may be difficult to establish a 'reference method' for comparative tests. In these cases, an alternative, including a polymerase chain reaction (PCR) method or a microscopical method, such as FISH (fluorescent *in situ* hybridisation), may be more reliable for detecting these pathogens. The alternatives are based on non-phenotypic characteristics involving DNA or rRNA.

To determine the most appropriate methods for diagnostic purposes and comparative studies, experts in research laboratories should be consulted and a consensus obtained. At present, within CEN TC275 WG 6, a working group is standardising the PCR system for detecting foodborne pathogens, such as verotoxin-producing *E. coli*. In addition, criteria for standardising diagnostic PCR and the attendant problems in harmonising PCR-based methods, including the possibilities for overcoming them, have been addressed in a EU research project, FOOD-PCR (Malorny *et al*., 2003). For example, PCR-assay development, analytical validation of the method and validation in a multi-centre collaborative trial have been carried out for detecting foodborne, thermotolerant campylobacters (Lübeck *et al.*, 2003a,b).

12.3.2 Number of food types to be tested and samples analysed

The results obtained from the method being evaluated are invariably affected by the type of food product being examined. The number of organisms present in the accompanying flora of the food product, as well as their character and biochemical activities, affect the behaviour of the target organism(s). In addition, the intrinsic and extrinsic properties of the food, including pH, a_w, temperature, storage atmosphere, naturally-occurring or added antimicrobials and the nature and severity of any processing to which the food was subjected are likely to cause sublethal injury to the target organism(s) and may reduce recovery by the proposed method (Struijk, 1996). Moreover, the composition of the food itself may complicate the assay further, because there are various constituents that can interfere directly with the assay procedure, as is often observed with PCR methods (Wilson, 1997).

The number of food types to be included depends upon the applicability of the method. If the method is to be validated for all foods, then five categories of food are usually included in the validation study. Food categories are mainly determined by the origin of individual products, e.g. meat products, poultry, fish and seafood, fruits and vegetables, dairy products, chocolate/bakery products and others, such as dressings and mayonnaise, egg products and cereals. Animal feed, veterinary and environmental samples should be regarded as a separate category. Also, the mode of processing to increase shelf-life, e.g. heat-processing, curing/salting, fermentation and freezing, may be used to select further types within a single food category (ISO, 2003). The types of food chosen should be relevant to the target organism(s) being sought. For example, validation studies for a *B. cereus* method might include samples of rice, spices, raw and heat-processed dairy products and heat-processed, vegetable-based products, whereas validation studies for *Campylobacter* might include raw poultry and red meat, raw seafood and raw milk. However, if the applicability of the method is restricted, then the number of food categories involved may be reduced. An example here would be the detection of *Vibrio parahaemolyticus* in fish and seafood products only. Thus, the outcome of a validation study, if successful, will be acceptance of the method for a particular type of food or otherwise for all foods.

Whenever a method is validated for detecting a particular target organism in all foods, questions arise as to whether the method is really effective for each individual type of food associated with that organism (Andrews, 1996). In the validation study, only a limited number of food types are included. If, for example, a validated method for *Listeria monocytogenes* produces reliable results for five food types from each of five categories of food, e.g. sliced, pre-packed cooked ham (heat-processed meat product), raw milk-based soft cheese (raw dairy product), smoked salmon (processed fish product), green salad (raw vegetable product) and pasta (other products), does this guarantee that the validated method will reliably detect the presence of *L. monocytogenes* in raw milk, pasteurised cheese, raw poultry, frozen fish, pâté, fermented meat, bakery yeast, etc? This is still an open question and therefore the validation report should state clearly and exactly the types of food that were involved in the validation study. Although official validation of a method is an indication that the method works well, the user should always demonstrate that the method produces reliable results, when applied to a particular type of food matrix in a limited, in-house validation study.

With regard to the number of samples to be included in the validation study, a sufficient number of identical samples should be analysed by both the reference method and the method under validation, in order to generate sufficient data for statistical analysis and correct interpretation. Examples of sample numbers included in the study protocols proposed by MicroVal and used by AOACI are shown in Table 12.1. It should be mentioned, however, that these figures are intended for an extended validation of a commercial test system. The validation of an internal method in a food control laboratory, industrial laboratory or applied research laboratory may include varying numbers of samples, depending upon the type and scope of the method. This is discussed further in the last part of the present chapter.

It is recommended that the food samples used in a comparative study come from as wide a distribution of sources as possible, in order to reduce any bias from local factors. It is desirable for qualitative methods to yield approximately equal proportions of positive and negative results for the same food type, although this may not always be feasible when analysing naturally-contaminated samples for foodborne pathogens. The reference method and the method under validation should utilise, as far as possible, exactly the same samples. For example, if the first stage of each method is the same, i.e. the same pre-enrichment broth or primary dilution, then subsamples can be taken after this first common step (ISO, 2003). Effective comparison of pathogen detection methods often requires testing at around the detection limit, where differences between methods are likely to be most apparent. In this case, it may be impossible to generate accurately paired samples, where it is known that the number of target organisms in one sample of the pair is the same as the other (Baylis *et al.*, 2001).

12.3.3 Naturally-contaminated versus artificially-contaminated food samples

Whenever possible, naturally-contaminated samples should be used in any comparison of methods, since these represent the real-life situation, with

Table 12.1 Comparison of validation schemes suggested by MicroVal and AOAC International

MicroVal[a]		AOAC International[b]	
Qualitative method	Quantitative method	Qualitative method	Quantitative method
Inclusivity 50 strains	*Inclusivity* 30 strains	*Inclusivity* 100–200 strains	*Inclusivity* 30–40 strains
Exclusivity 30 strains	*Exclusivity* 20 strains	*Exclusivity* 20–30 strains	*Exclusivity* 10–20 strains
Methods comparison study 5 categories of foods 60 samples per food category (minimum 3 food types)	*Methods comparison study* 5 categories of foods 5 levels of target organism(s) 2–10 replicates	*Pre-collaborative study* 20 food types 3 inoculum levels (single strain): – control (uninoculated) 5 replicates – low (near the detection limit) – high (10 × detection limit) 20 replicates each	*Pre-collaborative study* 20 food types 4 inoculum levels – control (uninoculated) – low – medium – high 5 replicates each
Detection limit 5 food types (5 food categories) 3 inoculum levels: – control (uninoculated) – low (near the detection limit) – high (10x detection limit) 6 replicates each	*Detection and quantification limit* 5 categories of foods 6–10 blank samples		
Inter-laboratory study At least 10 laboratories 1–3 food types 3 inoculum levels 8–10 replicates each	*Inter-laboratory study* At least 8 laboratories 1–3 food types 4 inoculum levels at least 2 replicates each	*Collaborative study* at least 15 laboratories 5 food types 3 inoculum levels 5 replicates each	*Collaborative study* at least 8 laboratories 5 food types 4 inoculum levels 2 replicates each

[a] EN/ISO 16140, Scotter *et al.* (2001a, b).
[b] Andrews *et al.* (1998).

the target organism(s) present as a minority (of pathogenic bacteria) among a large majority of other bacteria and in a non-optimal (stressed) condition, due to the intrinsic properties of the food and the processing and storage conditions. Naturally-contaminated samples can be collected from products being analysed routinely by the organising laboratory or other laboratories. Storage of the samples should be minimised to prevent changes in levels of the target organism(s) and further stress of those present. The presence of the required organism(s) should be confirmed by the reference method prior to or during the validation study (ISO, 2003).

If it is not possible to acquire a sufficient number of naturally-contaminated foods for each of the categories relevant to the validation study, artificially-contaminated (spiked) samples can be used. However, it is recommended that no more than 80 % of the samples should be of this kind (Andrews, 1998). When such samples are used, the inoculation levels should be similar to those expected in naturally contaminated foods, with no undue stress on the organisms. Where needed, a protocol for the preparation of stressed (sub-lethally injured) organisms should be established involving, for instance, chill storage, freeze stress and acid stress (Baylis *et al.*, 2001), and the degree of stress demonstrated at the time of inoculation by comparing the length of the lag phase and/or bacterial numbers in the stressed culture to those of a normal culture on a non-selective and/or a selective culture medium (Baylis *et al.*, 2001; Restaino *et al.*, 2001; Uyttendaele *et al.*, 2001). Also, the background flora (numbers, distribution and physiological state) in artificially-inoculated samples should be similar to that occuring naturally (Andrews, 1996).

12.3.4 Source, number of strains and inoculum level

Reference materials, containing appropriate, but well-defined levels of target organisms in a stable, but stressed state, may be used for spiking samples. However, their value is limited when only a few strains or serotypes of food origin are available in this form. For spiking purposes, strains that have been isolated from the same type of food product are preferred over clinical isolates. If this is not possible, then only fully characterised, reference strains should be used (Andrews, 1996; ISO, 2003).

In selecting strains to test the inclusivity and exclusivity of the method, the majority of strains should originate from the types of food used in the validation study and cover the recognised range of the target organism(s) with respect to geographical distribution, incidence and diversity of biochemical and physiological characteristics, as well as serotype, phage type, etc. If the proposed method can detect all species of a particular genus, a range of species from that genus and, if possible, all species of the genus should be included, as well as a number of representative species from a range of genera within the same family. When detection is only at species level, a range of strains from that species, originating from different sources, should be selected. It is also desirable to have representative strains from other species within the same genus. In testing the selectivity of the method, various non-target organisms known to be part of the background flora of the foods used in the validation study could be included (ISO, 2003).

Attention should be paid to the inoculum level, especially in the case of presence/absence testing. Inoculum levels should represent the levels that are normally encountered in the foods being analysed. In addition, tests should be made with inoculum levels near the detection limit of the method or the level that needs to be detected according to criteria set for the target organism. This will ensure the suitability of the test method for compliance testing.

12.4 Use of validation schemes for evaluating the performance of alternative methods

In the following sections, a few examples are given of the experimental design used in evaluating rapid methods to illustrate the above-mentioned principles. Because food manufacturers and regulators need quick and reliable information about the presence of foodborne pathogens in the food supply, new rapid or user-friendly methods for detecting foodborne pathogens are continually being introduced, particularly for *Salmonella*, *L. monocytogenes* and *E. coli* O157:H7. Thus, the majority of comparative studies are concerned with qualitative methods for these pathogenic bacteria. The examples mentioned hereafter include both immunological and molecular methods, which were chosen because of their wide application and/or potential in food microbiology laboratories throughout the world. These examples merely seek to illustrate the methodology applied in evaluating or validating the methods and no judgement on their actual performance should be implied, since further, more relevant information may be available.

12.4.1 Immunological methods for detecting foodborne pathogens

VIDAS® Listeria *enzyme-linked immunofluorescent assay*

The VIDAS® *Listeria* (LIS) (bioMérieux) is a qualitative, enzyme-linked, fluorescent immunoassay carried out in an automated system for the detection of *Listeria* spp. This method allows rapid screening for the presence of *Listeria* spp. in foods and environmental samples after a 44–48 hour prior enrichment step. Positive results must be confirmed by standard cultural methods.

The performance of the system was reviewed by the AOAC Research Institute's Performance Tested Methods Programme and found to meet the manufacturer's claims. Inclusivity and exclusivity were demonstrated with 206 strains of *Listeria* spp. and 50 strains of non-*Listeria* organisms. In a pre-collaborative study involving 980 samples of uninoculated, inoculated and naturally-contaminated samples, representing 17 different foods, the test was shown to be as good or better than cultural methods used on the same samples. The latter methods are described in the eighth edition of the US Food and Drug Administration – *Bacteriological Analytical Manual* (FDA, 1998). The collection of food matrices tested included dairy products, seafoods, vegetables, raw meats and poultry, as well as cooked and processed meats and poultry. A number of environmental surfaces were also sampled and subjected to testing.

Subsequently, the VIDAS® LIS method and the traditional culture method (FDA, 1998) were evaluated in a collaborative study. Nineteen laboratories participated in the study. Six food types (ice cream, cheese, green beans, fish, roast beef and ground turkey) were selected for inclusion in the study. Ice cream, green beans, and cheese were each inoculated with a different serovar of *L. monocytogenes*, roast beef was inoculated with a strain of *L. innocua*, and fish was inoculated with *L. welshimeri*. Ground turkey samples were naturally contaminated with *Listeria*. Each food type was divided into three portions; the first two were inoculated (1–5 cfu/25 g for low-level inoculation and 10–50 cfu/25 g for a high inoculum) and the third served as an uninoculated (negative) control. Cheese samples were stabilised by storing at 4 °C for five days. All other samples were stabilised by storage at –20 °C for five days. Each collaborator received a set of 15 samples for every food product (five replicates of each inoculum level and a negative control). Of 1558 samples tested, 935 were positive: 839 by the VIDAS® method and 809 by the standard culture method. Overall, false-negative rates were 10.3 and 13.5 % for the VIDAS® LIS and culture methods, respectively. The false-positive rate for the VIDAS® LIS assay was 1.4 %, based on nine VIDAS® LIS-positive assays that did not confirm positive by isolation of *Listeria*. For all samples tested, agreement between the two methods was 86 %. Results for each food type and contamination level were as good or better with VIDAS® LIS than those obtained with the traditional culture method (Gangar *et al.*, 2000). On the basis of these studies, the VIDAS® LIS method for detecting of *Listeria* spp. was recommended for Official First Action. The recommendation was approved by the Methods Committee on Microbiology and Extraneous Materials, and was adopted by the Official Methods Board of AOACI.

The VIDAS® LIS system also received an AFNOR validation certificate for the rapid detection of *Listeria* spp. in all food products. Inclusivity and exclusivity were described as 'specificity' in the AFNOR certificate and these attributes were demonstrated with 217 strains of *Listeria* (207 isolated from food and 10 from culture collections of *L. monocytogenes*) and 35 strains of non-*Listeria* organisms. The intrinsic detection limit, defined as the number of *Listeria* required to obtain a positive response with the VIDAS® LIS system, was 10^4–10^5 cfu/ml and was obtained with four pure strains of *Listeria*. The detection limit was determined using four different types of food (meat, vegetables, milk products and seafood), each artificially contaminated with four strains of *Listeria* at five different contamination levels: 0, 1–10, 2–20, 5–50 and 10–100 cfu/25 g. Agreement between the two methods was found to be 96.4 % (80/83). The three discrepant results involved levels of two or three cells/25 g. Accuracy was determined by a comparison of the VIDAS® LIS method with the reference method (ISO, 1996), using 204 samples of various products, of which 88 were naturally contaminated and 116 were not. All samples were tested in duplicate by both methods. Overall, the level of agreement between the two was described as 'good' by AFNOR (eight false-negatives were obtained with the VIDAS® LIS system and five with the reference method). Precision data were obtained from an inter-laboratory assay involving 13 different laboratories. Analyses were made on pasteurised milk samples, artificially contaminated with a strain of *L. monocytogenes* at four different levels: 0, 1–10, 5–50 and 10–100 cfu/25 g. All the results were as expected and the method was shown to be reliable. In 2002, an updated version of the VIDAS® LIS system (VIDAS LMO2) was launched, using two complementary monoclonal antibodies for the capture and detection processes. These are directed at different antigenic sites of a specific *L. monocytogenes* virulence protein. The updated test kit has also received AFNOR validation.

*Dynabeads ® anti-*Salmonella *system*

Dynabeads ® anti-Salmonella (Invitrogen–Dynal) employs magnetisable particles coated with specific antibodies to selectively concentrate all *Salmonella* serovariants from foods and environmental samples. The technique, which takes 15–20 minutes, may replace or enhance the performance of the 18–48 hour conventional selective enrichment. The protocol described originally included immunomagnetic separation (IMS) from buffered peptone water (BPW)-enriched food samples, followed by plating (IMS-Plating). The direct plating of bead-bacteria complexes onto solid media is suitable for processed foods or samples known to have low levels of microbial contamination. With raw food samples, such as raw poultry, IMS-Plating sometimes resulted in the overgrowth of target salmonellas by interfering enteric organisms on the plating media. As an alternative to the traditional culture method for detecting *Salmonella*, a modified IMS protocol can be used consisting of the standard pre-enrichment of samples in BPW, followed by IMS and subsequent selective enrichment of the bead–bacteria complexes in Rappaport–Vassiliadis soya peptone (RVS) broth, before plating on selective media (IMS-RVS-Plating). The performance of both the IMS-Plating and the IMS-RVS-Plating methods was compared with the conventional EN/ISO 6579 method (ISO, 2002), using ten food samples (powdered skimmed milk, mayonnaise, cake-mix, raw chicken meat, cooked sausage, cheese, pepper, meat balls, lasagne, casein) inoculated prior to pre-enrichment with 20 different *Salmonella* serotypes (two serotypes per sample) at low (1–5 cfu/25 g) and medium (10–50 cfu/25 g) levels. All inoculated samples were frozen for one month before being examined. A 100 % agreement was achieved between the direct IMS-Plating and IMS-RVS-Plating methods for the ten food samples, but the latter method favoured the formation of well-isolated, almost pure cultures of *Salmonella* on the plates. The IMS methods showed, respectively, a 90 % and 95 % agreement with the ISO method and isolated, respectively, two and one more *Salmonella*-positive samples at the low and medium inoculum level. In an evaluation of both IMS methods with 100 naturally-contaminated samples (50 poultry carcasses, 20 cloacal and/or faecal swabs, 15 chicken liver samples, 15 chicken breast-meat samples and 10 poultry feed samples), there was a clear advantage in the IMS-RVS-Plating method (39 positive samples) compared to the conventional ISO 6579 reference method with 31 positive samples and the IMS-Plating method (only 20 positive samples), as described by Cudjoe and Krona (1997).

The IMS separation method has been successfully combined with the ELISA technique as an end-detection method for the recovery of *S.* Enteritidis from eggs and skimmed milk powder. However, studies on raw chicken using a combined IMS and ELISA technique gave a significant number of false-negative results, because of high levels of competing organisms. A study was undertaken to evaluate a modified IMS-ELISA method involving re-suspension of the bead–bacteria complexes and incubation in Gram-negative (GN) broth at 42 °C for six hours prior to the ELISA. This study is, however, an example of one in which the experimental design was too limited. In the first part, there were only two food types, animal feed and raw chicken, artificially inoculated, not in the food matrix, but in BPW pre-enrichment broth, using rather high inoculum levels (2000, 200 and 20 cfu/ml). Also, only three serotypes of *Salmonella* were involved (18 samples in total) in demonstrating that direct application of IMS-ELISA failed most of the time, although RV-XLD (xylose–lysine–deoxycholate: conventional culture method) and IMS-XLD succeeded in detecting *Salmonella*. The second part was even more restricted, and evidence for improved detection of *Salmonella* using IMS-GN-ELISA compared to the conventional RV-XLD methodology was based on only 15 samples, involving three replicate pre-enrichment broths containing raw chicken and artificially inoculated with *S.* Enteritidis at five inoculum levels (Mansfield and Forsythe, 2001).

12.4.2 Molecular methods for detection/identification of foodborne pathogens

The Probelia®/iQ-Check™ Salmonella *PCR systems*

Recently, the PCR has emerged as an increasingly important diagnostic tool in food microbiology. The Probelia® *Salmonella* PCR amplification and detection systems (Bio-Rad Laboratories) have received AFNOR approval for detecting *Salmonella* in foodstuffs. The Probelia® *Salmonella* PCR system is based on amplification of the *Salmonella iagA* gene (involved in bacterial invasion), followed by probe hybridisation in a 96-well format for colorimetric detection. In an independent study (Wan *et al.*, 2000), the Probelia® PCR system was evaluated for rapid and specific detection of *Salmonella* in dairy products and compared to the Australian standard method. Using bacterial DNA preparations derived from ten-fold serial dilutions of a pure culture of *S.* Agona, the intrinsic detection limit of Probelia® was determined as being 8–79 cfu/ml, equivalent to 0.2–2 cfu/PCR reaction. Next, a comparative study was carried out using skim milk powder artificially inoculated with *S.* Agona at 5–10 cfu/g and subjected to analysis both immediately and after storage at 5, 15 or 25 °C for up to six weeks (five replicates at each sampling point). A second food type was included: artificially inoculated ricotta cheese at 1–2, 10–20 and 100–200 cfu/25 g (three replicates each). For all of the 40 milk powder samples and 12 ricotta cheese samples, the Probelia® results were consistent with those of the Australian standard method. The study was restricted to dairy products, two food types and only one *Salmonella* strain. No naturally-contaminated samples were examined.

In 2002, the Probelia® *Salmonella* PCR system was superseded by a real-time PCR-based system, iQ-Check™ *Salmonella*, that utilises a fluorescent probe (a molecular beacon), which hybridises with the amplified products generated and measures fluorescence directly during the PCR annealing step. The iQ-Check™ system has been evaluated for *Salmonella* detection with both artificially- and naturally-contaminated food and environmental samples (Uyttendaele *et al.*, 2003). The artificially-contaminated samples included poultry meat and ground red meat, subjected to chilling and freeze stress. The 120 naturally-contaminated samples included poultry neck skin and pork carcass swabs taken at various abattoirs, environmental samples taken at an egg-layer farm and retail samples of poultry and red meat. All were tested for *Salmonella* using the semi-solid DIASALM method, and the iQ-Check™ PCR assay after 24 hours enrichment in BPW. When the *Salmonella* cells were severely stressed, e.g. by freezing at −18 °C for seven days, inoculated samples gave false-negative results with the iQ-Check™ PCR assay. Stressed cells have an extended lag time and, especially when low numbers are present, the lag time may vary considerably (Stephens *et al.*, 1997). When cells are exposed to stress, the normal enrichment period may reduce the overall sensitivity of the combined enrichment-PCR assay (Uyttendaele *et al.* 1998). In total, 45 of the 120 naturally-contaminated samples were *Salmonella*-positive using the DIASALM method. The iQ-Check™ PCR showed 92 % agreement with the DIASALM system. In 2004, the PCR method for detecting *Salmonella* acquired AFNOR validation.

The TaqMan® Salmonella *PCR system*
The Applied Biosystems TaqMan ® assay is a fluorogenic PCR-based system enabling real-time detection of PCR products, based on Roche Molecular Systems TaqMan ® instrument. The assay utilises the 5'nuclease activity of *Taq* DNA polymerase to hydrolyse an internal fluorogenic probe for monitoring amplification of the DNA target (the *Salmonella inv*A gene). Studies were made to elucidate the specificity and sensitivity of the assay for pure cultures of *Salmonella* and for *Salmonella*-contaminated foods (Chen *et al.*, 1997). A total of 164 *Salmonella* strains, representing all the subspecies of *S. enterica* were detected, while 52 non-*Salmonella* strains were not. The intrinsic detection limit of the PCR assay was 2 cfu/PCR reaction, when 10-fold serial dilutions of a pure culture of *S.* Typhimurium were used (PCR assays conducted in duplicate and replicated on different days). An appropriate sample-preparation protocol was established for the isolation of PCR-amplifiable DNA from foods. A detection limit of 3–7 cfu/PCR reaction was obtained using post-enrichment, spiked food samples (ground beef and pork) inoculated with 10-fold dilutions of the *Salmonella* and *c.* 3 cfu/25 g were detected when foods (raw milk, ground beef, ground pork) were inoculated with two-fold dilutions of the same *Salmonella* strain and pre-enriched overnight. In both experiments, no replicates were involved. Finally, naturally-contaminated foods (50 chicken carcass rinses and 60 raw milk samples) were examined for *Salmonella* using both the fluorogenic TaqMan® assay and the MSRV culture method that was employed as the reference method. Agreement between the two methods was over 98 %. Two samples were *Salmonella*-positive by the PCR assay, but negative by the MSRV method.

A PCR-ELISA test for detecting Shiga toxin-producing E. coli *(STEC)*
Since the 1990s, STEC and, particularly, strains of serogroup O157, have emerged as important foodborne pathogens. A study was carried out to evaluate the specificity and sensitivity of an in-house PCR-ELISA test for detecting STEC in dairy products, using pure cultures, spiked and naturally-contaminated samples (Fach *et al.*, 2001). The specificity of the PCR-ELISA was determined with 94 STEC strains, including a large range of STEC serotypes isolated from humans and animals, and 84 non-STEC strains. The sensitivity of the test was determined with duplicates of three individual STEC strains diluted 10-fold in steps from *c.* 10^6 to 1 cfu/ml. The detection limit in dairy products was determined on five different pasteurised cheeses, artificially contaminated with three STEC strains at four levels (0, 10, 100 and 1000 cfu/10 g, *c.* 30 replicates each, except for the highest inoculum where *ca* 15 replicates were used) and immediately subjected to testing. In a comparative study of the PCR-ELISA and vero-cytotoxicity testing as the reference method, 527 naturally-contaminated samples were examined (raw milk samples, unpasteurised cheeses, pasteurised cheeses

and dairy environment samples). Of these, 30 yielded STEC by both the PCR-ELISA and the vero-cell assay. Only one sample, which had a cytotoxic effect on the vero cells, was PCR-negative. PCR-ELISA detected STEC in an additional 74 samples. Thus, results from the PCR-ELISA and the vero-cell assay were not in total agreement. The overall level of agreement (negative or positive by both methods) was 85.8 %.

PCR/restriction enzyme analysis (PCR/REA) for identifying thermophilic Campylobacter *spp.*
In industrialised countries, thermophilic *Campylobacter* spp. (*Camp. jejuni*, *Camp. coli*, *Camp. lari* and *Camp. upsaliensis*) are recognised as the most common bacterial agents responsible for gastroenteritis in humans. The use of traditional phenoytypic tests for differentiating and identifying campylobacters is often hampered by the fact that these bacteria are fastidious in their growth requirements, asaccharolytic and possess few distinguishing biochemical characteristics. Therefore, an in-house PCR/REA method was compared with standard phenotypic tests for the identification of these organisms. In total, 182 presumptive, thermophilic campylobacters from 12 different animal species were tested. By PCR/REA, 95 % of isolates were identified as either one of the four species or as not belonging to these organisms at all. By standard phenotyping, 174 of the 182 isolates were identified initially as one or other thermophilic species. The PCR/REA and standard phenotypic tests showed only 67 % agreement in species identification. However, for most of the 52 isolates that were identified differently by the two methods, additional tests could explain the discrepancies and prove the accuracy of the PCR/REA. For example, 19 hippurate-negative isolates initially identified as *Camp. coli* by phenotypic tests were shown to be *Camp. jejuni* by PCR/REA; these were verified as *Camp. jejuni* by being positive for the hippuricase gene (Engvall *et al.*, 2002).

12.5 Application of validated methods in accredited laboratories

Confidence in the results of a microbiological analytical procedure relies on the suitability of the analytical method (determined by a validation study) and also on the competence of the laboratory in question to provide accurate, reliable and repeatable test results under controlled conditions. Accreditation of laboratories carrying out microbiological food analyses provides an assurance to those relying on its services that the test results are always reliable. Accreditation of a laboratory is the formal approval given by a national, authorised body (linked in Europe by the European co-operation for Accreditation of Laboratories, EAL), indicating that the laboratory is competent to carry out specific methods of analysis and has a commitment to Quality Assurance (Bowles, 2000). A detailed account of the requirements is given in EN/ISO/IEC 17025 (ISO, 2005). This aspect is considered further in Chapter 13.

It is clear that, to comply with the requirements of EN/ISO/IEC 17025 (ISO, 2005), the analytical laboratory should only use validated methods. However, from past experience of technical audits carried out in microbiological laboratories concerned with food analysis, it is known that validation studies in these laboratories have not always been adequate. Such studies are needed in the accreditation process to demonstrate that the laboratory is competent to execute a test method and interpret the results correctly. Sometimes, however, there is no standardised operating procedure available for validation studies and the details are worked out on an *ad hoc* basis, without any systematic approach.

Also, validation may only include reference strains, with no testing of food matrices, or the number and/or type of food matrix is restricted, despite accreditation being claimed for a wide range of foodstuffs, with varying degrees of microbial contamination. On some occasions, foods are autoclaved before inoculation with the target organism(s) to eliminate interference from other organisms. This may improve the response of the test to a low-level inoculum, which could be below the critical level for detection, when other organisms are present. Often, non-target organisms are not included in validation studies or those used are not representative of the likely interfering flora. Such organisms should be those most likely to occur and be taxonomically related to the target strains. Thus, they would be expected to react in a similar way in the test method. At one extreme, validation tests sometimes include only blank samples that have tested negative for target organisms and have not been inoculated artificially. These show the competence of the laboratory in avoiding false-negative results, but give no guarantee that the target pathogen can be detected, especially when present only rarely as a natural contaminant. Sometimes, not all technicians involved in using a particular method have taken part in a validation study or have any recorded training or proven ability to use the method. These findings demonstrate that setting up validation studies is still not given sufficient attention in many laboratories and reveal the need for a more standardised approach based on appropriate guidelines. On the other hand, validation studies carried out for accreditation purposes or the in-house use of particular methods should not involve unnecessarily extensive, laborious and time consuming work, but should show evidence of technical competence, taking into account the cost–benefit aspect of the procedure.

Apart from the need to use validated methods that are properly documented in standard operating procedures for laboratory accreditation, there is also a need for a systematic approach to Quality Control in order to guarantee a consistently reliable test result from any validated method used in the laboratory (Bolton, 1998; Lightfoot and Maier, 1998). Examples of key requirements are as follows:

1. Test the performance of culture media before use. For solid media, the modified Miles-Misra method of inoculation can be used. Detailed information can be found in the recommendations of the IUMS-ICFMH Working Party on Culture Media (Corry, 1995) and the CEN/ISO/TS 111333 parts 1 and 2, dealing with guidelines on the preparation and production of culture media.
2. Check the performance of the relevant equipment by regular monitoring and establish schedules for routine cleaning and maintenance. This can include, for example, checking the temperature of an incubator with a calibrated thermometer. Recently, in the framework of the European FOOD-PCR project, a biochemical test was developed to check the efficiency of the thermocycler used in carrying out PCR reactions. This is described in ISO/TS 20836 (ISO, 2005).
3. Use defined, internal Quality Control procedures, including first, second and third lines of control, as follows. The first line of control is checks performed by the analysts themselves, during each series of tests being executed under similar conditions. These can include the analysis of blank samples (no test organism) or positive (target organism) and negative controls (non-target organism), using reference materials or spiked samples. The results obtained from the positive controls can be used to produce control charts that give an indication of the overall performance and consistency of the results over time.

   The second line of control is less frequently implemented. It includes checks initiated by the laboratory quality manager that need to be carried out by the various analysts involved, e.g. testing of an identical sample (naturally contaminated or artificially inoculated) or interpretation of incubated culture media by more than one analyst. Results can be used in determining intra-laboratory reproducibility.

   The third line of control involves the participation of the laboratory in approved proficiency-testing schemes and aims to compare overall performance with that of other laboratories engaged in the same activities. The results can be used by the laboratory to determine the degree of 'trueness' for an analytical procedure.

   The data obtained from internal Quality Control, along with that from in-house validation studies, can be used to estimate the 'uncertainty of measurement' of an analytical method.
4. Ensure that analytical staff are properly qualified. Of major importance for laboratory accreditation is the use of appropriately qualified staff, who are competent to carry out the required tests and encouraged to continue their professional development.

The introduction of a quality system as a requirement for laboratory accreditation ensures tight control over the laboratory's activities and enhances confidence in the results produced. Formal accreditation of laboratories carrying out microbiological testing for the food industry is increasingly demanded by food manufacturers and government bodies.

12.6 Future trends

Dialogue between international organisations that are involved in the development and validation of microbiological methods is of major importance in achieving uniformity of practices and performance and should ensure the acceptance of results by governmental inspection laboratories and laboratories in the food trade, thus facilitating international trade. International cooperation is occurring, as is illustrated by various international symposia and workshops, jointly organized by ISO, CEN, AOACI and IDF to create a forum for the exchange of ideas and experience between microbiologists. Meetings of ISO TC 34/SC9 and CEN TC275/WG 6 are held yearly in the same week and at the same location, in order to facilitate coordination between the two groups. Experts from AOACI and IDF also attend these meetings. This should result in the adoption of more harmonised standard protocols, although such an objective is not always easy to achieve, due to the wide variation in interests and experience (Andrews, 1996; Hitchins, 1996; Lahellec, 1998). For alternative methods, it is in the interests of both suppliers and users of novel test methods to harmonise protocols for validation purposes. In 2002, the European standard 'Protocol for the validation of alternative methods' was accepted by the CEN and has also been adopted as an ISO standard (ISO, 2003). Agreement has been reached with AOACI for mutual recognition of the different validation schemes. This should lead to greater harmonisation of validation procedures within Europe and, hopefully, acceptance in other countries. It will certainly help unification of the food trade in Europe. In 2005, the EN/ISO 16140 procedure was submitted for revision and a working group within ISO was established to make recommendations for in-house evaluation/validation of a method that had been optimised in a research laboratory, as well as for verification of a modified standard method used in gaining accreditation.

The use of validated methods is only part of providing a reliable result. Of equal importance is the execution of the method by qualified staff using validated equipment and utensils in a laboratory committed to Quality Assurance. These requirements need to be more widely recognised and applied. In addition, it should be clearly understood that a result obtained from an analytical method, particularly near the detection limit, is not an absolute indication of the presence or absence of the target organism(s) in the case of a qualitative method, or an absolute number of colony-forming units in a quantitative method. Inevitably, the result is subject to a degree of uncertainty that should be known in relation to the precision of the method, as determined from the validation study. The present ISO standard for the accreditation of testing laboratories EN/ISO/IEC 17025 (ISO, 2005a) refers explicitly to the necessity for determining the uncertainty of measurement in a microbiological analytical method. Recently, ISO/TS 19036 (ISO, 2006) was issued, and this deals with procedures for determining the uncertainty of measurement through the use of validation data, either external or in-house, and Quality Control data obtained as part of the accreditation scheme in establishing intra-laboratory reproducibility.

Many important decisions are based on the results of analytical procedures. Interpretation of these results, however, requires knowledge of the sampling procedure used, including primary sampling of the lot, transport of samples and secondary sub-sampling in the laboratory, appropriate definition of the goal or purpose of the analysis and adequate information about the food product involved and the technology used in processing. Only then is it possible to set realistic criteria for proper interpretation of the test results. The demand is likely to increase for personnel with good analytical skills to perform the analyses and a broad knowledge of microbiology, technology and food chemistry to interpret the results properly.

12.7 Sources of further information

Information concerning the definitions of technical performance characteristics, determination of these parameters and the measurement of uncertainty can be found in the Eurachem/CITAC guide (Ellison *et al.*, 2000), as well as in ISO 5725 (ISO, 1994).

For more details on the design of an appropriate validation protocol, the reader is referred to the CCFRA (Campden and Chorleywood Food Research Association) guidelines (Baylis *et al.*, 2001), the special issue of Food Control (1996, vol. 7, no. 1) on 'The validation of rapid methods in food microbiology' and EN/ISO 16140 (ISO, 2003).

Company websites

Applied Biosystems	www.appliedbiosystems.com
bioMérieux sa	www.biomerieux.com
Bio-Rad laboratories	www.bio-rad.com
bioTRADING	www.biotrading.com
Invitrogen Dynal	www.invitrogen.com
LabM	www.lab-m.com
Oxoid	www.oxoid.com
Remel Inc.	www.remelinc.com
Roche Molecular Systems	www.roche.com

12.8 References

ANDREWS W H (1996), 'Validation of modern methods in food microbiology by AOAC International: collaborative study', *Food Control*, **7**, 19–29.

ANDREWS W H (1998), 'Validation of rapid methods in microbiology', *De Ware(n) Chemicus*, **28**, 30–31.

ISO (1996), ISO 11290-1: 1996, 'Microbiology of food and animal feeding stuffs – Horizontal method for the detection and enumeration of *Listeria* monocytogenes – Part 1: Detection method', Geneva, Switzerland, International Organization for Standardization.

ISO (2002), EN/ISO 6579: 2002, 'Microbiology of food and animal feeding stuffs – Horizontal method for the detection of *Salmonella* spp.', Geneva, Switzerland, International Organization for Standardization.

ISO (2003), EN/ISO 16140: 2003, 'Microbiology of food and animal feeding stuffs – protocol for the validation of alternative methods', Geneva, Switzerland, International Organization for Standardization.

ISO (2004), EN/ISO 7932-2: 2004, 'Microbiology of food and animal feeding stuffs – Horizontal method for the enumeration of presumptive *Bacillus cereus* – colony count technique at 30 °C', Geneva, Switzerland, International Organization for Standardization.

ISO (2005a), EN/ISO/IEC 17025: 2005, 'General requirements for the competence of calibration and testing laboratories', Geneva, Switzerland, International Organization for Standardization.

ISO (2005b), ISO/TS 20836: 2005, 'Microbiology of food and animal feeding stuffs – Polymerase chain reaction (PCR) for the detection of food-borne pathogens – Performance testing for thermal cycles', Geneva, Switzerland, International Organization for Standardization.

ISO (2006), ISO/TS 19036: 2006, 'Microbiology of food and animal feeding stuffs – Guidelines for the estimation of measurement uncertainty for quantitative determinations', Geneva, Switzerland, International Organization for Standardization.

BAYLIS C L, JEWELL K, OSCROFT C A and BROOKS F L (2001), Guideline No. 29, *Guidelines for Establishing the Suitability of Food Microbiology Methods*, Chipping Campden, Gloucestershire, UK, Campden and Chorleywood Food Research Association.

BERTRAM-DROGATZ P A, WILBORN F, SCHEU P, PARDIGOL A, KOOB C, GRÖNEWALD C, FANDKE M, GASCHE A and BERGHOF K (2000), 'PCR-based commercial tests for pathogens', in *Encyclopaedia of Food Microbiology*, eds Robinson R K, Batt C A and Patel P D, London, UK, Academic Press, 1638–1640.

BOLTON F J (1998), 'Quality assurance in food microbiology – a novel approach', *International Journal of Food Microbiology*, **5**, 7–11.

BOWLES C (2000), 'Laboratory management – accreditation schemes', in *Encyclopaedia of Food Microbiology*, eds Robinson R K, Batt C A and Patel P D, London, UK, Academic Press, 1128–1133.

CHEN S, YEE A, GRIFFITHS M, LARKIN C, YAMASHIRO C T, BEHARI R, PASZKOKOLVA C, RAHN K and DEGRANDIS S A (1997), 'The evaluation of a fluorogenic polymerase chain reaction assay for the detection of *Salmonella* species in food commodities', *International Journal of Food Microbiology*, **35**, 239–250.

CORRY J E L (1995), 'Reviews from the sixth symposium of the Working Party for Culture Media', *International Journal of Food Microbiology*, **26**, 1–131.

CUDJOE K S and KRONA R (1997), 'Detection of *Salmonella* from raw food samples using Dynabeads anti-Salmonella and a conventional reference method', *International Journal of Food Microbiology*, **37**, 55–62.

DE SMEDT J M (1998), 'AOAC Validation of qualitative and quantitative methods for microbiology in foods', *International Journal of Food Microbiology*, **45**, 25–28.

DE ZUTTER L, DE SMEDT J M, ABRAMS R, BECKERS H, CATTEAU M, DE BORCHGRAVE J, DEBEVERE J, HOEKSTRA J, JONKERS F, LENGERS J, NOTERMANS S, VAN DAMME L, VANDERMEERSCH R, VERBRAEKEN R and WAES G (1991), 'Collaborative study on the use of motility enrichment on modified semisolid Rappaport–Vassiliadis medium for the detection of *Salmonella* from foods', *International Journal of Food Microbiology*, **13**, 11–20.

ELLISON S L R, ROSSLEIN M and WILLIAMS A (2000), *Quantifying Uncertainty in Analytical Measurement*, London, UK, Eurachem/CITAC.

ENGVALL E O, BRÄNDSTRÖM B, GUNNARSSON A, MÖRNER T, WAHLSTRÖM H and FERMER C (2002), 'Validation of a polymerase chain reaction/restriction enzyme analysis method for species identification of thermophilic campylobacters isolated from domestic and wild animals', *Journal of Applied Microbiology*, **92**, 47–54.

FACH P, PERELLE F, DILASSER F and GROUT J (2001), 'Comparison between a PCR-ELISA test and the vero cell assay for detecting Shiga toxin-producing *Escherichia coli* in dairy products and characterization of virulence traits of the isolated strains', *Journal of Applied Microbiology*, **90**, 809–818.

FDA (1998), Food and Drug Administration, *Bacteriological Analytical Manual*, 8th edition, revision A, Gaithersburg, MD, USA, AOAC International.

GANGAR V, CURIALE M S, D'ONORIO A and SCHULTZ A (2000), 'VIDAS® Enzyme-linked immunofluorescent assay for detection of *Listeria* in foods: collaborative study', *Journal of AOAC International*, **83**, 903–918.

HITCHINS A D (1996), 'The International Dairy Federation's procedure for the validation of microbiological analytical methods for dairy foods', *Food Control*, **7**, 13–18.

IN 'T VELD P H (1998), 'The use of reference materials in quality assurance programmes in food microbiology laboratories', *International Journal of Food Microbiology*, **45**, 35–41.

IN 'T VELD P H, HAVELAAR A H, VAN STRIJP-LOCKEFEER N G W M (1999), 'The certification of a reference material for the evaluation of methods for the enumeration of *Bacillus cereus*', *Journal of Applied Microbiology*, **86**, 266–274.

ISO (1994), ISO 5725: 1994 (Parts 1–6), 'Accuracy (trueness and precision) of measurement methods and results', Geneva, Switzerland, International Organization for Standardization.

LAHELLEC C (1998), 'Development of standard methods with special reference to Europe', *International Journal of Food Microbiology*, **45**, 13–16.

LIGHTFOOT N F and MAIER E A (1998), *Microbiological Analysis of Food and Water: Guidelines for Quality Assurance*, Amsterdam, The Netherlands, Elsevier Science.

LÜBECK P S, WOLFFS P, ON S L W, AHRENS P, RADSTRÖM P and HOORFAR J (2003a), 'Towards an international standard for PCR-based detection of food-borne thermotolerant campylobacters: assay development and analytical validation', *Applied and Environmental Microbiology*, **69**, 5664–5669.

LÜBECK P S, COOK N, WAGNER M, FACH P and HOORFAR J (2003b), 'Towards an international standard for PCR-based detection of food-borne thermotolerant campylobacters: validation in a multicenter collaborative trial', *Applied and Environmental Microbiology*, **69**, 5670–5672.

MALORNY B, TASSIOS P T, RADSTRÖM P, COOK N, WAGNER M and HOORFAR J (2003), 'Standardisation of diagnostic PCR for the detection of foodborne pathogens', *International Journal of Food Microbiology*, **83**, 39–48.

MANSFIELD P L and FORSYTHE S J (2001), 'The detection of *Salmonella* serovars from animal feed and raw chicken using a combined immunomagnetic separation and ELISA method', *Food Microbiology*, **18**, 361–366.

MICROVAL SECRETARIAT (1998), 'MicroVal: a European approach to the certification of new microbiological methods', *International Journal of Food Microbiology*, **45**, 17–24.

NOTERMANS S, BEUMER R and ROMBOUTS F (1997), 'Detecting foodborne pathogens and their toxins: conventional versus rapid and automated methods', in *Food Microbiology Fundamentals and Frontiers*, eds Doyle M P, Beuchat L R and Montville T J, Washington, DC, USA, ASM Press, 705–708.

POPPE C and DUNCAN C L (1996), 'Comparison of detection of *Salmonella* by the Tecra® Unique Salmonella test and the modified Rappaport Vassiliadis medium', *Food Microbiology*, **13**, 75–81.

RENTENAAR I M F (1996), 'MicroVal, a challenging Eureka project', *Food Control*, **7**, 31–36.

RESTAINO L, FRAMPTON E W and SPITZ H (2001), 'Repair and growth of heat- and freeze-injured *Escherichia coli* O157 : H7 in selective enrichment broths', *Food Microbiology*, **18**, 617–629.

SCHULTEN S M, IN'T VELD P H, NAGELKERKE N J D, SCOTTER S, DE BUYSER M L, ROLLIER P and LAHELLEC C (2000), 'Evaluation of the ISO 7932 standard for the enumeration of *Bacillus cereus* in foods', *International Journal of Food Microbiology*, **57**, 53–61.

SCOTTER S L, LANGTON S, LOMBARD B, SCHULTEN S, NAGELKERKE N, IN 'T VELD P H, ROLLIER P and LAHELLEC C (2001a), 'Validation of ISO method 11290 Part 1 – Detection of *Listeria monocytogenes* in foods', *International Journal of Food Microbiology*, **64**, 295–306.

SCOTTER S L, LANGTON S, LOMBARD B, LAHELLEC C, SCHULTEN S, NAGELKERKE N, IN 'TVELD P H and ROLLIER P (2001b), 'Validation of ISO method 11290 Part 2. Enumeration of *Listeria monocytogenes* in foods', *International Journal of Food Microbiology*, **70**, 121–129.

STRUIJK C B (1996), 'Guidelines for method validation techniques used in the microbiological examination of food samples', *Food Control*, **7**, 53–58.

STEPHENS P J, JOYNSON J A, DAVIES K W, HOLBROOK R, LAPPIN-SCOTT H M and HUMPHREY T J (1997), 'The use of an automated growth analyser to measure recovery times of single heat-injured *Salmonella* cells', *Journal of Applied Microbiology*, **83**, 445–455.

UYTTENDAELE M R, GRANGETTE C, ROGERIE F, PASTEAU S, DEBEVERE J and LANGE M (1998), 'Influence of cold stress on the prior enrichment time needed for detection of entero-haemorrhagic *E. coli* in ground beef using the polymerase chain reaction', *Applied and Environmental Microbiology*, **64**, 1640–1643.

UYTTENDAELE M, BAGAMBOULA C F, DE SMET E, VAN WILDER S and DEBEVERE J (2001), 'The evaluation of culture media for enrichment and isolation of *Shigella sonnei* and *S. flexneri*', *International Journal of Food Microbiology*, **70**, 255–265.

UYTTENDAELE M, VANWILDEMEERSCH K and DEBEVERE J (2003), 'Evaluation of the iQ PCR system versus the Vidas ELISA and Diassalm for rapid detection of *Salmonella* in meat and poultry', *Letters in Applied Microbiology*, **37**, 386–391.

VAN DER ZEE H, WIT B and DE BOER E (2002), 'Validation of the semi-solid media MSRV and Diassalm vs ISO 6579 combined with the use of incubated pre-enrichment after refrigeration for 24 and 48 hours', *Proceedings, International Symposium on Salmonella and Salmonellosis*, Ploufragan, France, 29–31 May, 129–130.

WAN J, KING K, CRAVEN H and MCAULEY C, TAN S E and COVENTRY M J (2000), 'Probelia™ PCR system for rapid detection of *Salmonella* in milk powder and ricotta cheese', *Letters in Applied Microbiology*, **30**, 267–271.

WILSON I G (1997), 'Inhibition and facilitation of nucleic acid amplification', *Applied and Environmental Microbiology*, **63**, 3741–3751.

13

Quality Assurance in the food microbiology laboratory

R. Wood, Food Standards Agency, UK

13.1 Introduction

It is universally recognised that a laboratory must be able to produce and report data that are fit-for-purpose, i.e. suitable for the customer's intended use. For a laboratory to produce consistently reliable data, it must implement an appropriate programme of Quality Assurance measures. Such measures are now required in the European Union (EU) by virtue of legislation on food control, by the Codex Alimentarius Commission (CAC) for laboratories involved in the import/export of foodstuffs and, in the case of the United Kingdom (UK) Food Standards Agency, for contractors undertaking survey work. Thus, customers now demand that analytical data meet established quality requirements, which are described below. The present chapter also considers the significance of the measures identified and then discusses the future of analytical methods within the food laboratory.

'Food analysis' is generally understood to refer to chemical or physical tests, assays or measurements, and could include a wide variety of analyses, such as determination of water, fat, fibre, nitrite or nitrate content, and measurement of mycotoxins and pesticide or herbicide residues. Microbiological tests, on the other hand, usually involve the presence or absence of pathogenic microbes, total viable counts or estimation of numbers of pathogenic or indicator bacteria, or various types of spoilage organisms. Analysis of the microbiological quality of food is sometimes referred to as 'food examination' and has been treated separately from food analysis.

In the UK, a distinction is made between chemical analysis and microbiological examination for the purposes of the 1990 Food Safety Act

(Anonymous, 1990) and Regulations made under the Act. This is unusual in that most countries do not make such a distinction and, for them, 'analysis' embraces both chemical and microbiological analyses. Thus, in the EU, legislation on analysis is generally taken to refer to both chemistry and microbiology. It is important that this is appreciated when non-UK analytical documents are being considered. One reason for the distinction made in the UK is the assumption that it is more difficult to apply Quality Control systems to microbiological tests. For example, microbes are distributed heterogeneously in many foods, it is much more difficult to prepare stable control samples and also, perhaps, microbiology has been considered traditionally to be partly an art – the results depending to some extent on the particular skill of the individual microbiologist. This has been particularly true for detecting/isolating pathogens, such as salmonellas, which has depended on the ability of the microbiologist to detect one or a few suspect salmonella colonies among many others growing on the same agar plate and which contrasts with analytical (chemical) methods, that invariably rely on a more objective measurement, such as weight, volume or absorbance.

Although most official laboratories still use some traditional (colony-count type) microbiological methods, more rapid methods, often partly or wholly mechanised and more akin to those used in chemical analyses, are gradually gaining in popularity. Development of improved traditional-type methods, including selective media with better indicator systems (e.g. chromogenic substrates), or immunological or polymerase chain reaction-(PCR)-type tests applied after sample enrichment, have also made microbiological testing less subjective than in the past. In addition, it is becoming increasingly clear that Quality Control systems, similar to those used for analytical work, can and should be applied to microbiological tests, even though the results of the latter can be much more variable. Thus, it will be appropriate to outline the legislative aspects of assuring food-laboratory performance, applied in the first instance to chemical analyses, and highlight the differences with respect to microbiological examinations.

13.2 Legislation and codes of practice

Since the UK acceded to the European Community in 1972, methods of analysis have been prescribed by legislation for a number of foodstuffs. However, the Community now recognises that the quality of the results provided by a laboratory is just as important as the choice of method used to obtain the results. Whilst it is generally recognised that every laboratory should use Quality Assurance measures, the feed and food control laboratories in the EU are actually required to do so by law. The requirements are described below and they provide good guidance on the measures that all laboratories should follow to produce reliable data. Other international organisations, most notably the CAC, have taken a similar approach. The

Codex system is therefore included in this chapter. Because of these developments, there is a move away from the need to specify the exact details of an analytical method in favour of a requirement for a general quality system within which the laboratory must operate. This gives greater flexibility to the laboratory, without detracting from the quality of the results it will produce.

Although such an approach is relatively easy to understand in relation to analytical chemistry, it is less obvious when applied to microbiology, where the test result frequently depends on the method of analysis. It is all the more surprising, therefore, that, until comparatively recently, the EU did not define precisely the methods to be used for the microbiological criteria laid down in Directives for various foods, e.g. egg products, live and cooked shellfish, milk and milk products, including cheeses of various types, meat preparations and minced meat. In the last few years, however, policy on this has changed and the Member States of the EU have taken the lead in developing new standard methods and revising old ones *via* the European Committee for Standardisation (CEN), the International Organisation for Standardisation (ISO) and the International Dairy Federation (IDF). New and revised standard methods include sections on their repeatability and reproducibility, and there are standards, both completed and in preparation, that are concerned with general quality systems, such as ISO (1996) and ISO (2000). An EU Regulation on microbiological criteria for foodstuffs (EU, 2005) is being implemented and this specifies certain ISO standard methods. It also allows the use of alternative methods that have been validated according to the relevant EN/ISO protocol.

13.2.1 The EU Feed and Food Control Regulation

Methods of analysis have been prescribed by legislation for a number of foodstuffs, ever since the formation of the European Community, now the EU. However, the Union now recognises the importance of Quality Control in relation to laboratory data. This is best illustrated by considering Council Regulation 882/2004 on official controls performed to ensure the verification of compliance with feed and food law, animal health and animal welfare rules (EU, 2004a). Most of the provisions of this Regulation came into force on 1 January 2006. The essential elements relating to Quality Control in the laboratory are given in Article 11 (and the associated Annex III) and in Article 12:

Article 11 (Methods of sampling and analysis)

1. Sampling and analysis methods used in the context of official controls shall comply with relevant Community rules or,
   (a) if no such rules exist, with internationally recognised rules or protocols, for example those that the European Committee for Standardisation (CEN) has accepted or those agreed in national legislation; or,

(b) in the absence of the above, with other methods fit for the intended purpose or developed in accordance with scientific protocols.

2. Where paragraph 1 does not apply, validation of methods of analysis may take place within a single laboratory according to an internationally accepted protocol.
3. Wherever possible, methods of analysis shall be characterised by the appropriate criteria set out in Annex III.
4. The following implementing measures may be taken in accordance with the procedure referred to in Article 62(3):
   (a) methods of sampling and analysis, including the confirmatory or reference methods to be used in the event of a dispute;
   (b) performance criteria, analysis parameters, measurement uncertainty and procedures for the validation of the methods referred to in (a); and
   (c) rules on the interpretation of results.
5. The competent authorities shall establish adequate procedures in order to guarantee the right of feed and food business operators whose products are subject to sampling and analysis to apply for a supplementary expert opinion, without prejudice to the obligation of competent authorities to take prompt action in case of emergency.
6. In particular, they shall ensure that feed and food business operators can obtain sufficient numbers of samples for a supplementary expert opinion, unless impossible in case of highly perishable products or very low quantity of available substrate.
7. Samples must be handled and labelled in such a way as to guarantee both their legal and analytical validity.

ANNEX III (Characterisation of methods of analysis)

1. Methods of analysis should be characterised by the following criteria:
   (a) accuracy;
   (b) applicability (matrix and concentration range);
   (c) limit of detection;
   (d) limit of determination;
   (e) precision;
   (f) repeatability;
   (g) reproducibility;
   (h) recovery;
   (i) selectivity;
   (j) sensitivity;
   (k) linearity;
   (l) measurement uncertainty;
   (m) other criteria that may be selected as required.
2. The precision values referred to in 1(e) shall either be obtained from a collaborative trial which has been conducted in accordance with an internationally recognised protocol on collaborative trials (e.g. ISO 5725:1994 or the IUPAC International Harmonised Protocol) or, where performance criteria for analytical methods have been established, be based on criteria compliance tests. The repeatability and reproducibility values shall be expressed in an internationally recognised form (e.g. the 95 % confidence intervals as defined by ISO 5725:1994 or IUPAC). The results from the collaborative trial shall be published or freely available.
3. Methods of analysis which are applicable uniformly to various groups of commodities should be given preference over methods which apply only to individual commodities.

4. In situations where methods of analysis can only be validated within a single laboratory then they should be validated in accordance with e.g. IUPAC Harmonised Guidelines, or where performance criteria for analytical methods have been established, be based on criteria compliance tests.
5. Methods of analysis adopted under this Regulation should be edited in the standard layout for methods of analysis recommended by the ISO.

Article 12 (Official laboratories)

1. The competent authority shall designate laboratories that may carry out the analysis of samples taken during official controls.
2. However, competent authorities may only designate laboratories that operate and are assessed and accredited in accordance with the following European Standards:
   (a) EN ISO/IEC 17025 on 'General requirements for the competence of testing and calibration laboratories';
   (b) EN 45002 on 'General criteria for the assessment of testing laboratories';
   (c) EN 45003 on 'Calibration and testing laboratory accreditation system–General requirements for operation and recognition', taking into account criteria for different testing methods laid down in Community feed and food law.
3. The accreditation and assessment of testing laboratories referred to in paragraph 2 may relate to individual tests or groups of tests.
4. The competent authority may cancel the designation referred to in paragraph 1 when the conditions referred to in paragraph 2 are no longer fulfilled.

There is further information on the role and duties of Community and National Reference Laboratories. Although such laboratories have a limited role in the EU at the present time, it may be anticipated that the Commission will wish to expand both the areas covered and scope of such laboratories.

As a result of the adoption of the above Regulations, legislation is now in place to ensure that there is confidence in the performance of food analysis laboratories. The requirements of the legislation serve as a valid model for the Quality Assurance measures that laboratories undertaking microbiological analysis should follow.

13.2.2 Codex Alimentarius Commission

The decisions of the CAC are becoming increasingly important, because of the acceptance of Codex Standards in World Trade Organisation (WTO) agreements. They can be regarded as semi-legal in status. Thus, on a worldwide basis, the establishment of the WTO and formal acceptance of the Agreements on the Application of Sanitary and Phytosanitary Measures (SPS Agreement) and Technical Barriers to Trade (TBT Agreement) have dramatically raised the status of Codex itself. As a result, Codex Standards are now seen as *de facto* international standards and, increasingly, are being adopted by reference into the food laws of both developed and developing countries.

Because of the current status of the CAC, the work that it has carried out on laboratory Quality Assurance must be considered carefully. One of the CAC Committees, the Codex Committee on Methods of Analysis and Sampling, has developed criteria for assessing the competence of testing laboratories involved in the official import and export control of foods (CAC, 1997a, b). The criteria are intended to assist countries in fair-trading of foodstuffs and to protect consumers. They mirror the EU recommendations for laboratory quality standards and methods of analysis. The criteria for laboratories involved in the import and export control of foods, now adopted by the CAC, are:

- to comply with the general criteria for testing laboratories laid down in ISO (2005c); this is a form of accreditation;
- to participate in appropriate proficiency testing schemes for food analysis, which conform to the requirements laid down in Thompson and Wood (1993);
- to use, whenever available, methods of analysis that have been validated according to the principles laid down by the CAC, and
- to use internal Quality Control procedures, such as the harmonised guidelines described in Thompson and Wood (1995).

In addition, the bodies assessing the laboratories should comply with the general criteria for laboratory accreditation, such as those laid down in ISO (1993a).

Thus, in the EU, the requirements have been based on accreditation, proficiency testing, the use of validated methods of analysis and, in addition, the formal requirement to use internal Quality Control procedures that comply with the harmonised guidelines. Although the EU and CAC referred initially to different sets of accreditation standards, these have effectively been replaced by the ISO/IEC Standard 17025 (ISO, 2005c).

The outcome of the Official Feed and Food Control Regulation (OFFC) and the Codex requirements is that food control laboratories must consider the following aspects within the laboratory:

- the organisation of the laboratory;
- how well the laboratory carries out analyses; and
- the methods of analysis used in the laboratory.

All these aspects are inter-related, but in simple terms may be thought of as:

- becoming accredited to an internationally recognised standard – such accreditation is aided by the use of internal Quality Control procedures;
- participating in proficiency schemes; and
- using validated methods.

These considerations will be addressed in turn, as will other Quality Assurance measures that a laboratory should take into account.

13.2.3 UK Food Standards Agency

Surveys

The Food Standards Agency undertakes food survey exercises. Therefore, it has produced information for potential contractors on the analytical Quality Assurance requirements for chemical surveillance. These requirements are outlined below and they emphasise the need for a laboratory to produce and report data of appropriate quality. The requirements are divided into three parts.

Part A: Quality Assurance requirements for project proposals provided by potential contractors at the time tender documents are completed and when commissioning a survey. Here, information is sought on:

- the formal Quality Control system used in the laboratory when assessed by a third party (e.g. UK Accreditation Service (UKAS)-accredited or Good Laboratory Practice (GLP)-compliant);
- the quality system, if not assessed by a third party;
- proficiency testing;
- internal Quality Control;
- validation of methods.

Part B: information required by the FSA customer once the contract has been awarded and to be agreed with the contractor, e.g. sample storage conditions, proposed methods and a copy of the Standard Operating Procedures (SOPs), where accredited, internal Quality Control (IQC) procedures, measurement limits (i.e. limit of detection: limit of determination/quantification; reporting limits and measurement uncertainty). Although clearly defined for chemical analyses, such aspects are less well established for microbiological examinations, despite the need to address similar issues.

Part C: information to be provided by the contractor on an on-going basis, once the contract has been awarded. This should be agreed with the customer to ensure that the contractor remains in 'analytical control'.

Contractual research

The procedures employed by the UK Food Standards Agency have been considered recently by a working group (FSA, 2001). The report recommends that the quality systems employed by the Agency's research contractors be reviewed, with a view to introducing a formal, third-party-assessed system during 2006.

Thus, all participating research organisations must consider the following aspects: organisation of the laboratory; how well it actually carries out analyses; and the analytical methods used.

13.3 Accreditation

Although formal accreditation is not essential to ensure that a laboratory will produce 'quality data', an accredited laboratory is able to state that it has been third-party assessed and this does give additional assurance to its 'customers'.

The OFFC Regulation requires that all feed and food control laboratories should be accredited to ISO Standard 17025 (ISO, 2005c). In the EU Member States, governments nominate their own accreditation services to carry out this function under the Regulation. For example, in the UK, UKAS is the body in question. However, since the accreditation agency will also be required to comply with EN 45003 Standard and to carry out assessments in accordance with EN 45002, any other accreditation agency that is a member of the European Co-operation for Accreditation of Laboratories (EA) may also be nominated for the purpose. A similar procedure is followed in the other Member States, all of which have developed, or are developing, organisations equivalent to UKAS. Laboratories carrying out official microbiological analyses for control purposes will have to be accredited. Other (non-control) laboratories may wish to reach the same standard to inspire confidence in their work.

In the UK, it has been the normal practice for UKAS to define the scope of an individual laboratory on a method-by-method basis. However, in the case of official food-control laboratories undertaking non-routine or investigative chemical analyses, it is accepted that accredited, fully-documented methods are not practicable, i.e. those that specify each sample type and analyte. Nevertheless, such laboratories must have a protocol defining the approach to be adopted, including requirements for validation of methods and IQC. Full details of the procedures used, including instrumental parameters, must be recorded at the time of each analysis, to allow the procedure to be repeated in the same manner at a later date. Thus, appropriate methods may be accredited on a generic basis, with such accreditation being underpinned, where necessary, by specific accreditation of the test method. This approach is particularly appropriate for laboratories using methods that undergo continuous development and where a laboratory may wish for any investigative activities to be covered by accreditation, as well as defined methods that are used on a routine basis.

In the UK, official food control laboratories undertaking microbiological examinations are accredited on a method-by-method basis for the detection and/or enumeration of pathogens indicators and organisms routinely determined in food. The tests cover the aerobic colony count, Enterobacteriaceae, coliform bacteria, *Escherichia coli* (including serotype 0157), *Staphylococcus aureus*, *Bacillus cereus*, *Clostridium perfringens*, *Salmonella*, *Listeria monocytogenes* and *Campylobacter* spp. Where legislation prescribes the methods, any official laboratory that intends to use them must be accredited to do so. If not prescribed by statute, then methods published by ISO, CEN, AOACI or others that have been validated may be used.

Microbiological examinations for which there are no approved standard methods may be undertaken, if the laboratory has in place a series of accredited methods or generic protocols dealing with, for example: sample preparation, colony counting, impedimetric techniques, immunological procedures, gene probe methods, PCR and electron or other microscopical techniques. It will be necessary for laboratories to demonstrate Quality Control procedures to ensure compliance with the ISO/IEC 17025 Standard, an example of which would be meeting the ISO/AOACI/IUPAC (International Union of Pure and Applied Chemistry) guidelines (Thompson and Wood, 1995).

ISO Standard 17025 (ISO, 2005c) is similar in intent to the previous ISO/IEC Guide 25 (ISO, 1990) and the equivalent EN Standards, but gives more emphasis to method validation, measurement uncertainty and traceability than did previous standards/guides. Requirements for the accreditation of microbiological laboratories are summarised by the EA (2002).

13.4 Internal Quality Control

Internal Quality Control (IQC) is one of a number of concerted measures that analytical chemists can take to ensure that the data produced in the laboratory are of known quality and certainty. In practice, these attributes are determined by comparing the results achieved at a given time with a standard. Thus, IQC comprises the routine practical procedures that enable the analyst to accept a result or group of results, or reject the results and repeat the analysis. The process involves the inclusion of particular reference materials, known as 'control materials', in the analytical sequence and duplicate analyses. The application of IQC in food microbiology is discussed elsewhere. ISO, IUPAC and AOAC INTERNATIONAL have co-operated to produce the relevant protocols (Horwitz, 1988, (updated by Horwitz, 1995); Thompson and Wood, 1993). The working group that produced these protocols has prepared a further one on the IQC of data produced in analytical laboratories (Thompson and Wood, 1995). Use of the procedures outlined in the protocol should facilitate compliance with the accreditation requirements specified above.

IQC in microbiology laboratories differs somewhat from that used in analytical laboratories, mostly because the analyte is less stable. However, 'standard' or 'reference' materials can still be used, as can replicate testing and evaluation of test results, and the spiking of samples with appropriate, standard microbial strains.

13.4.1 Basic concepts

The above-mentioned protocol sets out guidelines for implementing IQC in analytical laboratories. IQC helps to ensure that the data produced are fit for their intended purpose. In practice, fitness for purpose is determined by comparing the accuracy achieved in the test at a given time with a required level of accuracy. The results are either accepted as fit-for-purpose or they are rejected and the analysis repeated. Therefore, IQC is an important determinant of the quality of analytical data, and is recognised as such by accreditation agencies.

Wherever possible, the control materials used in IQC should be representative of those being tested with respect to matrix composition, physical state and the concentration range of the analyte. Since control materials are treated in exactly the same way as those undergoing the test, they are regarded as surrogates that can be used to characterise the performance of the analytical system, both at a specific time and over longer intervals. IQC is a final check on the correct execution of all procedures (including calibration) that are prescribed in the analytical protocol, and every other QA measure that underlies good analytical practice. In this way, it is necessarily retrospective. As far as possible, it is also required to be independent of the analytical protocol, especially the calibration that it is designed to test.

Ideally, both the control materials and those used to create the calibration should be traceable to appropriate certified reference materials or a recognised empirical reference method. When this is not possible, control materials should be traceable at least to a material of guaranteed purity or other, well-characterised material. However, the two traceability pathways must not become coincident at too late a stage in the analytical process. For instance, if control materials and calibration standards were prepared from a single stock solution of analyte, IQC would not detect any inaccuracy stemming from the incorrect preparation of the stock solution.

In a typical analytical situation, several, or perhaps many, similar test materials will be analysed together, and control materials will be included in the group. Frequently, determinations will be duplicated by analysing separate test portions of the same material. Such a group of materials is referred to as an analytical 'run'. (The terms 'set', 'series' and 'batch' have also been used as synonyms for run.) Runs are regarded as being analysed under essentially constant conditions. The batches of reagents used, the instrument settings, the analyst and the laboratory environment will, under ideal conditions, remain unchanged during the analysis of a run. Therefore, systematic errors should remain constant, as should values of the parameters that describe random errors. Since the monitoring of these errors is of concern, the run is the basic operational unit of IQC.

Thus, a run is seen to be tested under conditions that demonstrate repeatability, i.e. the random measurement errors are of a magnitude that would be encountered in a 'short' period of time. In practice, the analysis of a run may occupy sufficient time for small systematic changes to occur. For example, reagents may degrade, instruments may drift, minor adjustments to instrumental settings may be called for, or the laboratory temperature may rise. For the purposes of IQC, however, these systematic effects are subsumed into the repeatability variations. Sorting the materials of a run into a randomised order converts the effects of any drift into random errors.

Although spiked samples of the food being examined microbiologically can be used, there are a number of difficulties/uncertainties with this approach. Firstly, unless the food has been sterilised, it is impossible to guarantee that it does not already contain the target organism. Secondly, if the food is sterile (e.g. by autoclaving or, preferably, irradiating), then there will be no competitive flora. Thirdly, the wild strains being sought may have different properties from those of the control strain used to spike the food, and they may also differ in their physiological state. The first difficulty is fairly easy to overcome by using a relatively rare strain that can be readily recognised on isolation. This is also a useful precaution in case of accidental cross-contamination from the 'positive control' to the test culture(s). For instance, cultural methods for isolating *Salmonella* are extremely sensitive, such that even one organism per 25 g sample can be detected with relative ease. The consequences of reporting a sample of processed food positive for *Salmonella* can be extremely serious and may involve the recall of large quantities of product, with potential losses of millions of pounds sterling.

13.4.2 Recommendations made in the guidelines

Specific recommendations given in the guidelines represent integrated approaches to IQC that are suitable for many types of analysis and areas of application. Managers of laboratory quality systems will have to adapt these recommendations to the demands of their own particular situations. For example, such adaptations could involve adjusting the number of duplicates and control materials inserted into a run, or including any additional measures favoured in the particular area of application. The procedure finally chosen and its accompanying decision criteria must be specified in an IQC protocol that is separate from the one relating to the analytical system.

The practical approach to Quality Control is determined by the frequency with which the measurement is carried out and the size and nature of each run. The use of control charts and decision rules are covered in Appendix 1 of the guidelines.

By following the guidelines, laboratories would introduce IQC measures that are essential in ensuring the reliability of their data. The guidelines stress, however, that IQC is not foolproof, even when properly executed. Obviously, it is subject to 'errors of both kinds', i.e. runs that are in control will occasionally be rejected and runs that are out of control occasionally be accepted. Of more importance, IQC cannot usually identify sporadic gross errors or short-term disturbances in the analytical system that affect the results obtained for individual test materials. Moreover, inferences based on IQC results are applicable only to test materials that fall within the scope of the analytical method validation. Despite these limitations, which professional experience and diligence can alleviate to a degree, IQC is the principal means of ensuring that only data of appropriate quality are released from a laboratory. When properly executed, it is very successful.

The guidelines also stress that the perfunctory execution of any quality system will not guarantee the production of data of adequate quality. The correct procedures for feedback, remedial action and staff motivation must also be documented and acted upon. In other words, there must be a genuine commitment to quality within a laboratory for an IQC programme to succeed, i.e. IQC must be part of a complete quality management system.

13.4.3 Quality control of media

Almost all microbiological tests require the use of culture media, most of which are not chemically defined, but contain mixtures of nutrients, and frequently selective agents, designed to inhibit unwanted microbes, as well as indicator systems for identifying colonies of the microbes being sought. The proper performance of these media is therefore essential, if a laboratory is to obtain reliable test results. Although recipes with detailed lists of ingredients and instructions for preparation are provided in standard test protocols, few laboratories prepare their media from basic ingredients. Most buy them in a dehydrated form which needs only to be mixed with the correct quantity of water and sterilised. Heat-labile ingredients are added after sterilisation of the basal media and are also available commercially. Medium manufacturers test the functioning of the ingredients of their media, e.g. gelling properties of agar, composition of peptones, inhibitory effects of bile salts, brilliant green, and also the complete medium. Nevertheless, laboratories should check the functioning of each new lot of medium they buy. This is normally done by using test inocula of target and (in the case of selective media) unwanted (competitive) organisms (Corry *et al.*, 2003). The choice of test organisms may include recent isolates that reflect those most likely to be encountered, as well as standard strains, and possibly strains known to be particularly sensitive to sub-optimal media. Methods of testing can be quantitative (comparison of colony counts on control versus test media) or semi-quantitative (standardised streaking or 'ecometry'). The appearance and size, as well as the number of colonies, should be checked. Methods have also been devised for liquid media. Standard methods are in preparation on this topic (ISO, 2000, 2003). Less detailed tests, e.g. a qualitative streak-plate, should also be set up for each batch of prepared medium.

13.5 Proficiency testing

Participation in proficiency testing schemes provides laboratories with an objective means of assessing and documenting the reliability of the data they are producing. Although there are several types of scheme, they all share a common feature: test results obtained by one laboratory are compared with those obtained by one other or more. The proficiency-testing schemes must provide a transparent interpretation and assessment of results. Laboratories wishing to demonstrate their proficiency should seek and participate in schemes that are relevant to their area of work. A proficiency-testing scheme is defined as a system for objectively checking laboratory results by an external agency. It includes comparison of a laboratory's results at intervals with those of other laboratories, the main objective being the establishment of trueness.

In addition, although various protocols for proficiency testing have been produced, the need now is for a harmonised protocol that will be accepted universally; progress towards the preparation and adoption of an internationally-recognised protocol is described below. Various terms have been used for schemes conforming to the draft protocol, e.g. 'external quality assessment', 'performance scheme', but the preferred term is 'proficiency testing'.

Proficiency testing schemes are based on regular circulation of homogeneous samples by a co-ordinator, analysis of the samples, normally by the laboratory's method of choice, and an assessment of the results. However, although many organisations follow such a scheme, there has been no international agreement on how this should be done – in contrast to the collaborative trial situation. In order to rectify this, the same international group that drew up collaborative trial protocols was invited to prepare one for proficiency schemes and the first meeting to do so was held in April 1989. The first protocol was published in 1993 and subsequently revised (Thompson *et al.*, 2006).

13.5.1 Microbiological proficiency testing schemes

Currently, there are no internationally or nationally recognised standards for proficiency testing in relation to the microbiological examination of food. Therefore, the available schemes for food examination will be recognised by the FSA on a case-by-case basis. Schemes that satisfy the requirements will be accepted and food laboratories wishing to be recognised as official control laboratories will be required to participate in the relevant parts of one or more of the recognised schemes. The FSA requires that schemes recognised for the purposes of the OFFC must comply with the general principles of the international harmonised protocol (Thompson and Wood, 1993), in so far as they are appropriate.

Proficiency test samples should mirror routine situations that are likely to be encountered when examining foods in the UK under the OFFC.

There should be at least 12 distributions per year. Each distribution may contain a number of test materials. Each test material may contain a single organism, a mixture of organisms or may be devoid of organisms of significance. Detection and/or determination of specific pathogens and indicators are required at least once each year.

Where quantitative determinations are assessed, schemes should treat the results statistically to determine whether performance is satisfactory, for example by converting counts to $\log_{10}$ values and then applying the procedures that have been developed in the international harmonised protocol (Thompson and Wood, 1993). Recognised proficiency testing schemes for the microbiological examination of food should also include a procedure for the recognition of unsatisfactory qualitative results. Currently, there are no nationally or internationally recognised protocols for assessing satisfactory performance in qualitative (presence/absence) tests on food. Nevertheless, it is proposed that, in assessing performance, schemes should take due account of false-positive and false-negative results.

Proficiency test results that fall outside acceptable confidence intervals prescribed for the schemes are unsatisfactory. In such cases, it will be necessary for laboratories to demonstrate to UKAS that appropriate remedial action has been taken. The performance of laboratories in proficiency testing schemes recognised as suitable for official control laboratories will be monitored by UKAS on behalf of the FSA. Therefore, these laboratories must consent to UKAS reporting to the FSA on their performance.

13.5.2 Why proficiency testing is important

Participation in proficiency testing schemes provides laboratories with a means of objectively assessing and demonstrating the reliability of the data they produce. Although there are several types of scheme, they all share a common feature of comparing test results obtained by one laboratory with those of others. Schemes may be 'open' to any laboratory or participation may be invited. Schemes may set out to assess the competence of laboratories undertaking a very specific analysis, e.g. lead in blood, or more general analyses, such as food analysis. Although accreditation and proficiency testing are separate exercises, it is anticipated that accreditation assessments will increasingly use proficiency testing data. The format of a typical, quantitative proficiency testing scheme is outlined in Appendix I (see page 329).

13.5.3 Attitude of accreditation agencies to proficiency testing

It is now recommended in ISO/IEC Standard 17025 (ISO, 2005c), the prime standard to which accreditation agencies now operate, that such agencies require laboratories seeking accreditation to participate in an appropriate proficiency testing scheme before accreditation can be gained. There is now an internationally recognised protocol to which proficiency testing schemes should comply; this is the IUPAC/AOAC/ISO harmonised protocol. Because of the importance of proficiency testing, the protocol is outlined in an appendix to this chapter. The elements of the protocol apply equally well to microbiological and chemical measurements.

13.5.4 Blind proficiency testing schemes

It should be recognised by laboratories that the use of blind proficiency testing, i.e. where the laboratory receives a sample for analysis from a customer who knows the characteristics of the sample, but does not inform the laboratory, is becoming more frequent. This is because some customers wish to assess for themselves the effectiveness of their contractors.

13.6 Quality Assurance requirements: analytical methods

Methods should be validated as being fit-for-purpose before use by a laboratory. Laboratories should ensure that, as a minimum, the methods they use are fully documented, laboratory staff are trained in their use and control mechanisms are established to ensure that the procedures are under statistical control. The development of methods of analysis for incorporation into international standards or into foodstuffs legislation was, until comparatively recently, not systematic. However, the EU and Codex have certain requirements for methods of analysis and these are outlined below. They are followed by other international standardising organizations, e.g. AOACI and CEN.

13.6.1 Codex Alimentarius requirements

The Codex Alimentarius Commission was the first international organisation working at government level in the food sector to lay down principles for the establishment of its methods. That it was necessary for guidelines and principles to be laid down reflects the confused and unsatisfactory situation in the development of legislative methods of analysis that existed in the food sector until the early 1980s. The principles (CAC, 1997c) are given below; other organisations that subsequently established procedures for developing analytical methods in their particular sector have followed these principles to a significant degree. They require that preference be given to methods of analysis for which reliability has been established with respect to the following criteria, selected as appropriate:

- specificity;
- accuracy;
- precision – laboratory, reproducibility within the laboratory and between laboratories;

- limit of detection;
- sensitivity;
- practicability and applicability under normal laboratory conditions;
- other criteria that may be selected, as required.

13.6.2 EU requirements

The EU has harmonised requirements for sampling and analysis in an attempt to meet the current demands of national and international enforcement agencies and the likely increases in problems that the open market will bring. To aid this process, the EU has adopted the requirements for methods of analysis described in the OFFC Regulation (see Section 13.2.1 above).

The criteria to which EU methods of analysis for foodstuffs should now conform are as stringent as those recommended by any international organisation that follows the Directive. The requirements are in line with those described above for Codex, and are given in the Annex to the Directive.

The precision values referred to in the EU Regulation must be obtained from a collaborative trial conducted in accordance with an internationally recognised protocol on such trials, e.g. ISO 5725 (ISO, 1994) or the IUPAC harmonised protocol. The repeatability and reproducibility values must be expressed in an internationally recognised form, e.g. the 95 % confidence intervals, as defined by ISO 5725 or IUPAC. The results from the collaborative trial shall be published formally or otherwise made freely available. Methods of analysis that are applicable uniformly to various groups of commodities should be given preference over those that apply only to individual commodities.

In situations where analytical methods can only be validated within a single laboratory, they should be validated in accordance with IUPAC harmonised guidelines. Methods adopted under this Regulation should be described according to the standard layout for such methods recommended by ISO. The above provisions apply equally well to microbiological testing as to chemical analyses, for which they were developed originally.

13.6.3 Other organisations – CEN and AOACI

There are other international organisations, most notably CEN and AOACI, which have requirements similar to those already discussed. Although CEN methods are not prescribed by legislation, the European Commission does place considerable importance on the work that CEN carries out in developing specific methods for the food sector; CEN has been given direct mandates by the Commission to publish particular methods, e.g. those for detecting food irradiation. Because of this, some of the methods being developed by CEN are described below. CEN, like the other organisations described above, has adopted a set of guidelines to which its Methods Technical Committees should conform, when developing an analytical method. The guidelines are:

Details of the inter-laboratory test on the precision of the method are to be summarised in an annex to the method. It should be noted that the values derived from the inter-laboratory test may not be applicable to analyte concentrations and matrices other than those given in the annex.

The precision clauses must be worded as follows:

Repeatability: 'The absolute difference between two single test results found on identical test materials by one operator using the same apparatus within the shortest feasible time interval will exceed the repeatability value *r* in not more than 5 % of the cases.

The value(s) is (are): . . .'

Reproducibility: 'The absolute difference between two single test results on identical test material reported by two laboratories will exceed the reproducibility value R in not more than 5 % of the cases.

The value(s) is (are): . . .'

There shall be certain minimum requirements regarding the information to be given in an informative annex, these being:

- year of inter-laboratory test and reference to the test report (if available);
- number of samples;
- number of laboratories retained after eliminating outliers;
- number of outliers;
- number of accepted results;
- mean value (with the respective unit);
- repeatability standard deviation (*sr*) (with the respective unit);
- repeatability relative standard deviation (*RSDr*) (%);
- repeatability limit (*r*) with the respective unit;
- reproducibility relative standard deviation (*sR*) (with the respective unit);
- reproducibility relative standard deviation (*RSDR*) (%);
- reproducibility limit (*R*) (with the respective unit);
- sample types clearly described;
- notes, if further information is to be given.

13.6.4 Validation requirements of official bodies

Consideration of the above requirements confirms that, in future, all methods must be fully validated if at all possible, i.e. subjected to a collaborative trial conforming to an internationally recognised protocol. This is now a legislative requirement in the food sector of the EU. The concept of a valid analytical method in the food sector and its requirements are described below.

13.7 Criteria for valid methods of analysis

It would be simple to say that any new method should be tested fully for the criteria given above. However, the most 'difficult' of these is obtaining performance values for accuracy and precision.

13.7.1 Accuracy

Accuracy is defined as the closeness of the agreement between the result of a measurement and its true value (ISO, 1993b). It may be assessed with the use of reference materials. However, with microbiological analyses, there is a particular problem. In many instances, the numerical value of a characteristic (or criterion) in a standard, or whether the organism is present or not, depends on the test procedures used. This highlights the need for sampling and analytical provisions in a standard to be developed at the same time that the specification in the standard is negotiated, to ensure that the latter is related to the analytical procedures prescribed.

13.7.2 Precision

Precision is defined as the closeness of agreement between independent test results obtained under prescribed conditions (ISO, 1992). In a standard method, the precision characteristics are obtained from a properly organised collaborative trial, i.e. a trial conforming to the requirements of an international standard (the AOAC/ISO/IUPAC harmonised protocol or ISO 5725). Because of the importance of collaborative trials, and the resources that are now being devoted to the assessment of precision characteristics of analytical methods before their acceptance, they are described in detail below:

13.7.3 Collaborative trials

As mentioned above, all 'official' methods of analysis are required to include precision data. These may be obtained by subjecting the method to a collaborative trial conforming to an internationally agreed protocol. A collaborative trial is a procedure whereby the precision of an analytical method may be assessed and quantified. The precision of a method is usually expressed in terms of repeatability and reproducibility values. Accuracy is not the objective here.

Recently, there has been progress towards universal acceptance of collaboratively tested methods and collaborative trial results, no matter by whom these trials are organised. This has been facilitated by the publication of the IUPAC/ISO/AOAC harmonisation protocol (Horwitz, 1988). The protocol was developed under the auspices of IUPAC, aided by representatives from the major organisations interested in conducting collaborative studies. In particular, from the food sector, AOACI, ISO, IDF, the

Collaborative International Analytical Council for Pesticides (CIPAC), the Nordic Analytical Committee (NMKL), the Codex Committee on Methods of Analysis and Sampling and the International Office of Cocoa and Chocolate were involved. The protocol gives a series of 11 recommendations dealing with:

- the components that make up a collaborative trial;
- participants;
- sample type;
- sample homogeneity;
- sample plan;
- method(s) to be tested;
- pilot study/pre-trial;
- the trial proper.

13.7.4 Statistical analysis

It is important to appreciate that the statistical significance of the results is wholly dependent on the quality of the data obtained from the trial. Data which contain obvious gross errors should be removed prior to statistical analysis. It is essential that participants inform the trial co-ordinator of any gross error known to occur during the analysis and also if any deviation from the written method has taken place. The statistical parameters calculated, and the outlier tests performed are those used in the internationally agreed protocol (Horwitz, 1988).

13.7.5 Alternative validation procedures

In the microbiology sector, there will be an interest in alternative validation procedures, most notably for 'test kits'. Such procedures are currently being prepared by both AOACI and CEN.

13.7.6 Single method validation

There is concern among the analytical community that, although methods should be validated where possible by a collaborative trial, this is not always feasible for economic or practical reasons. As a result, IUPAC guidelines are being developed for in-house method validation to inform food analysts about an acceptable procedure. These guidelines have been published recently (Thompson *et al.*, 2002) and point readers to relevant protocols.

13.8 Method validation through proficiency testing

The prime objective of proficiency testing is to assess the 'quality' of the laboratory. However, in some proficiency testing schemes, a significant number of laboratories will use the same method of analysis. This is particularly so for schemes involving microbiological laboratories. Consequently, there are initiatives to develop procedures for validating analytical methods using the results from proficiency testing schemes, when this situation occurs.

13.9 Measurement uncertainty for the microbiologist

13.9.1 General

It is increasingly being recognised by both laboratories and their customers that any reported analytical result is only an estimate and the 'true value' will lie within a range around the reported result. The extent of the range may be derived in a number of different ways, e.g. by using the results from method validation studies or determining the inherent variation from different components of the method, i.e. estimating these variances as standard deviations and developing an overall standard deviation for the method. There is some concern within the analytical community as to the most appropriate way to estimate this variability in food analysis.

Guidelines on measurement uncertainty have now been adopted by Codex: these are given in Appendix II of this chapter (see page 000). In addition, the EU has published a report to the Standing Committee on the Food Chain and Animal Health on the relationship between analytical results, measurement uncertainty, recovery factors and provisions in EU food and feed legislation (EU, 2004b). This report covers the following topics:

1. Introduction
2. Issues involved
3. number of significant figures taken into account when reporting results and interpreting them against statutory limits
3.1 Introduction
3.2 Solution
4. Reporting of results with respect to their measurement uncertainty
4.1 Introduction
4.2 Reporting of results by food and feed control analysts
4.3 Consequences of reporting results in different ways
4.4 Solution
4.5 Procedures for estimating measurement uncertainty
4.6 Value of the measurement uncertainty
5. The use of recovery information in analytical measurement
6. Other legislation
7. Recommendations
8. Future

Annex I:	Diagrammatic illustration of the effect of measurement uncertainty and the limit
Annex II:	Procedures for the estimation of measurement uncertainty
Annex II.1:	ISO guide to the expression of uncertainty in measurement
Annex II.2:	EURACHEM guide to quantifying uncertainty in analytical measurement

	A. Component-by-component approach
	B. Use of collaborative trial data
Annex II.3:	Use of collaborative trial: data – ISO 5725 critical differences
Annex II.4:	Draft ISO TS 21748 – guide to the use of repeatability, reproducibility and trueness estimates in measurement uncertainty estimation
Annex II.5:	Concept set by Commission Decision 2002/657/EC implementing Council Directive 96/23/EC concerning the performance of analytical methods and the interpretation of results
Annex II.6:	AOAC INTERNATIONAL approach
Annex II 7:	Internal Quality Control approach
Annex II.8:	NMKL (Nordic Committee On Food Analysis) approach
Annex II.9:	Microbiological analyses
Annex II.10:	Useful general measurement uncertainty references

13.9.2 Microbiology laboratories

Until recently, few laboratories indicated their uncertainty of measurement, when quoting test results, even when these were expressed as numbers of colony-forming units. The reasons given included the fact that the distribution of microbes in the substrates examined, particularly solids, such as foods, was inherently heterogeneous. Also, microbes are often present in clumps that break up to varying degrees during sampling, mixing, diluting and plating, while reference materials cannot be prepared with exactly known numbers of microbes. This attitude to uncertainty of measurement has changed recently, partly because the results of standard tests are sometimes used to assess whether a food complies with statutory microbial limits, and partly as a result of the widespread introduction of Quality Assurance and accreditation systems in microbiological laboratories. As with chemical analysis, overall errors can be estimated by investigating individual errors within the method (weighing, pipetting, bias from different individuals counting the colonies, etc.), but the method generally favoured is to estimate overall uncertainty by determining the repeatability and reproducibility of the method concerned. Uncertainty is minimised by Quality Assurance systems that minimise errors within the method, e.g. temperature, time of incubation, weighing, measurement of pH, productivity and selectivity of culture media, accuracy of volume measurement. However, these errors cannot be eliminated completely and other sources of uncertainty are inherent. For example, the numbers of microbes in replicate samples generally follow the Poisson distribution, so there is inherently greater uncertainty than that found in chemical testing. Uncertainty of measurement can be estimated by replicate testing within the laboratory, as well as from results obtained by participation in proficiency testing schemes. Uncertainty will be affected by factors such as the food substrate being tested and the analytical method used, so it needs to be determined for each food/method combination. In general, selective colony-count methods have greater uncertainty than those involving non-selective 'total' colony counts and counts from liquids are less uncertain than those obtained from solid foods. There are a number of useful publications and draft standards on this topic (NMKL, 1999; Voysey and Jewell, 1999; UKAS, 2000; ISO, 2005a, b; 2002, 2003, 2006; Niemelä, 2002).

13.10 Future trends

For the microbiological laboratory, as for all laboratories, it is likely that the most significant development will be the need to demonstrate better the quality of their work. For survey work, that objective is readily achieved through accreditation. However, the requirement to demonstrate quality through a third-party assessment is likely to be adopted by the major funding agencies in the UK. Such a requirement would have a major impact on the work of the laboratory.

13.11 References

ANONYMOUS (1990), *Food Safety Act, 1990*, London, UK, HMSO, p. 16.

CAC (1997a), 'Report of the 21st session of the Codex committee on methods of analysis and sampling – ALINORM 97/23A', Rome Italy, Codex Alimentarius Commission.

CAC (1997b), 'Report of the 22nd session of the Codex Alimentarius Commission – ALINORM 97/37', Rome Italy, Codex Alimentarius Commission.

CAC (1997c), *Procedural Manual of the Codex Alimentarius Commission*, 10th edition, Rome, Italy, Codex Alimentarius Commission.

CORRY J E L, CURTIS G D W and BAIRD R M (eds) (2003), *Handbook of Culture Media for Food Microbiology*, 2nd edition, Progress in Industrial Microbiology, vol 37, Amsterdam, The Netherlands, Elsevier.

EA (2002), 'European Co-operation for Accreditation: Accreditation for microbiological laboratories – EA-04/10 2002, http://www.european-accreditation.org.

EU (2004a), 'Regulation (EC) NO 882/2004 of the European Parliament and of the Council of 29 April 2004 on official controls performed to ensure the verification of compliance with feed and food law, animal health and animal welfare rules', *Official Journal of the European Union*, **L165**, 1–141.

EU (2004b), 'Report to the standing committee on the food chain and animal health on the relationship between analytical results, the measurement uncertainty, recovery factors and the provisions in EU food and feed legislation', http://europa.eu.int/comm/food/food/chemicalsafety/contaminants/report-sampling_analysis_2004_en.pdf.

EU (2005), 'Commission Regulation (EC) No 2073/2005 of 15 November 2005 on microbiological criteria for foodstuffs, *Official Journal of the European Union*, **L338**, 1–26.

EURACHEM (2000), *EURACHEM/CITAC Guide quantifying uncertainty in analytical measurement*, 2nd edition, Berlin, Germany, EURACHEM Secretariat, http://www.eurachem.ul.pt/.

FSA (2001), 'A review of the Food Standards Agency's research portfolio and research management systems', London, UK, Food Standards Agency.

HORWITZ W (1982), Evaluation of analytical methods used for regulation of food and drugs, *Analytical Chemistry*, **54**, 67A–76A.
HORWITZ W (1988), 'Protocol for the design, conduct and interpretation of method performance studies, *Pure and Applied Chemistry*, **60**, 855–864.
HORWITZ W (ed.) (1995), 'Protocol for the design, conduct and interpretation of method performance studies, *Pure and Applied Chemistry*, **67**, 331–343.
ISO (1990), 'General requirements for the competence of calibration and testing laboratories – ISO/IEC Guide 25, Geneva, Switzerland, International Organization for Standardization.
ISO (1992), ISO Guide 30:1992, 'Terms and definitions used in connection with reference materials', Geneva, Switzerland, International Organization for Standardization.
ISO (1993a), ISO/IEC Guide 58:1993, 'Calibration and testing laboratory accreditation systems – General requirements for operation and recognition', Geneva, Switzerland, International Organization for Standardization.
ISO (1993b), *International vocabulary of basic and general terms in metrology* – 2nd edition, Geneva, Switzerland, International Organization for Standardization.
ISO (1993c), *Guide to the expression of uncertainty in measurement*, Geneva, Switzerland, International Organization for Standardization.
ISO (1994), ISO 5725:1994, 'Accuracy (trueness and precision) of measurement methods and results – Parts 1–6', Geneva, Switzerland, International Organization for Standardization.
ISO (1996), BS 5763-0:1996, ISO 7218:1996, 'Methods for microbiological examination of food and animal feeding stuffs – general laboratory practices', Geneva, Switzerland, International Organization for Standardization.
ISO (2000), BS DD ENV ISO 11133-1:2000, 'Microbiology of food and animal feeding stuffs – guidelines on quality assurance and performance testing of culture media in the laboratory', Geneva, Switzerland, International Organization for Standardization.
ISO (2002), ISO/TC34/SC 9N543:2002, 'Draft: Statistics of analytical data. Protocol for the establishment of precision characteristics of microbiological quantitative methods by interlaboratory studies', Geneva, Switzerland, International Organization for Standardization.
ISO (2003), ISO/TS 11133-2:2003, 'Microbiology of food and animal feeding stuffs – Guidelines on preparation and production of culture media – Part 2: Practical guidelines on performance testing of culture media', Geneva, Switzerland, International Organization for Standardization.
ISO (2005a), ISO/CD 14461-1 (IDF 169-1):2001, 'Milk and milk products – Quality control in microbiological laboratories – Part 1: Analyst performance assessment for colony counts', Geneva, Switzerland, International Organization for Standardization.
ISO (2005b), ISO/CD 14461-2 (IDF 169-2):2001, 'Milk and milk products – Quality control in microbiological laboratories – Part 2: Determination of the reliability of colony counts of parallel plates and subsequent dilution steps', Geneva, Switzerland, International Organization for Standardization.
ISO (2005c), EN/ISO/IEC 17025:2005, 'General requirements for the competence of testing and calibration laboratories', Geneva, Switzerland, International Organization for Standardization.
ISO (2006), ISO/TS 19036: 'Microbiology of food and animal feeding stuffs – Guidelines for the estimation of measurement uncertainty for quantitative determinations', Geneva, Switzerland, International Organization for Standardization.
NIEMELÄ S I (2002), 'Uncertainty of quantitative determinations derived by cultivation of microorganisms, 2nd edition', Centre for Metrology and Accreditation,

Advisory Commission for Metrology, Chemistry Section, Expert Group for Microbiology, Helsinki, Finland, Publication J3/2002.

NMKL (1999), 'Procedure No. 8: Measurement of uncertainty in microbiological examination of foods', Oslo, Norway, Nordic Committee on Food Analysis, www.nmkl.org.

THOMPSON M and WOOD R (eds) (1993), 'International Union of Pure and Applied Chemistry: The international harmonised protocol for the proficiency testing of (chemical) analytical laboratories', *Pure and Applied Chemistry*, **65**, 2123–2144.

THOMPSON M and WOOD R (eds) (1995), 'International Union of Pure and Applied Chemistry: Guidelines on internal quality control in analytical chemistry laboratories', *Pure and Applied Chemistry*, **67**, 649–666.

THOMPSON M, ELLISON S L R and WOOD R (eds) (2002), 'Harmonised guidelines for single-laboratory validation of methods of analysis', *Pure and Applied Chemistry*, **74**, 835–855.

THOMPSON M, ELLISON S L R and WOOD R (eds) (2006), 'International Union of Pure and Applied Chemistry: The international harmonised protocol for the proficiency testing of analytical chemistry laboratories', *Pure and Applied Chemistry*, **78**, 145–196.

UKAS (2000), 'The expression of uncertainty in testing, edition 1', UKAS publication ref. LAB 12, London, UK, United Kingdom Accreditation Service.

VOYSEY P A and JEWELL K (1999), 'Uncertainty associated with microbiological measurement, Project no. 29732, Review no. 15', Chipping Campden, UK, Campden and Chorleywood Food Research Association.

Appendix I: The ISO/IUPAC/AOAC international harmonised protocol for proficiency testing of (chemical) analytical laboratories

The protocol (Thompson and Wood, 1993) is recognised within the food sector of the EU and also by the CAC. It makes the following recommendations about the organisation of proficiency testing, all of which are important:

1. Framework

Samples must be distributed regularly to participants who are required to return the results within a given time. The results will be analysed statistically by the organiser and participants will be notified of their performance. Advice will be available to poor performers, and participants will be kept fully informed of the scheme's progress. Participants will be identified by code only, to preserve confidentiality.

The structure of the scheme for any one analyte or round in a series should be:

- samples prepared;
- samples distributed regularly;
- participants analyse samples and report results;
- results analysed and performance assessed;
- participants notified of their performance;
- advice available for poor performers, on request;
- co-ordinator reviews performance of scheme;
- next round commences.

2. Organisation

The running of the scheme will be the responsibility of a co-ordinating laboratory/organisation. Sample preparation will either be contracted out or carried out in house. The co-ordinating laboratory must be of high reputation in relation to the type of analysis being undertaken. Overall management of the scheme should be in the hands of a small steering committee (advisory panel) having representatives from the co-ordinating laboratory (who should be practising laboratory scientists), contract laboratories (if any), appropriate professional bodies and ordinary participants.

3. Samples

The samples to be distributed must be generally similar in matrix to the unknown samples that are analysed routinely (in respect of matrix composition and analyte concentration range). It is essential they are of acceptable homogeneity and stability. The bulk material prepared must be effectively homogeneous so that all laboratories will receive samples that do not differ significantly in analyte concentration. The co-ordinating laboratory should also be able to demonstrate that the bulk sample is sufficiently stable to ensure it will not undergo significant change during the proficiency test. Thus, prior to sample distribution, matrix and analyte stability must be determined by analysis after appropriate storage. Ideally, the quality checks on samples should be performed by a different laboratory from the one preparing the bulk sample, although it is recognised that this may cause considerable difficulty for the co-ordinating laboratory. The number of samples to be distributed per round for each analyte should be no more than five.

4. Frequency of sample distribution

Sample distribution frequency in any one series should not be more than every two weeks and not less than every four months. A frequency greater than once every two weeks could lead to problems in turn-round of samples and obtaining results. If the period between distributions extends much beyond four months, there will be unacceptable delays in identifying any analytical problems and the impact of the scheme on participants will be small. The frequency also relates to the field of application and extent of IQC that is required for that field. Thus, although the frequency range stated above should be adhered to, there may be circumstances where it is acceptable to have a longer time-scale for sample distribution, e.g. if sample throughput per annum is very low. Advice on this respect would be a function for the advisory panel.

5. Estimating the assigned value (the 'true' result)

There are a number of possible approaches to determining the nominally 'true' result for a sample, but only three are normally considered. The result may be established from the amount of analyte added to the bulk sample by the laboratory preparing it; alternatively, a 'reference' laboratory, or group of such expert laboratories, may be asked to measure the analyte concentration using definitive methods or, thirdly, the results obtained by the participating laboratories or a substantial sub-group of these may be used as the basis for obtaining the nominal 'true' result. The organisers of the scheme should provide the participants with a clear statement showing the basis for assigning reference values, and these should take into account the views of the advisory panel.

6. Choice of analytical method

Participants can use the analytical method of their choice, except when otherwise instructed to adopt a specified method. It is recommended that all methods should be properly validated before use. In situations where the analytical result is method-dependent, the true value will be assessed using results obtained with a defined procedure. If participants use a method that is not 'equivalent' to the defined method, then there will be an automatic bias in their results, when performance is assessed.

7. Performance criteria

For each analyte in a round, a criterion for the performance score may be set, against which the score obtained by a laboratory can be judged. A 'running score' could be calculated to give an assessment of performance spread over a longer period of time.

8. Reporting results

Reports issued to participants should include data on the results from all laboratories, together with the participant's own performance score. The original results should be presented to enable participants to check correct data entry. Reports should be made available before the next sample distribution. Although all results should be reported, it may not be possible to do this in very extensive schemes (800 participants undertaking 15 analyses in a round). Participants should, therefore, receive at least a clear report, with the results of all laboratories in histogram form.

9. Liaison with participants

Participants should be provided with a detailed information pack on joining the scheme. Communication with participants should be by newsletter or annual report, together with a periodic open meeting; participants should be advised of any changes in scheme design. Advice should be available to poor performers. Feedback from laboratories should be encouraged so that participants contribute to the scheme's development. Participants should view it as *their* scheme rather than one imposed by a distant bureaucracy.

10. Collusion and falsification of results

Collusion might take place between laboratories, so that independent data are not submitted. Proficiency testing schemes should be designed to ensure that there is as little collusion and falsification as possible. For example, alternative samples could be distributed within a round. Also instructions should make it clear that collusion is contrary to professional scientific conduct and serves only to nullify the benefits of proficiency testing.

11. Statistical procedure for analysis of results

The first stage in producing a score from a result x (a single measurement of analyte concentration in a test material) is to obtain an estimate of the bias, thus:

$$\text{bias} = x - X$$

where X is the true concentration or amount of analyte.

The efficacy of any proficiency test depends on using a reliable value for X. Several methods are available for establishing a working estimate of $\hat{X}$, i.e. the assigned value, outlined above.

Formation of a z-score

Most proficiency testing schemes compare bias with a standard error. An obvious approach is to form the z-score given by:

$$z = (x - \hat{X})/\sigma$$

where σ is a standard deviation. σ could be either an estimate of the actual variation encountered in a particular round ($\tilde{s}$), obtained from the laboratories' results after outlier elimination, or a target representing the maximum permitted variation that is consistent with valid data. This is the procedure recommended in the international protocol and hence that to be followed in the EU.

A fixed target value for σ is preferable and can be arrived at in several ways. It could be fixed arbitrarily, with a value based on a perception of how laboratories should perform. It could be an estimate of the precision required for a specific task of data interpretation. σ could be derived from a model of precision, such as the 'Horwitz Curve' (Horwitz, 1982). However, while this model provides a general picture of reproducibility, substantial deviation from it may be experienced for particular methods. In the case of toxicants, data on the target deviations will have been drawn from both collaborative trials and predictions using the Horwitz Curve.

Interpretation of z-scores
If $\hat{X}$ and σ are good estimates of the population mean and standard deviation, then z will approximate a normal distribution, with a mean of zero and unit standard deviation. An analytical result is described as 'well behaved' when it complies with this condition.

An absolute value of $z(|z|)$ greater than three suggests poor performance in terms of accuracy. This judgement depends on assuming a normal distribution which, outliers apart, seems to be justified in practice.

As z is standardised, it is comparable for all analytes and methods. Thus, values for z can be combined to give a composite score for a laboratory in one round of a proficiency test.

Therefore, the z-scores can be interpreted as follows:

$\|z\| < 2$	Satisfactory:	will occur in 95 % cases produced by well-behaved results
$2 < \|z\| < 3$	Questionable:	but will occur in ≈5 % of cases produced by well-behaved results
$\|z\| > 3$	Unsatisfactory:	will only occur in ≈0.1 % of cases produced by well-behaved results

Appendix II: Codex guidelines on measurement uncertainty

Introduction

It is important and required by ISO/IEC 17025:2005 that analysts are aware of the uncertainty associated with each analytical result and that they estimate the uncertainty. The measurement uncertainty may be derived by a number of procedures. Food analysis laboratories are required, for Codex purposes, to be in control (as outlined in Codex GL 27-1997 'Guidelines for the assessment of the competence of testing laboratories involved in the import and export of food'), use collaboratively tested or validated methods, when available, and verify their application before taking them into routine use. Such laboratories therefore have available to them a range of analytical data that can be used to estimate measurement uncertainty.

These guidelines only apply to quantitative analysis.

Most quantitative analytical results take the form of '$a \pm 2u$ or $a \pm U$', where 'a' is the best estimate of the true value of the concentration of analyte (the analytical result), 'u' is the standard uncertainty and 'U' (equal to $2u$) is the expanded uncertainty. The range '$a \pm 2u$' represents a 95 % level of confidence, where the true value would be found. The value of 'U' or '$2u$' is the value that is normally used and reported by analysts, and is hereafter referred to as 'measurement uncertainty'. It may be estimated in a number of different ways.

Terminology

The international definition for measurement uncertainty is:

> 'Parameter, associated with the result of a measurement, that characterises the dispersion of the values that could reasonably be attributed to the measurand' (see International vocabulary of basic and general terms in metrology, ISO 1993b, 2nd edition).

Notes:

1. The parameter may be, for example, a standard deviation (or a given multiple of it), or the half-width of an interval having a stated level of confidence.
2. Uncertainty of measurement comprises, in general, many components. Some of these components may be evaluated from the statistical distribution of results of a series of measurements and can be characterised by experimental standard deviations. The other components, which can also be characterised by standard deviations, are evaluated from assumed probability distributions, based on experience or other information.
3. It is understood that the result of a measurement is the best estimate of the value of a measurand, and that all components of uncertainty, including those arising from systematic effects, such as components associated with corrections and reference standards, contribute to the dispersion.

Recommendations

1. The measurement uncertainty associated with all analytical results is to be estimated.

2. The measurement uncertainty of an analytical result may be estimated by a number of procedures, notably those described by ISO (1993c) and EURACHEM (2000). These documents recommend procedures based on a component-by-component approach, method validation data, internal quality control data and proficiency test data. The need to undertake an estimation of the measurement uncertainty using the ISO component-by-component approach is not necessary, if the other forms of data are available and used to estimate the uncertainty. In many cases, the overall uncertainty may be determined by an inter-laboratory (collaborative) study by a number of laboratories and a number of matrices (Horowitz, 1995) or by the ISO 5725 protocols (ISO, 1994).
3. The measurement uncertainty and its level of confidence must, on request, be made available to the user (customer) of the results.

Index